HANDBOOK OF THERMOCHEMICAL DATA
FOR COMPOUNDS AND AQUEOUS SPECIES

HANDBOOK OF THERMOCHEMICAL DATA
FOR COMPOUNDS AND AQUEOUS SPECIES

HERBERT E. BARNER
Ledgemont Laboratory
Kennecott Cooper Corporation

RICARD V. SCHEUERMAN
Dow Chemical Company
formerly of
Ledgemont Laboratory
Kennecott Cooper Corporation

A Wiley-Interscience Publication

JOHN WILEY & SONS, New York • Chichester • Brisbane • Toronto

Library of Congress Cataloging in Publication Data

Barner, Herbert E   1936-
   Handbook of thermochemical data for compounds and
aqueous species.

   "A Wiley-Interscience Publication."
   1. Thermodynamics—Tables, calculations, etc.
I. Scheuerman, Ricard V., 1946-   joint author.
II. Title.

QD511.8.B38   541'.369'0212   77-20244
ISBN 0-471-03238-7

Printed in the United States of America

10 9 8 7 6 5 4 3 2 1

PREFACE

The purpose of this handbook is to present the thermodynamic properties of a wide variety of ions and complexes in aqueous solution over a range of temperatures. To make the tables as useful as possible, data on a large number of compounds and minerals are also included. This permits the use of a single set of tables for most aqueous solution applications.

The thermodynamic properties of ions and complexes are needed for analyzing the behavior of aqueous electrolyte systems. Although several compilations report extensive data at 25°C, no comprehensive tabulation has heretofore been available for <u>aqueous</u> species at <u>elevated</u> temperatures.

The high-temperature properties of ions and complexes were developed from 25°C data by means of generalized correlations. It has been shown that the selected correlations provide good estimates of the effect of temperature on the thermodynamic properties of these substances. The high-temperature data for compounds and minerals were developed by standard, rigorous methods.

The free energy of formation, heat of formation, enthalpy, and entropy are tabulated for compounds, minerals, and ions over the temperature range of 25°C to 300°C. Similar data are given for neutral complexes up to 200°C.

Any compilation of data suffers from being incomplete. Occasionally, users may find coverage for all but one or two species needed in an analysis. We have, therefore, included Fortran listings of the computer programs that were used to generate the tables of data. Detailed instructions for the use of these programs are given, including sample calculations. (Card copies of the computer programs are available from the authors upon request.)

Thus, providing that the basic data at 25°C are available or can be estimated, users can readily supplement this compilation with additional substances as the need arises.

We would like to give special acknowledgment to C. M. Criss and J. W. Cobble, and to H. C. Helgeson, whose correlations have been applied in developing the data for aqueous species. These investigators have made monumental contributions to the thermochemistry of aqueous solutions, and without their pionerring efforts this compilation would not have been possible. We are also grateful to Ledgemont Laboratory, Kennecott Copper Corporation, for the encouragement and support of this work, and for permission to publish the results.

Finally, we wish to express our sincere thanks to Claudette Devol of the Ledgemont Laboratory staff for her patience and skill in preparing the manuscript.

Herbert E. Barner
Ricard V. Scheuerman

Lexington, Massachusetts
Pittsburgh, California
October 1977

CONTENTS

1. Introduction,   1

2. Calculation Methods,   15

3. Tabulated Data,   25

4. Prediction of Data for Additional Substances,   93

SYMBOLS,   117

APPENDIX.   FORTRAN LISTINGS OF COMPUTER PROGRAMS,   119

HANDBOOK OF THERMOCHEMICAL DATA
FOR COMPOUNDS AND AQUEOUS SPECIES

INTRODUCTION

The purpose of this book is to provide convenient tables of thermodynamic data for a wide variety of compounds, ions, and neutral complexes. The data are intended primarily for use in calculating equilibrium constants and energy balances for chemical and metallurgical processes carried out in aqueous solutions.

A summary of definitions and important thermodynamic relationships is given in this introductory chapter. No attempt is made to provide a complete background on solution equilibria, the calculation of activities and activity coefficients in aqueous solutions, and other related topics. The reader is referred to standard texts (Refs. 1, 3, 4) that deal with these topics in detail.

CONVENTIONS FOR PURE SUBSTANCES

The values of the thermodynamic properties of pure substances given in Chapter 3 are for the substances in their standard states. These standard states are defined here.

The standard state for a pure solid or liquid is the substance at the specified temperature under a pressure of 1 atm.

For a gas, the standard state is the hypothetical ideal gas at the specified temperature and a fugacity of 1 atm. In this state the enthalpy of the ideal gas is equal to the enthalpy of the real gas at the same temperature and at zero pressure. Since the behavior of most gases is nearly ideal at low pressure, this state approximates that of the real gas at the specified temperature and 1 atm. pressure.

The enthalpy and free energy of formation data given in Chapter 3 represent the change in the appropriate thermodynamic property when 1 mole

1

of the substance in its standard state is formed at the specified temperature from the elements, each in its standard reference state. With two
exceptions, the standard reference state for each element has been chosen
to be the standard state that is thermodynamically stable at the specified
temperature and 1 atm. pressure. The exceptions are

1. Sulfur. The ideal diatomic gas, $S_2(g)$, was selected as the standard
    reference state at all temperatures.
2. Phosphorus. The following standard reference states were selected:

    (a)   25- 44°C--The crystalline white form of phosphorus.

    (b)   44-274°C--Liquid phosphorus.

    (c)   274-300°C--The ideal tetratomic gas, $P_4(g)$.

CONVENTIONS FOR DISSOLVED SPECIES

The standard state for a dissolved species in aqueous solution, such as
$Cu^{2+}(aq)$ or $AgCl(aq)$, is defined as the hypothetical ideal solution at a
molality of 1.0. This state is arrived at by extrapolation from the
behavior in infinitely dilute solutions where behavior is ideal.

The tabulated values of the enthalpy and free energy of formation for
individual ions are based on the convention that the enthalpy and free
energy of formation of the hydrogen ion (aq; standard state, molality =
1.0) are zero at all temperatures. This convention has two consequences.
First, the values of ΔH_T^O and ΔF_T^O calculated for any half-cell reaction
(i.e., any reaction in which electrons appear in the balanced chemical
equation) are relative to the hydrogen ion formation reaction at the given
temperature. Second, the thermodynamic relation,

$$\Delta F_T^O = \Delta H_T^O - T\Delta S_T^O$$

will not hold for individual half-cell reactions. These consequences are
of no concern when complete reactions are considered, that is, when no
electrons appear in the balanced chemical equations.

CHEMICAL EQUILIBRIUM

Standard Free Energy of Reaction and the Equilibrium Constant

A chemical reaction between substances, B_i, may be written by

$$\nu_1 B_1 + \nu_2 B_2 + \ldots = \nu_j B_j + \ldots \tag{1.1}$$

where ν_i denotes the stoichiometric coefficients. The thermodynamic equilibrium constant is represented by

$$K = \Sigma \ a_{B_i}^{\nu_i} \tag{1.2}$$

where the stoichiometric coefficients on the right-hand side of (1.1) are positive and those on the left-hand side are negative. The activity of a reactant or product, a_{B_i}, may be thought of as a measure of the "effective concentration" of the substance in the chemical reaction.

The standard free-energy change for reaction (1.1) at temperature T is the sum of the free energies of formation of the products in their standard states minus the free energies of formation of the reactants in their standard states,

$$\Delta F_T^O \ (reaction) = \Sigma \ \nu_i \ (\Delta F_T^O)_i \tag{1.3}$$

The standard free-energy change of the reaction is related to the thermodynamic equilibrium constant by the relation,

$$\Delta F_T^O \ (reaction) = - \ RT \ \ln K \tag{1.4}$$

where R is the gas constant and T the absolute temperature.

The free-energy data tabulated in Chapter 3 may be used to calculate equilibrium constants by means of Eqs. (1.3) and (1.4). If ΔF_T^O is expressed in cal/mole, consistent units are obtained by using,

R = 1.987 cal/deg K-mole

T = absolute temperature in ^{O}K.

ACTIVITY-CONCENTRATION RELATIONSHIPS

Equilibrium constants calculated from thermochemical tabulations are expressed in terms of activities. The use of equilibrium constants in applications, on the other hand, usually requires that results be expressed in concentrations. Thus the analysis of equilibrium problems through thermochemical calculations requires translation from activities to concentrations and vice versa. This translation is readily made for gases and pure solids

and liquids. The translation for solutions, however, is considerably more complex unless concentrations are sufficiently low that ideal solution behavior is approached.

Activity of Gases

The activity of a gas at any state is given by the ratio of its fugacity in that state to its fugacity in its standard state. Since the fugacity is equal to unity in the standard state ($f_i^0=1$), the activity of any gas i becomes equal to its fugacity,

$$a_i = f_i/f_i^0 = f_i/1 = f_i \tag{1.5}$$

Furthermore, for a perfect gas (for which $pV = nRT$), fugacity is equal to pressure. Since most pure gases tend to follow perfect gas behavior up to 1 atm. pressure, the activity of a gas at low pressure may be approximated as numerically equal to the pressure in atmospheres.

At most temperatures of interest, gas mixtures at total pressures up to 1 atm. also behave as perfect gas mixtures. For such mixtures the activity of each gas may be approximated by its partial pressure.

The exact pressure up to which the previous approximations are permissible cannot be specified because it will depend on the nature of the gas(es) in question, the temperature, and the accuracy required. However, well-developed methods are available for estimating the fugacities of nonpolar gases in their pure states or in mixtures. These methods are based on generalized equations of state and generally require only a knowledge of the critical constants of the gases in question. Two particularly useful approaches are based on the virial equation of state or the equation due to Redlich and Kwong.[11] An excellent review of these approaches is given by Prausnitz.[9]

Activity of Pure Solids and Liquids

The activity of a pure solid or pure liquid at any temperature and at 1 atm. pressure is unity. Furthermore, because the free energies of solids and liquids are essentially independent of pressure, the activities of pure solids and liquids are unity to a very near approximation for wide ranges of pressures.

Nonelectrolyte Solutions

The data in this book are directed toward aqueous electrolyte solutions and not toward nonelectrolyte solutions such as those formed by mixtures of hydrocarbon liquids. Therefore, the treatment of activities in nonelectrolyte systems is not discussed here. A basic treatment with geological orientation is given by Garrels and Christ[1]; a more extensive discussion from a chemical engineering viewpoint is given by Prausnitz.[9]

Aqueous Electrolyte Solutions

Based on the conventions adopted in this book, the activity of a dissolved species in aqueous solution is given by

$$a_i = \gamma_i m_i \tag{1.6}$$

where m_i is the concentration of the species expressed on a molality basis, and γ_i is its activity coefficient. In the infinitely dilute solution $\gamma_i = 1$, and the activity becomes equal to the concentration (m_i).

Activity Coefficients for Ions

It is usually necessary to estimate the activity coefficients of ions to make reasonable predictions of equilibrium in aqueous solutions. Although many procedures have been proposed for this purpose, most are based on the Debye-Hückel equation or some variation or extension thereof. According to the Debye-Hückel expression, the activity coefficient for an individual ion is given by

$$- \log \gamma_i = \frac{A \, z_i^2 \, \sqrt{I}}{1 + \mathring{a}_i B \, \sqrt{I}} \tag{1.7}$$

The coefficients A and B are constants characteristics of the solvent (i.e., water) at the specified temperature and pressure. The quantity $\mathring{a}_i$ has a value dependent on the effective diameter of the ion. The quantity z_i is the charge of the ion, and the I is the ionic strength of the solution. The ionic strength is defined by

$$I = \frac{1}{2} \Sigma m_i z_i^2 \tag{1.8}$$

where m_i is the molality and z_i the charge of the ith ion in the solution.

Values of A and B are given in Figures 1.1 and 1.2 as a function of temperature (at 1 atm. pressure). The values of $\overset{\circ}{a}_i$ are established largely from experiment; values for common ions are presented in Table I.1.

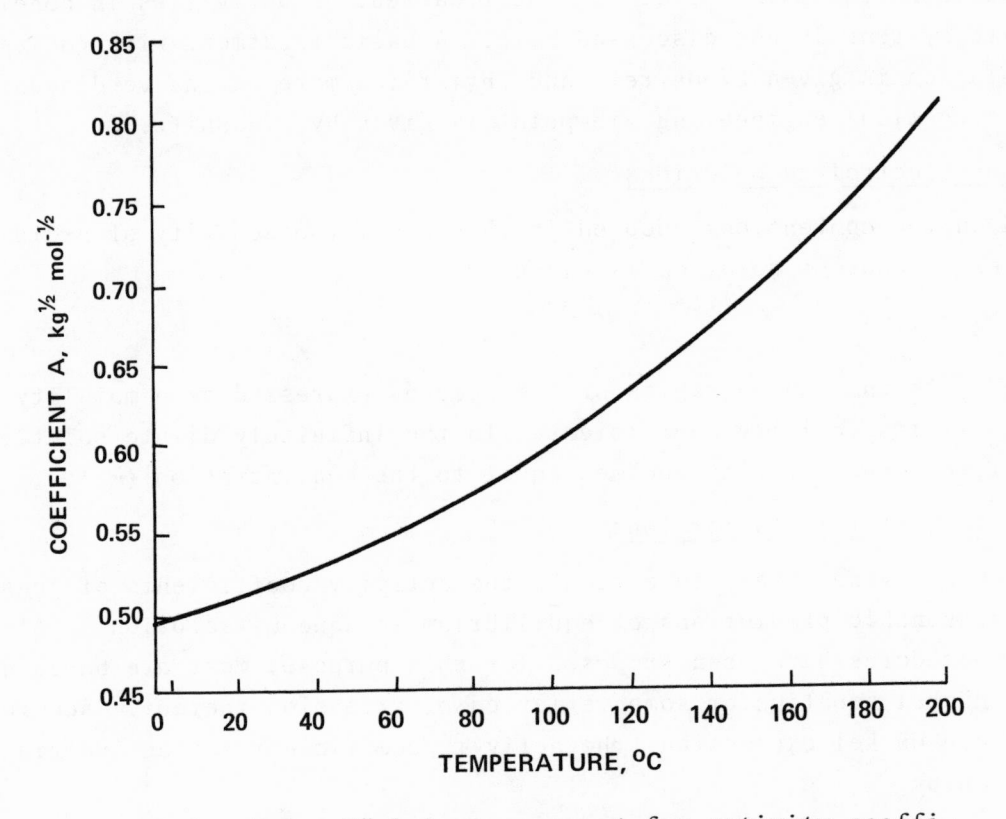

Figure 1.1.   Debye-Hückel parameter A for activity coeffi-
cient in aqueous solutions [Ref., W. J. Hamer,
Natl. Stand. Ref. Data Series, NSRDS-NBS-24,
Washington, D.C. (1968) and W. L. Marshall,
R. Slusher, and E. V. Jones, *J. Chem. Eng. Data*,
9, 187 (1964)]

Reasonable estimates of activity coefficients can often be made with Eq. (1.7) at ionic strengths extended up to approximately 0.1. Unfortunately, in many industrial processes higher ionic strengths are encountered. A method for estimating ionic activity coefficients in concentrated aqueous solutions has been developed by Helgeson (Ref. 4, page 23). This procedure is based on a modification of the Debye-Hückel model.

It is often convenient to consider the "mean activity coefficient" of an electrolyte, since individual ionic coefficients cannot be extracted

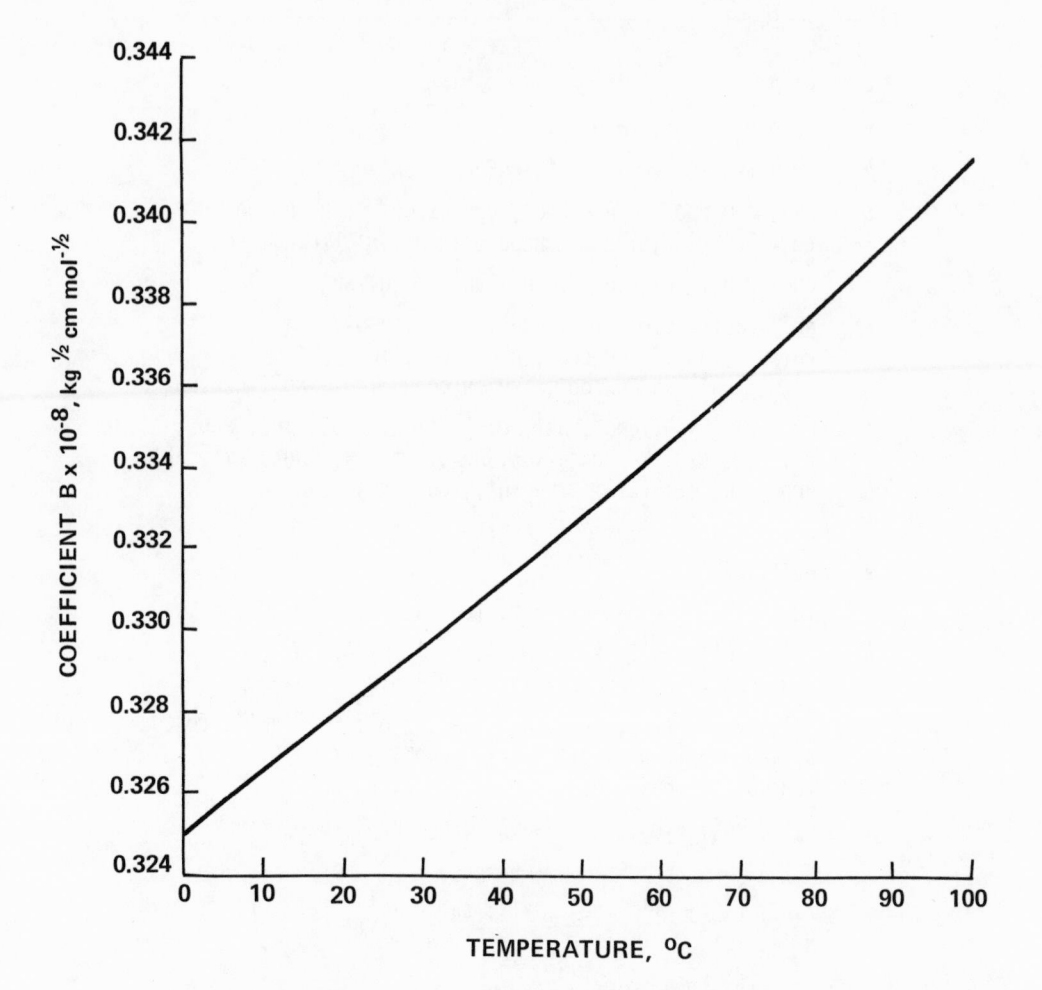

Figure 1.2.   Debye-Hückel parameter B for activity coeffi-
cient in aqueous solutions [Ref., W. J. Hamer,
Natl. Stand. Ref. Data Series, NSRDS-NBS-24,
Washington, D.C. (1968)]

from experimental data.   For an electrolyte $A_x B_y$ that dissociates into x
ions of A and y ions of B, the mean activity coefficient is defined as

$$\gamma = (\gamma_+^x \; \gamma_-^y)^{1/(x+y)} \tag{1.9}$$

where γ_+^x and γ_-^y are the activity coefficients of the two ions.   Meissner
and co-workers[6-8] have developed a generalized graphical correlation for
the mean activity coefficient that appears to be best suited to high ionic
strengths ($I>2$).   This procedure requires that at least one value of the

Table I.1. Values of the Parameter $\mathring{a} \times 10^8$ for Selected Ions

---

Charge 1

9   H^+

8    $(C_6H_5)_2CHCOO^-$, $(C_3H_7)_4N^+$

7    $OC_6H_2(NO_3)_3^-$, $(C_3H_7)_3NH^+$, $CH_3OC_6H_4COO^-$

6   Li^+, $C_6H_5COO^-$, $C_6H_4OHCOO^-$, $C_6H_4ClCOO^-$, $C_6H_5CH_2COO^-$, $CH_2=CHCH_2COO^-$, $(CH_3)_2CCHCOO^-$, $(C_2H_5)_4N^+$, $(C_3H_7)_2NH_2^+$

5    $CHCl_2COO^-$, CCl_3COO^-, $(C_2H_5)_3NH^+$, $(C_3H_7)NH_3^+$

4   Na^+, $CdCl^+$, ClO_2^-, IO_3^-, HCO_3^-, $H_2PO_4^-$, HSO_3^-, $H_2AsO_4^-$, $Co(NH_3)_4(NO_2)_2^+$, CH_3COO^-, CH_2ClCOO^-, $(CH_3)_4N^+$, $(C_2H_5)_2NH_2^+$, $NH_2CH_2COO^-$, $^+NH_3CH_2COOH$, $(CH_3)_3NH^+$, $C_2H_5NH_3^+$

3   OH^-, F^-, CNS^-, CNO^-, HS^-, ClO_3^-, ClO_4^-, BrO_3^-, IO_4^-, MnO_4^-, K^+, Cl^-, Br^-, I^-, CN^-, NO_2^-, NO_3^-, Rb^+, Cs^+, NH_4^+, Tl^+, Ag^+, $HCOO^-$, $H_2(citrate)^-$, $CH_3NH_3^+$, $(CH_3)_2NH_2^+$

Charge 2

8   Mg^{++}, Be^{++}

7    $(CH_2)_5(COO)_2^=$, $(CH_2)_6(COO)_2^=$, $(congo\ red)^=$

6   Ca^{++}, Cu^{++}, Zn^{++}, Sn^{++}, Mn^{++}, Fe^{++}, Ni^{++}, Co^{++}, $C_6H_4(COO)_2^=$, $H_2C(CH_2COO)_2^=$, $(CH_2CH_2COO)_2^=$

5   Sr^{++}, Ba^{++}, Ra^{++}, Cd^{++}, Hg^{++}, $S^=$, $S_2O_4^=$, $WO_4^=$, Pb^{++}, $CO_3^=$, $SO_3^=$, $MoO_4^=$, $Co(NH_3)_5Cl^{++}$, $Fe(CN)_5NO^=$, $H_2C(COO)_2^=$, $(CH_2COO)_2^=$, $(CHOHCOO)_2^=$, $(COO)_2^=$, $H(citrate)^=$

4   Hg_2^{++}, $SO_4^=$, $S_2O_3^=$, $S_2O_8^=$, $SeO_4^=$, $CrO_4^=$, $HPO_4^=$, $S_2O_6^=$

Charge 3

9   Al^{+3}, Fe^{+3}, Cr^{+3}, Sc^{+3}, Y^{+3}, La^{+3}, In^{+3}, Ce^{+3}, Pr^{+3}, Nd^{+3}, Sm^{+3}

6    $Co\ (ethylenediamine)_3^{+3}$

5    $Citrate^{-3}$

4   PO_4^{-3}, $Fe(CN)_6^{-3}$, $Cr(NH_3)_6^{+3}$, $Co(NH_3)_6^{+3}$, $Co(NH_3)_5H_2O^{+3}$

Charge 4

11   Th^{+4}, Zn^{+4}, Ce^{+4}, Sn^{+4}

6    $Co(S_2O_3)(CN)_5^{-4}$

5    $Fe(CN)_6^{-4}$

Charge 5

9    $Co(S_2O_3)_2(CN)_4^{-5}$

Source: J. N. Butler, *Ionic Equilibrium, A Mathematical Approach*, Reading, Mass.: Addison-Wesley, 1964, after J. Kielland, *J. Am. Chem. Soc.*, 59, 1675 (1937).

activity coefficient at a given ionic strength (preferably at $I>2$) be known or otherwise estimated.

The use of the Debye-Hückel expression, Eq. (1.7), together with Eq. (1.9) indicates that at a given temperature the mean activity coefficient of an electrolyte is determined by the properties of the electrolyte's ions (the z_i and $\mathring{a}_i$ values) and the total ionic strength. This estimating procedure is a useful first approximation in mixtures of electrolytes. However, in general, the mean activity coefficient is also affected by specific interactions with other ions present in solution. For example, the mean activity coefficient of HCl in an HCl-water solution is not identical to the mean activity coefficient of HCl in an HCl-BaCl-water solution even though the total ionic strength may be the same. Methods of approaching the correlation and calculation of activity coefficients in mixed electrolyte systems have been proposed by Harned and Owen (Chapter 13 of Ref. 3) and by Meissner and Kusik.[6] A review of the latter method with examples of applications in calculating chemical equilibrium and solubility limits has recently been published by Kusik and Meissner.[5]

Activity of Water in Electrolyte Solutions

Since the activity of a pure liquid at 1 atm. pressure and at a given temperature is unity, the activity of pure water is taken as unity. It is often assumed that the activity of water in electrolyte solutions is also unity. As shown later, this is a reasonable assumption for dilute solutions. However, the assumption becomes progressively poorer as the concentration of solutes increases. For concentrated solutions, experimental data should be used to obtain accurate values of water activity.

The activity of water in an electrolytic solution is related to its fugacity by the expression

$$a_{H_2O} = f_{H_2O}/f_{H_2O}^* \qquad (1.10)$$

where $f_{H_2O}^*$ is the fugacity of pure water. At low pressures we may replace the fugacity by partial pressure

$$a_{H_2O} = p_{H_2O}/p_{H_2O}^* \qquad (1.11)$$

where p_{H_2O} is the partial pressure of water above the solution and $p_{H_2O}^*$ is the vapor pressure of pure water. The partial pressure of water is given by

$$p_{H_2O} = p_{H_2O}^* \, \gamma_{H_2O} \, x_{H_2O} \tag{1.12}$$

where x_{H_2O} is the mole fraction of water in the liquid and γ_{H_2O} is the activity coefficient based on the mole fraction concentration scale. Hence the activity is given by

$$a_{H_2O} = \gamma_{H_2O} \, x_{H_2O} \tag{1.13}$$

The activity of water in an electrolyte solution can be determined from Eq. (1.11) by measuring the partial pressure of water above the solution. Typical data for several electrolytes[12] are presented in Figure 1.3. The activity data are plotted as a function of concentration units that take into account the total number of solute particles. Thus a 1 molal $CaCl_2$ solution is taken to have a total solute concentration of 3 molal (i.e., 1 molal Ca^{2+} plus 2 molal Cl^-).

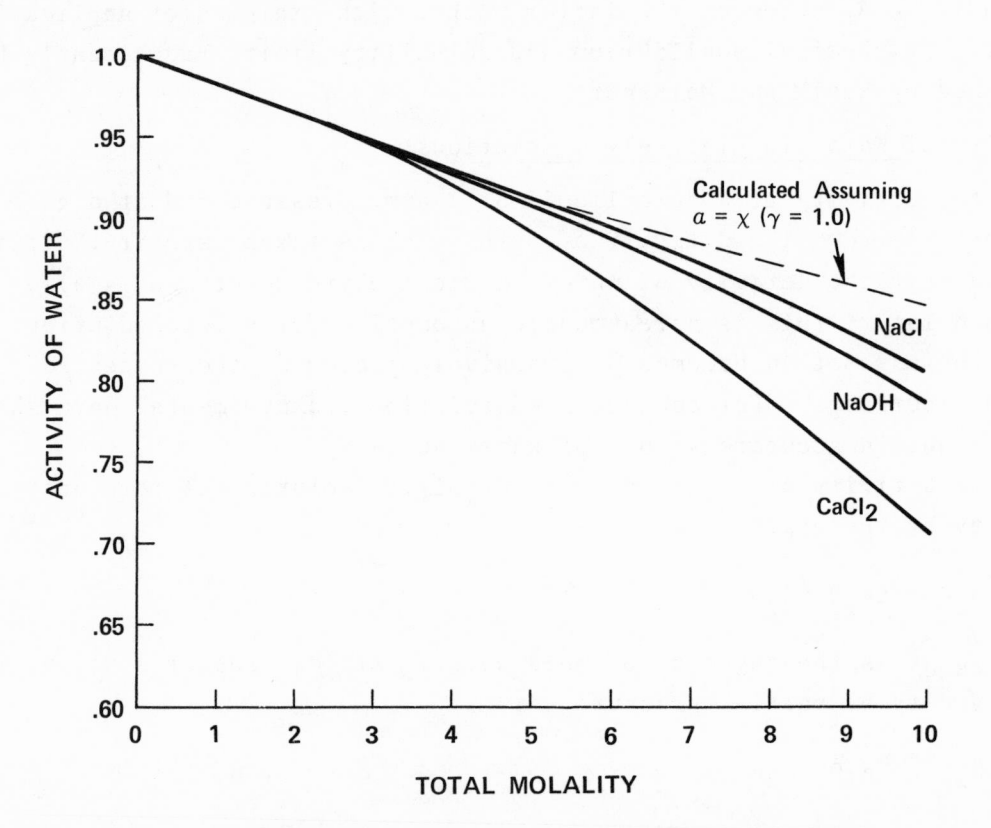

Figure 1.3. Activity of water in the presence of various salts (Source of data: R. A. Robinson and R. H. Stokes, *Electrolyte Solutions*, London: Butterworths, 1959).

Figure 1.3 indicates that the assumption $\gamma_{H_2O} = 1$ generally does not introduce errors greater than 5% provided that the total solute concentration is below 4-5 molal.

Data for the partial pressure of water in various electrolyte solutions have been assembled.[2] This information can be used to estimate the magnitude of γ_{H_2O} in specific systems by means of Eqs. (1.11) and (1.13).

In general, the activity of water in an electrolyte solution is also a function of temperature. Garrels and Christ[1] indicate that for moderate concentrations of the dissolved electrolyte this variation is not appreciable over the range $0°$-$100°C$. However, as the concentration of the electrolyte increases, the temperature dependence can be expected to become greater.

Activity of Neutral Molecules in Electrolyte Solutions

Activity coefficients for dissolved gases (such as N_2, O_2, and H_2) in electrolytes generally must be established by experiment. The activity coefficient is usually correlated with the ionic strength of the solution. Data for various gases and neutral organic molecules in a large variety of electrolyte solutions are available.[3,10]

In the present state of the art it is generally not possible to estimate the activity coefficient of neutral complexes such as $NiSO_4(aq)$. The usual recourse is to take the activity coefficient as unity.

ENERGY BALANCES

A process involving chemical reaction between substances, B_i, may be represented by

$$\nu_1 B_1(T_1) + \nu_2 B_2(T_2) + \ldots \rightarrow \nu_j B_j(T_j) + \ldots \qquad (1.14)$$

where ν_i denotes the moles of substance i. The temperatures are designated in Eq. (1.14) because feed and product substances may appear at different temperatures (e.g., a gaseous reactant may enter a process at a different temperature than other reactants). The heat requirement Q for the process is given by

$$Q = \Sigma \nu_i (H_T^O)_i + \Delta H (mixing) \qquad (1.15)$$

where $(H_T^O)_i$ is the enthalpy of the pure or dissolved substance i at temperature T, and $\Delta H (mixing)$ represents the enthalpy of mixing for the process.

In this compilation the enthalpy of a substance is defined by

$$H_T^O = \Delta H_{298}^O + \Sigma \int_{298}^{T} C_p(T)\,dT + \Sigma \Delta H_{T_t}^{t} \tag{1.16}$$

where ΔH_{298}^O is the heat of formation at 298.15^OK. The integral accounts for sensible heat effects of all phases of the substance between 298.15^OK and temperature T, and the last term accounts for all transitions of the substance between 298.15^OK and T.

The heat of mixing term is often disregarded in a first approximation. However, for many processes that involve electrolytes in aqueous solutions, and in particular processes involving dilution of concentration of strong electrolytes, the mixing term is substantial and cannot be neglected. Data relating to mixing effects is beyond the scope of this compilation. (A large body of data on the heat of dilution at 25^OC is given by the National Bureau of Standards in their Technical Note 270 series.[13])

REFERENCES

1. R. M. Garrels and C. L. Christ, *Solution, Minerals, and Equilibria*. New York: Harper & Row, 1965.

2. *Handbook of Chemistry and Physics*, 52nd ed. Cleveland: Chemical Rubber, 1971.

3. H. S. Harned and B. B. Owen, *The Physical Chemistry of Electrolytic Solutions*. New York: Reinhold, 1958.

4. H. C. Helgeson, *Complexing and Hydrothermal Ore Deposition*. New York: The MacMillan Company, 1964.

5. C. L. Kusik and H. P. Meissner, *Intl. J. Mineral Processing*, 2, 105 (1975).

6. H. P. Meissner and C. L. Kusik, *AIChE J.*, 18, 294 (1972).

7. H. P. Meissner and N. A. Peppas, *AIChE J.*, 19, 4806 (1973).

8. H. P. Meissner and J. W. Tester, *Ind. Eng. Chem. Process Des. Devel.*, 11, 1128 (1972).

9. J. M. Prausnitz, *Molecular Thermodynamics of Fluid-Phase Equilibria*. Englewood Cliffs, N.J.: Prentice-Hall, 1969.

10. M. Randall and C. F. Failey, *Chem. Rev.*, 4, 271 (1927).

11. O. Redlich and J. N. S. Kwong, *Chem. Rev.*, 44, 233 (1949).

12. R. A. Robinson and R. H. Stokes, *Electrolyte Solutions*. London:
     Butterworths, 1959.

13. *Selected Values of Chemical Thermodynamic Properties*. National Bureau of
     Standards, Technical Notes 270-3 (1968), 270-4 (1969), 270-5 (1971),
     270-6 (1971), 270-7 (1973).

CALCULATION METHODS

FUNCTIONS TABULATED

These functions are tabulated as a function of temperature:

ΔH_T^O--Enthalpy of formation.

ΔF_T^O--Gibbs free energy of formation

H_T^O--Enthalpy. It represents the sum of (1) the enthalpy of formation at 25^OC and (2) the enthalpy at temperature T less the enthalpy at 25^OC.

S_T^O--Entropy.

REFERENCE DATA USED FOR THE ELEMENTS

Data for the elements in their reference states must be used to calculate the data for compounds and other species. The entropy of elements at 25^OC was taken from the National Bureau of Standards 270 Technical Note series.[7] Heat capacity data (in the form of polynomial equations relating C_p to temperature) and phase transition data (temperature and enthalpy and entropy of transition) were taken from Barin and Knacke.[1] The reference sources used for each element are listed in the tables in Chapter 3.

COMPOUNDS

The thermodynamic properties of compounds were calculated rigorously using standard procedures. Thus the enthalpy and entropy at a given temperature were computed from these relationships:

15

$$H_T^o = \Delta H_{298}^o + \Sigma \int_{298}^{T} C_p^o(T)\,dT + \Sigma \Delta H_{T_t}^{t} \tag{2.1}$$

$$S_T^o = S_{298}^o + \Sigma \int_{298}^{T} (C_p^o(T)/T)\,dT + \Sigma \Delta S_{T_t}^{t} \tag{2.2}$$

where

H_T^o is the enthalpy at 1 atm. pressure and temperature T.

ΔH_{298}^o is the heat of formation of the compound from the elements in their reference states at 298.15^oK.

$C_p^o(T)$ is the heat capacity of the compound at constant pressure. The summations include all phases between 298.15 and T^oK.

$\Delta H_{T_t}^{t}$ and $\Delta S_{T_t}^{t}$ are the heat and entropy of transition of the compounds, respectively, at temperature T_t. The summations include all transformations between 298.15^o and T^oK.

S_T^o and S_{298}^o are the entropies of the compound at 1 atm. pressure and temperatures T and 298.15^oK, respectively.

As far as possible, the 298.15^oK data for the compounds were taken from the National Bureau of Standards.[7] However, it was necessary to supplement this source with other references for the compounds and minerals not covered by the Bureau of Standards.

The heat capacity for compounds is expressed in polynomial form as a function of temperature. The principle source for the heat capacity data is Barin and Knacke.[1]

Equations (2.1) and (2.2) were also used to calculate the enthalpy and entropy for the elements. The heat of formation and entropy of formation for compounds were subsequently calculated from the expressions

$$\Delta H_T^o = H_T^o \ (compound) - \Sigma H_T^o \ (elements) \tag{2.3}$$

$$\Delta S_T^o = S_T^o \ (compound)) - \Sigma S_T^o \ (elements) \tag{2.4}$$

The summations in Eqs. (2.3) and (2.4) include, of course, the appropriate stoichiometric coefficients needed to form the compound from the elements.

Finally, the Gibbs free energy for compounds was computed using the identity

$$\Delta F_T^O = \Delta H_T^O - T\Delta S_T^O \tag{2.5}$$

IONS IN AQUEOUS SOLUTIONS

Few thermodynamic data are available for ions in aqueous solution over a wide range of temperatures. Therefore, the thermodynamic properties of ions at elevated temperatures were estimated* from known 25°C data using the entropy correspondence principle developed by Criss and Cobble.[2] The National Bureau of Standards[7] and Latimer[6] were the principle sources for the 25°C data. The correspondence principle forms a framework for relating the entropy of individual ions at different temperatures to the entropy at a reference temperature (taken to be 25°C). Once the entropy of an ion can be satisfactorily predicted by this means, the average value of the heat capacity of the ion can also be calculated. From this, the temperature dependence of the free energy and the heat of formation of the ion can be predicted. The calculation procedure used in this book is now described.

Temperature Dependence of Free Energy of Formation

The fundamental thermodynamic equation that relates the Gibbs free energy of a chemical reaction at temperature T to that at a reference temperature T_r is given by

$$\Delta F_T^O = \Delta F_{T_r}^O - \Delta S_{T_r}^O \Delta T + \int_{T_r}^{T} \Delta C_p^O dT - T\int_{T_r}^{T} \Delta C_p^O d(\ln T) \tag{2.6}$$

where

ΔF^O = the Gibbs free energy for the reaction.

ΔS^O = the entropy change for the reaction.

ΔC_p^O = the change in heat capacity for the reaction (products minus reactants, including appropriate stoichiometric coefficients).

ΔT = $T - T_r$.

---

* Several exceptions were made for complex metal ammine ions for which requisite 25°C data were not available. In these cases, free energy of formation values were determined from equilibrium constant data reported by Sillen and Martell.[8] In addition, data for several dissolved gases were taken from Himmelblau.[5]

The subscripts T and T_r designate the temperature and reference temperature, respectively.

If $\Delta \overline{C}_p^O \big]_{T_r}^T$ is taken as the "average" value of ΔC_p^O between the two temperatures, Eq. (2.6) can be written as

$$\Delta F_T^O = \Delta F_{T_r}^O - \Delta S_{T_r}^O \Delta T + \Delta \overline{C}_p^O \Big]_{T_r}^T [\Delta T - T \ln T/T_r] \qquad (2.7)$$

Since ΔF^O and ΔS^O for ions are often available at a reference temperature (usually 25°C), the problem reduces to that of estimating $\Delta \overline{C}_p^O \big]_{T_r}^T$ and more directly the average heat capacity of ions.

Heat Capacity and Entropy of Ions

The partial molal entropy of an ion is related to its partial molal heat capacity by

$$\overline{S}_T^O - \overline{S}_{T_r}^O = \int_{T_r}^T \overline{C}_p^O \, d \ln T \qquad (2.8)$$

Denoting the average value of the heat capacity of the ion between temperatures T and T_r as $\overline{C}_p^O \big]_{T_r}^T$, it is now assumed that

$$\overline{S}_T^O - \overline{S}_{T_r}^O = \overline{C}_p^O \Big]_{T_r}^T \cdot \int_{T_r}^T d \ln T$$

which leads to

$$\overline{C}_p^O \Big]_{T_r}^T = (\overline{S}_T^O - \overline{S}_{T_r}^O)/\ln T/T_r \qquad (2.9)$$

Having evaluated the average heat capacity of an ion by Eq. (2.9), $\Delta \overline{C}_p^O \big]_{T_r}^T$ for the chemical reaction can be obtained by merely summing the average heat capacities of products minus reactants. (The average heat capacities of nonionic species are calculated from existing C_p data.) Thus, using Eq. (2.7), the free energy of the reaction can be evaluated at any temperature.

Values of $\overline{S}_T^O$ and $\overline{S}_{T_r}^O$ must be known to evaluate the average heat capacity of an ion by Eq. (2.9). $\overline{S}_{T_r}^O$ is chosen as the entropy of the ion in its standard state at 25°C ($\overline{S}_{298}^O$) and is usually known. $\overline{S}_T^O$ is generally unknown

but can be estimated from the entropy correspondence principle.[2] According to this principle $\bar{S}_T^O$ is linearly related to $\bar{S}_{298}^O$,

$$S_T^O = a(T) + b(T) \, S_{298}^O \tag{2.10}$$

where the coefficients a and b are functions of temperature and type of ion. Criss and Cobble have shown that Eq. (2.10) is satisfied if the entropies of ions are expressed on an "absolute" scale rather than on the conventional scale. The conversion of scales for any ion is accomplished through the relationship

$$\bar{S}_{298}^O \; (absolute) = \bar{S}_{298}^O \; (conventional) - 5.0 \, z \tag{2.11}$$

where z is the ionic charge.

The values of the coefficients a and b recommended by Criss and Cobble[2,3] are given in Table II.1. The "absolute" entropy values assigned to the hydrogen ion by Criss and Cobble are also listed.

Free Energy of Formation of Ions

The procedure used to calculate the free energy of ions is described using the $Al^{3+}(aq)$ ion as an example.

We may write the following reaction for the formation of $Al^{3+}(aq)$ from its element:

$$Al(c) \rightarrow Al^{3+}(aq) + 3e \tag{2.12}$$

To avoid ambiguities in defining the properties of electrons we may combine Eq. (2.12) with the hydrogen half-cell reaction

$$H^+(aq) + e \rightarrow \tfrac{1}{2} H_2(g) \tag{2.13}$$

to give

$$Al(c) + 3H^+(aq) \rightarrow Al^{3+}(aq) + \tfrac{3}{2} H(g) \tag{2.14}$$

The calculation sequence for a given temperature T may now be summarized:

1. The reference temperature T_r is set equal to $25^O C$ ($298.15^O K$).

2. The conventional entropy of Al^{3+} at $25^O C$ (such as found in the NBS Technical Series 270[7]) is converted to the "absolute" scale using Eq. (2.11).

Table II.1. Values of Coefficients $a(T)$ and $b(T)$ for Entropy of Ions

Temperature (°C)	Simple cations		Simple anions and OH⁻		Oxy anions (XO_n^{-m})		Acid oxy anions $(XO_n(OH)_i^{-m})$		Absolute entropy of $H^+(aq)$
	$a(T)$	$b(T)$	$a(T)$	$b(T)$	$a(T)$	$b(T)$	$a(T)$	$b(T)$	
25	0	1.000	0	1.000	0	1.000	0	1.000	-5.0
60	3.9	0.955	- 5.1	0.969	- 14.0	1.217	- 13.5	1.380	-2.5
100	10.3	0.876	-13.0	1.000	- 31.0	1.476	- 30.3	1.894	2.0
150	16.2	0.792	-21.3	0.989	- 46.4	1.687	- 50.0	2.381	6.5
200	23.3	0.711	-30.2	0.981	- 67.0	2.020	- 70.0	2.960	11.1
250	29.9	0.630	-38.7	0.978	- 86.5	2.320	- 90.0	3.530	16.1
300	36.6	0.548	-49.2	0.972	-106.0	2.618	-110.0[a]	4.1[a]	20.7

[a] Extrapolated

Source: C. M. Criss and J. W. Cobble, *J. Am. Chem. Soc.*, 86, 5385 and 5390 (1964).

3. The properties of $Al(c)$, $H^+(aq)$, $Al^{3+}(aq)$, and $H_2(g)$ at $25^{\circ}C$ are used to calculate ΔF^O_{298} and ΔS^O_{298} for Eq. (2.14). Since the free energy of formation of $Al(c)$, $H^+(aq)$, and $H_2(g)$ are zero by definition, ΔF^O_{298} for the reaction is simply the free energy of formation of $Al^{3+}(aq)$ at $25^{\circ}C$.

4. The entropy of $Al^{3+}(aq)$ at temperature T is estimated by Eq. (2.10) using the a and b parameters given in Table II.1; the average heat capacity is then evaluated from Eq. (2.9).

5. The average heat capacities for $Al(c)$ and $H_2(g)$ between T and $298.15^{\circ}K$ are determined from existing C_p data. The average heat capacity for $H^+(aq)$ is specified in Table II.1.

6. The average heat capacity change, $\Delta C^O_p\big]^T_{T_r}$, for Eq. (2.14) is obtained by summing the average heat capacities of products minus reactants.

7. The free energy of formation for the Al^{3+} ion at temperature T is calculated from Eq. (2.7),

$$\Delta F^O_T = \Delta F^O_{298} - \Delta S^O_{298}(T-298.15) + \Delta \overline{C}^O_p\big]^T_{T_r} [T-298.15 - T \ln T/298.15]$$

Since the free energy of formation of $Al(c)$, $H^+(aq)$ and $H_2(g)$ are, by definition, zero at all temperatures, this equation gives directly the free energy of formation of the Al^{3+} ion.

In making the transition from Eq. (2.6) to Eq. (2.7), it was assumed that the linear ΔT and $\Delta \log T$ averages of $\overline{C}^O_p$ are identical. This is an acceptable assumption over small temperature intervals. However, to minimize errors, Criss and Cobble (ref. 3, p. 5393) suggest that the calculations above $200^{\circ}C$ be carried out in $50^{\circ}C$ intervals. Therefore, above $200^{\circ}C$ we use a reference temperature of $200^{\circ}C$, and likewise, above $250^{\circ}C$ we use a reference temperature of $250^{\circ}C$. In addition, whenever a phase transition in the forming element (for example, $58^{\circ}C$ for bromine) is exceeded, the transition temperature is used as the reference temperature. The calculation sequence is similar to that just described, with the exception that the revised reference temperature replaces the $25^{\circ}C$ temperature.

Heat of Formation and Enthalpy of Ions

The heat of formation of an ion at temperature T is calculated from the relationship

$$\Delta H_T^O = \Delta F_T^O + T \Delta S_T^O \tag{2.15}$$

Using the example of $Al^{3+}(aq)$ cited previously, the ΔF_T^O and ΔS_T^O values relate to the free energy and entropy changes for the full-cell reaction, Eq. (2.14).

The enthalpy of an ion is computed from the expression[*]

$$H_T^O = H_{298}^O + \Delta \overline{C}_p^O(ion) \Big]_{298}^T [T - 298.15] \tag{2.16}$$

where

ΔH_{298}^O is the enthalpy of formation of the ion at $298.15\,^O K$, and $\overline{C}_p^O(ion) \Big]_{298}^T$ is the mean heat capacity of the ion between temperatures $298.15\,^O K$ and T.

UNDISSOCIATED NEUTRAL COMPLEXES

This section describes the methods used to generate the properties of undissociated complexes in aqueous solution.

In an attempt to preserve internal consistency as much as possible, the properties of complexes at $25\,^O C$, when available, were taken from the National Bureau of Standards.[7] These data were subsequently extended to higher temperatures using the correlations of Helgeson[4] and Criss and Cobble.[2,3]

The approach may be summarized by reference to the complex $AgCl(aq)$. The equilibrium dissociation reaction for this complex is

$$AgCl(aq) \rightleftharpoons Aq^+(aq) + Cl^-(aq) \tag{2.17}$$

Helgeson[4] has developed correlations that may be used to estimate the temperature variation of ΔF_T^O and ΔS_T^O for Eq. (2.17). Thus, assuming that data for all species are available at $25\,^O C$, ΔF_T^O, ΔS_T^O, and hence also ΔH_T^O for

---

[*] Above either $200\,^O C$ or a phase transition temperature for one of the forming elements, the enthalpy is calculated from

$$H_T^O = H_{T_r}^O + \overline{C}_p^O(ion) \Big]_{T_r}^T [T - T_r]$$

where T_r is the reference temperature and $\overline{C}_p^O \Big]_{T_r}^T$ the mean heat capacity between T and T_r.

Eq. (2.17) can be defined as a function of temperature. Furthermore, the properties of the dissociation product ions, $Ag^+(aq)$ and $Cl^-(aq)$ can be estimated at higher temperatures using the method of Criss and Cobble,[2] as previously described. The properties of the complex $AgCl(aq)$ can then be obtained by difference. The specific equations are described below.

The Gibbs free energy of dissociation for a complex at temperature T is given by Helgeson[4] as

$$\Delta F_T^O(dissociation) =$$
$$-\Delta S_{T_r}^O(dissociation)\left[T_r - \frac{\theta}{\omega}\left\{1 - \exp[\exp(b-aT) - c + (T-T_r)/\theta]\right\}\right] +$$
$$\Delta H_{T_r}^O(dissociation) \tag{2.18}$$

where

$a = 0.01875$

$b = -12.741$

$c = \exp(b + aT_r)$

$\theta = 219.0$

$\omega = 1 + ac\theta$.

Similarly, the entropy of dissociation for a complex at temperature T is given by:[4]

$$\Delta S_T^O(dissociation) =$$
$$\frac{\Delta S_{T_r}^O(dissociation)\left\{\exp[\exp(b+aT) - c + (T-T_r)/\theta]\right\}[1+\phi\exp(b+aT)]}{\omega} \tag{2.19}$$

where

$\phi = 4.106$.

The reference temperature T_r in these equations is taken as $298.15^O K$.

The free energy of formation and the entropy of the neutral complex are subsequently calculated by difference, for example,

$$\Delta F_T^O(complex) = \Sigma\Delta F_T^O(ion) - \Delta F_T^O(dissociation) \tag{2.20}$$

where the free energies of formation of the ions are calculated by the method of Criss and Cobble[2] as previously described.

The mean heat capacity of the complex is evaluated by the relation

$$\overline{C_p^O}\Big]_{T_r}^T = (S_T^O - S_{298}^O)/\ln(T/298.15) \tag{2.21}$$

The enthalpy of reaction for the dissociation reaction is calculated from

$$\Delta H_T^O(dissociation) = \Delta F_T^O(dissociation) + T\Delta S_T^O(dissociation) \tag{2.22}$$

The enthalpy of formation of the complex is then calculated by difference in a manner analogous to Eq. (2.20). Finally, the enthalpy of the complex is estimated from the relationship

$$H_T^O = \Delta H_{298}^O(complex) + \overline{C_p^O}(complex)\Big]_{298}^T [T-298.15] \tag{2.23}$$

REFERENCES

1. I. Barin and O. Knacke, *Thermochemical Properties of Inorganic Substances.* Berlin: Springer-Verlag, 1973.

2. C. M. Criss and J. W. Cobble, *J. Am. Chem. Soc.*, 86, 5385 (1964).

3. C. M. Criss and J. W. Cobble, *J. Am. Chem. Soc.*, 86, 5390 (1964).

4. H. C. Helgeson, *J. Phy. Chem.*, 71, 3121 (1967).

5. D. M. Himmelblau, *J. Phy. Chem.*, 63, 1803 (1959).

6. W. M. Latimer, *The Oxidation States of the Elements and their Potentials in Aqueous Solutions.* Englewood Cliffs, N.J.: Prentice Hall, 1952.

7. *Selected Values of Chemical Thermodynamic Properties.* National Bureau of Standards Technical Notes 270-3 (1968), 270-4, (1969), 270-5 (1971), 270-6 (1971), 270-7 (1973).

8. L. G. Sillen and A. E. Martell, *Stability Constants of Metal-Ion Complexes.* London: The Chemical Society, 1964.

TABULATED DATA

The properties tabulated as a function of temperature (in $^{\circ}$C) are defined thus:

DELTA H. This column presents values of ΔH_T°, the enthalpy of formation in kcal/g-mole. It represents the enthalpy change when one gram-formula weight of the substance in its standard state is formed at the indicated temperature from the elements, each in its standard reference state.

DELTA F. This column presents values of ΔF_T°, the Gibbs free energy of formation in kcal/g-mole. It represents the free energy change when one gram-formula weight of the substance in its standard state is formed at the indicated temperature from the elements, each in its standard reference state.

ENTHALPY. This column presents values of $\Delta H_{298}^{\circ} + (H_T^{\circ} - H_{298}^{\circ})$ in kcal/g-mole. It represents the sum of (1) the enthalpy of formation at 298°K(25°C), and (2) the enthalpy in the standard state at temperature T less the enthalpy in the standard state at 298°K.

ENTROPY. This column presents values of entropy (S°) in cal/($^{\circ}$K)(g-mole). For ions, the tabulated entropies are on an "absolute" scale[3] rather than on the scale conventionally used. The "absolute" scale was adopted because it forms the basis for estimating the temperature dependence of ionic entropies as discussed in Chapter 2.

REF. This column indicates the source of the data.

PRECAUTIONS

Although discussed elsewhere, we reiterate the following precautions in using the tabulated data.

1. <u>Standard States for Elements</u>. Generally, the standard reference state
    for each element is chosen to be the state that is thermodynamically
    stable at the given temperature and 1 atm. pressure. The two excep-
    tions are sulfur and phosphorus (see Chapter 2).

2. <u>Standard States for Substances</u>. The standard state for a solid or pure
    liquid is the substance at the specified temperature and 1 atm. pres-
    sure. For a gas, the standard state is the hypothetical ideal gas at
    the given temperature and a fugacity of 1 atm. Solutes in water are
    designated as aqueous (aq). The standard state for a solute is taken
    as the hypothetical ideal solution at a molality of 1.0.

3. <u>Half-Cell Reactions</u>. The data tabulated for individual ions in aqueous
    solution are based on the convention that ΔH_T^O and ΔF_T^O for the hydrogen
    ion (aq; standard state, molality = 1.0) are zero at all temperatures.
    As a result, the values of ΔH_T^O and ΔF_T^O calculated for any half-cell
    reaction are relative to the hydrogen ion formation reaction at the
    given temperature. In addition, the thermodynamic relation $\Delta F_T^O =$
    $\Delta H_T^O - T\Delta S_T^O$ will not hold for individual half-cell reactions. These
    consequences are of no concern when complete reactions are considered,
    that is, where no electrons appear in the balanced chemical equations.

SOURCES OF DATA

The sources of data indicated in the table refer to the numbered data
references:

1. *Selected Values of Chemical Thermodynamic Properties*. National Bureau of
    Standards, Technical Notes 270-3 (1968), 270-4 (1969), 270-5 (1971),
    270-6 (1971), 270-7 (1973).

2. I. Barin and O. Knacke, *Thermochemical Properties of Inorganic Substances*.
    Berlin: Springer-Verlag, 1973.

3. C. M. Criss and J. W. Cobble, *J. Am. Chem. Soc.*, 86, 5385 (1964).

4. E. G. King, A. D. Mah, and L. B. Pankratz, *Thermodynamic Properties of
    Copper and Its Inorganic Compounds*. The International Copper Research
    Association and the U. S. Bureau of Mines, INCRA Monograph II, 1973.

5. W. M. Latimer, *The Oxidation States of the Elements and their Potentials in
    Aqueous Solutions*. Englewood Cliffs, N.J.: Prentice-Hall, 1952.

6. H. C. Helgeson, *J. Phys. Chem.*, 71, 3121 (1967).

7. Y. S. Touloukian and T. Makita, *Thermophysical Properties of Matter*, Vol. 6. New York: IFI/Plenum, 1970.

8. L. B. Pankratz and E. G. King, *High Temperature Enthalpies and Entropies of Chalcopyrite and Bornite*, U. S. Dept. of the Interior, Bureau of Mines ROI 7435 (1970).

9. R. Barany, L. B. Pankratz, and W. W. Weller, *Thermodynamic Properties of Cuprous and Cupric Ferrites*, U. S. Dept. of the Interior, Bureau of Mines ROI 6513 (1974).

10. H. C. Helgeson, *Am. J. Sci.*, 267, 729 (1969).

11. R. A. Robie and D. R. Waldbaum, *Thermodynamic Properties of Minerals and Related Substances at 298.15°K and One Atmosphere Pressure and at Higher Temperatures*, U. S. Dept. of the Interior, Geological Survey Bulletin 1259 (1968).

12. K. K. Kelley, *Contributions to the Data on Theoretical Metallurgy. XIII. High-Temperature Heat-Content, Heat-Capacity, and Entropy Data for the Elements and Inorganic Compounds*, U. S. Dept. of the Interior, Bureau of Mines Bulletin 584 (1960).

13. Internal sources, Kennecott Copper Corporation.

14. D. M. Himmelblau, *J. Phys. Chem.*, 63, 1803 (1959).

15. L. G. Sillen and A. E. Martell, *Stability Constants of Metal-Ion Complexes*. London: The Chemical Society, 1964.

Table III.3. Thermochemical Data for Compounds
and Aqueous Species

AG (C), SILVER

T(C)	DELTA H	DELTA F	ENTHALPY	ENTROPY	REF
25	0.0	0.0	0.0	10.2	1
50	0.0	0.0	0.2	10.7	2
75	0.0	0.0	0.3	11.1	2
100	0.0	0.0	0.5	11.5	2
150	0.0	0.0	0.8	12.3	2
200	0.0	0.0	1.1	13.0	2
250	0.0	0.0	1.4	13.6	2
300	0.0	0.0	1.7	14.2	2

AG(CN)2- (AQ)
(EXTRAPOLATED AS CRISS & COBBLE TYPE 3)

T(C)	DELTA H	DELTA F	ENTHALPY	ENTROPY	REF
25	64.6	73.0	64.6	51.0	1
50	63.9	73.7	63.9	48.9	3
75	63.5	74.5	63.2	46.7	3
100	63.1	75.3	62.3	44.3	3
150	61.9	77.0	60.5	39.6	3
200	61.2	78.8	58.9	36.0	3
250	60.3	80.7	56.8	31.8	3
300	59.3	82.7	54.4	27.5	3

AG(NH3)2+ (AQ)

T(C)	DELTA H	DELTA F	ENTHALPY	ENTROPY	REF
25	-26.6	-4.1	-26.6	53.6	1
50	-27.6	-2.2	-26.3	54.7	3
75	-28.7	-0.2	-25.8	55.9	3
100	-30.0	1.9	-25.4	57.3	3
150	-32.7	6.4	-24.8	58.7	3
200	-35.1	11.2	-23.6	61.4	3
250	-38.0	16.2	-22.5	63.7	3
300	-40.8	21.5	-21.2	66.0	3

AG(S2O3)2--- (AQ)
(EXTRAPOLATED AS CRISS & COBBLE TYPE 3)

T(C)	DELTA H	DELTA F	ENTHALPY	ENTROPY	REF
25	-346.9	-285.5	-346.9	122.3	5
50	-343.7	-280.5	-344.1	131.2	3
75	-339.7	-275.8	-341.0	140.3	3
100	-334.8	-271.5	-337.8	149.5	3
150	-327.9	-263.3	-333.4	159.9	3
200	-316.2	-257.1	-325.0	180.0	3
250	-303.0	-251.6	-316.4	197.2	3
300	-288.9	-247.3	-307.2	214.1	3

AG+ (AQ)

T(C)	DELTA H	DELTA F	ENTHALPY	ENTROPY	REF
25	25.2	18.4	25.2	12.4	1
50	25.3	17.9	26.0	14.8	3
75	25.5	17.3	27.0	17.8	3
100	25.6	16.7	28.2	21.1	3
150	25.7	15.5	30.1	26.0	3
200	26.1	14.2	32.7	32.1	3
250	26.3	13.0	35.5	37.7	3
300	26.8	11.6	38.6	43.4	3

AG2CO3 (C), SILVER CARBONATE

T(C)	DELTA H	DELTA F	ENTHALPY	ENTROPY	REF
25	-120.9	-104.4	-120.9	40.0	1
50	-120.9	-103.1	-120.2	42.2	2
75	-120.8	-101.7	-119.5	44.2	2
100	-120.7	-100.3	-118.8	46.2	2
150	-120.6	-97.6	-117.4	49.9	2
200	-120.4	-94.9	-115.8	53.3	2

AG2O (C), SILVER OXIDE

T(C)	DELTA H	DELTA F	ENTHALPY	ENTROPY	REF
25	-7.4	-2.7	-7.4	29.0	1
50	-7.4	-2.3	-7.0	30.3	2
75	-7.4	-1.9	-6.6	31.5	2
100	-7.4	-1.5	-6.2	32.7	2
150	-7.3	-0.7	-5.3	34.9	2
200	-7.2	0.0	-4.4	36.9	2

AG2S (C), SILVER SULFIDE
PHASE TRANSITION OF AG2S AT 176C

T(C)	DELTA H	DELTA F	ENTHALPY	ENTROPY	REF
25	-23.1	-19.2	-23.1	34.4	1
50	-23.1	-18.9	-22.7	35.9	2
75	-23.0	-18.5	-22.2	37.3	2
100	-22.9	-18.2	-21.7	38.7	2
150	-22.7	-17.6	-20.7	41.3	2
176	-22.6	-17.3	-20.1	42.6	2
176	-21.2	-17.3	-18.7	45.7	2
200	-21.0	-17.1	-18.2	46.8	2
250	-20.8	-16.7	-17.1	49.0	2
300	-20.6	-16.3	-16.0	50.9	2

AG2SO4 (C), SILVER SULFATE

T(C)	DELTA H	DELTA F	ENTHALPY	ENTROPY	REF
25	-186.4	-157.3	-186.4	47.9	1
50	-186.4	-154.9	-185.6	50.5	2
75	-186.4	-152.4	-184.8	52.9	2
100	-186.3	-150.0	-184.0	55.2	2
150	-186.1	-145.1	-182.3	59.5	2
200	-185.9	-140.3	-180.5	63.4	2
250	-185.6	-135.5	-178.7	67.2	2
300	-185.3	-130.7	-176.7	70.7	2

AGBR (C), SILVER BROMIDE

T(C)	DELTA H	DELTA F	ENTHALPY	ENTROPY	REF
25	-24.0	-23.2	-24.0	25.6	1
50	-24.0	-23.1	-23.7	26.6	2
58	-24.0	-23.1	-23.6	27.0	2
58	-27.6	-23.1	-23.6	27.0	2
75	-27.6	-22.9	-23.3	27.6	2
100	-27.5	-22.5	-23.0	28.5	2
150	-27.3	-21.9	-22.3	30.3	2
200	-27.1	-21.2	-21.6	32.0	2
250	-26.8	-20.6	-20.8	33.5	2
300	-26.6	-20.0	-20.0	35.0	2

AGCL (C), SILVER CHLORIDE

T(C)	DELTA H	DELTA F	ENTHALPY	ENTROPY	REF
25	-30.4	-26.3	-30.4	23.0	1
50	-30.3	-25.9	-30.1	24.0	2
75	-30.3	-25.6	-29.7	25.0	2
100	-30.2	-25.2	-29.4	25.9	2
150	-30.0	-24.6	-28.7	27.6	2
200	-29.8	-24.0	-28.0	29.1	2
250	-29.7	-23.3	-27.3	30.6	2
300	-29.5	-22.8	-26.6	31.9	2

AGCL (AQ), UNDISSOCIATED COMPLEX

T(C)	DELTA H	DELTA F	ENTHALPY	ENTROPY	REF
25	-17.4	-17.4	-17.4	36.8	1
50	-18.0	-17.4	-17.7	35.9	6
75	-18.4	-17.3	-17.9	35.3	6
100	-18.7	-17.2	-18.0	35.1	6
150	-19.7	-16.9	-18.4	33.9	6
200	-19.8	-16.5	-18.2	34.6	6

AGF (C), SILVER FLUORIDE

T(C)	DELTA H	DELTA F	ENTHALPY	ENTROPY	REF
25	-48.5	-44.2	-48.5	20.0	2
50	-48.4	-43.8	-48.2	21.0	2
75	-48.4	-43.5	-47.9	21.9	2
100	-48.3	-43.2	-47.6	22.8	2
150	-48.2	-42.5	-46.9	24.5	2
200	-48.0	-41.8	-46.3	25.9	2
250	-47.9	-41.2	-45.6	27.3	2
300	-47.7	-40.5	-44.9	28.5	2

AGF (AQ), UNDISSOCIATED COMPLEX

T(C)	DELTA H	DELTA F	ENTHALPY	ENTROPY	REF
25	-57.1	-48.7	-57.1	6.2	1
50	-58.1	-48.0	-57.8	3.9	6
75	-59.2	-47.1	-58.7	1.4	6
100	-60.4	-46.2	-59.6	-1.4	6
150	-63.5	-44.1	-62.2	-9.1	6
200	-67.5	-41.5	-65.4	-15.8	6

AGI (C), SILVER IODIDE
PHASE TRANSITION OF AGI AT 147C

T(C)	DELTA H	DELTA F	ENTHALPY	ENTROPY	REF
25	-14.8	-15.8	-14.8	27.6	1
50	-14.8	-15.9	-14.5	28.7	2
75	-14.7	-16.0	-14.1	29.7	2
100	-14.7	-16.1	-13.7	30.7	2
114	-14.7	-16.2	-13.5	31.3	2
114	-16.5	-16.2	-13.5	31.3	2
147	-16.5	-16.1	-13.0	32.5	2
147	-15.1	-16.1	-11.5	36.0	2
150	-15.1	-16.1	-11.5	36.1	2
185	-15.2	-16.2	-11.0	37.2	2
185	-20.2	-16.2	-11.0	37.2	2
200	-20.1	-16.1	-10.8	37.6	2
250	-20.0	-15.7	-10.1	39.0	2
300	-19.9	-15.3	-9.5	40.2	2

AGNO3 (C,L), SILVER NITRATE SOL PHASE TRANSI-
TION AT 160C,
SOL-LIQ TRANS AT 210C

T(C)	DELTA H	DELTA F	ENTHALPY	ENTROPY	REF
25	-29.7	-8.0	-29.7	33.7	1
50	-29.7	-6.2	-29.2	35.5	2
75	-29.6	-4.4	-28.6	37.3	2
100	-29.5	-2.6	-27.9	39.0	2
150	-29.2	1.0	-26.6	42.4	2
160	-29.1	1.7	-26.3	43.1	2
160	-28.5	1.7	-25.7	44.4	2
200	-28.1	4.5	-24.5	47.0	2
210	-28.0	5.2	-24.2	47.7	2
210	-25.3	5.2	-21.4	53.4	2
250	-24.9	7.7	-20.2	55.9	2
300	-24.4	10.8	-18.7	58.7	2

AGSO4- (AQ)
(EXTRAPOLATED AS CRISS & COBBLE TYPE 4)

T(C)	DELTA H	DELTA F	ENTHALPY	ENTROPY	REF
25	-206.3	-170.8	-206.3	37.0	1
50	-206.3	-167.8	-206.1	37.4	3
75	-205.9	-164.9	-205.8	38.4	3
100	-205.1	-162.0	-205.3	39.8	3
150	-205.3	-156.1	-205.9	38.1	3
200	-204.1	-150.5	-205.3	39.5	3
250	-202.5	-144.9	-204.8	40.6	3
300	-200.9	-139.5	-204.2	41.7	3

AL (C), ALUMINUM

T(C)	DELTA H	DELTA F	ENTHALPY	ENTROPY	REF
25	0.0	0.0	0.0	6.8	1
50	0.0	0.0	0.2	7.2	2
75	0.0	0.0	0.3	7.7	2
100	0.0	0.0	0.4	8.1	2
150	0.0	0.0	0.8	8.9	2
200	0.0	0.0	1.1	9.6	2
250	0.0	0.0	1.4	10.2	2
300	0.0	0.0	1.7	10.8	2

AL(OH)3 (C), ALUMINUM HYDROXIDE

T(C)	DELTA H	DELTA F	ENTHALPY	ENTROPY	REF
25	-307.0	-274.2	-307.0	17.0	2
50	-307.1	-271.4	-306.4	18.8	2
75	-307.2	-268.7	-305.8	20.6	2
100	-307.2	-265.9	-305.2	22.4	2
150	-307.2	-260.3	-303.8	25.8	2
200	-307.1	-254.8	-302.3	29.2	2
250	-306.9	-249.3	-300.7	32.4	2
300	-306.6	-243.8	-299.0	35.6	2

AL(OH)4- (AQ)
(EXTRAPOLATED AS CRISS & COBBLE TYPE 2)

T(C)	DELTA H	DELTA F	ENTHALPY	ENTROPY	REF
25	-356.3	-310.2	-356.3	33.0	1
50	-358.1	-306.3	-357.7	28.6	3
75	-359.7	-302.2	-359.2	24.3	3
100	-361.2	-298.0	-360.7	20.0	3
150	-364.7	-289.3	-364.1	11.3	3
200	-368.5	-280.1	-368.0	2.2	3
250	-372.3	-270.5	-372.3	-6.4	3
300	-377.6	-260.6	-378.2	-17.1	3

AL+++ (AQ)

T(C)	DELTA H	DELTA F	ENTHALPY	ENTROPY	REF
25	-127.0	-116.0	-127.0	-91.9	1
50	-126.8	-115.1	-125.2	-86.2	3
75	-126.6	-114.2	-122.8	-78.7	3
100	-126.4	-113.3	-119.7	-70.2	3
150	-126.1	-111.6	-114.4	-56.6	3
200	-125.6	-109.9	-108.1	-42.0	3
250	-125.9	-108.2	-101.1	-28.0	3
300	-125.5	-106.5	-93.3	-13.8	3

AL2(SO4)3 (C), ALUMINUM SULFATE

T(C)	DELTA H	DELTA F	ENTHALPY	ENTROPY	REF
25	-868.4	-769.4	-868.4	57.2	1
50	-868.4	-761.1	-866.8	62.4	2
75	-868.4	-752.8	-865.1	67.5	2
100	-868.3	-744.5	-863.3	72.5	2
150	-867.8	-727.9	-859.4	82.2	2
200	-867.3	-711.4	-855.4	91.2	2
250	-866.6	-695.0	-851.2	99.7	2
300	-865.8	-678.6	-846.8	107.6	2

AL203 (C), ALUMINUM OXIDE

T(C)	DELTA H	DELTA F	ENTHALPY	ENTROPY	REF
25	-400.5	-378.2	-400.5	12.2	1
50	-400.6	-376.3	-400.0	13.8	2
75	-400.6	-374.4	-399.5	15.3	2
100	-400.6	-372.5	-398.9	16.8	2
150	-400.6	-368.8	-397.8	19.8	2
200	-400.5	-365.0	-396.5	22.5	2
250	-400.5	-361.3	-395.3	25.1	2
300	-400.4	-357.5	-394.0	27.5	2

AL203.H2O (C), ALUMINUM OXIDE MONOHYDRATE
(BOEHMITE)

T(C)	DELTA H	DELTA F	ENTHALPY	ENTROPY	REF
25	-472.0	-436.3	-472.0	23.2	1
50	-472.1	-433.3	-471.2	25.6	2
75	-472.1	-430.3	-470.5	27.9	2
100	-472.2	-427.3	-469.7	30.0	2
150	-472.3	-421.3	-468.1	33.9	2
200	-472.4	-415.3	-466.6	37.5	2
250	-472.6	-409.3	-465.0	40.7	2
300	-472.7	-403.2	-463.3	43.7	2

AL2O3.3H2O (C), ALUMINUM OXIDE TRIHYDRATE (GIBBSITE)

T(C)	DELTA H	DELTA F	ENTHALPY	ENTROPY	REF
25	-612.5	-546.7	-612.5	33.5	1
50	-612.7	-541.2	-611.4	37.2	2
75	-612.8	-535.7	-610.2	40.8	2
100	-612.9	-530.1	-608.9	44.3	2
150	-612.9	-519.0	-606.2	51.2	2
200	-612.7	-507.9	-603.2	57.8	2
250	-612.3	-496.8	-599.9	64.3	2
300	-611.7	-485.8	-596.5	70.6	2

AL2SI2O5(OH)4 (C), KAOLINITE

T(C)	DELTA H	DELTA F	ENTHALPY	ENTROPY	REF
25	-979.6	-903.0	-979.6	48.5	1
50	-979.8	-896.6	-978.1	53.3	10
75	-979.9	-890.1	-976.6	57.9	10
100	-980.0	-883.7	-974.9	62.4	10
150	-980.2	-870.7	-971.6	70.8	10
200	-980.2	-857.8	-968.1	78.6	10
250	-980.2	-844.9	-964.5	85.8	10
300	-980.1	-831.9	-960.8	92.5	10

AL2SIO5 (C), KYANITE

T(C)	DELTA H	DELTA F	ENTHALPY	ENTROPY	REF
25	-656.4	-620.5	-656.4	20.0	1
50	-656.5	-617.4	-655.7	22.4	12
75	-656.6	-614.4	-654.9	24.8	12
100	-656.6	-611.4	-654.0	27.1	12
150	-656.6	-605.3	-652.2	31.7	12
200	-656.5	-599.3	-650.3	35.9	12
250	-656.4	-593.2	-648.3	40.0	12
300	-656.2	-587.2	-646.2	43.7	12

AL2SIO5 (C), ANDALUSITE

T(C)	DELTA H	DELTA F	ENTHALPY	ENTROPY	REF
25	-655.9	-620.6	-655.9	22.3	1
50	-655.9	-617.7	-655.1	25.0	12
75	-655.9	-614.7	-654.2	27.6	12
100	-655.9	-611.8	-653.3	30.1	12
150	-655.8	-605.9	-651.4	34.9	12
200	-655.6	-600.0	-649.4	39.4	12
250	-655.4	-594.1	-647.3	43.5	12
300	-655.2	-588.3	-645.2	47.4	12

AL2SIO5 (C), SILLIMANITE

T(C)	DELTA H	DELTA F	ENTHALPY	ENTROPY	REF
25	-662.6	-627.6	-662.6	23.0	1
50	-662.7	-624.6	-661.8	25.5	12
75	-662.7	-621.7	-661.0	28.0	12
100	-662.7	-618.7	-660.1	30.4	12
150	-662.7	-612.8	-658.3	34.9	12
200	-662.7	-606.9	-656.4	39.1	12
250	-662.6	-601.0	-654.5	43.0	12
300	-662.5	-595.2	-652.5	46.7	12

ALBR3 (C,L), ALUMINUM BROMIDE

MELT PT = 97C
BOIL PT = 255C

T(C)	DELTA H	DELTA F	ENTHALPY	ENTROPY	REF
25	-126.0	-120.6	-126.0	43.1	2
50	-126.2	-120.1	-125.4	45.1	2
58	-126.2	-119.9	-125.2	45.7	2
58	-136.9	-119.9	-125.2	45.7	2
75	-136.8	-119.1	-124.8	46.9	2
97	-136.6	-118.0	-124.2	48.5	2
97	-134.0	-118.0	-121.5	55.8	2
100	-133.9	-117.9	-121.4	56.0	2
150	-133.4	-115.8	-119.9	59.8	2
200	-132.9	-113.7	-118.4	63.1	2
250	-132.4	-111.7	-116.9	66.1	2
255	-132.3	-111.5	-116.8	66.4	2

ALCL3 (C,L), ALUMINUM CHLORIDE

MELT PT OF ALCL3 = 193C

T(C)	DELTA H	DELTA F	ENTHALPY	ENTROPY	REF
25	-168.3	-150.3	-168.3	26.5	1
50	-168.2	-148.8	-167.8	28.2	2
75	-168.1	-147.3	-167.2	29.9	2
100	-168.0	-145.9	-166.6	31.4	2
150	-167.8	-142.9	-165.5	34.3	2
193	-167.6	-140.4	-164.5	36.6	2
193	-159.2	-140.4	-156.0	54.7	2
200	-159.1	-140.1	-155.8	55.1	2
250	-158.7	-138.1	-154.5	579.0	0
300	-158.4	-136.2	-153.2	60.2	2

ALF3 (C), ALUMINUM FLUORIDE

T(C)	DELTA H	DELTA F	ENTHALPY	ENTROPY	REF
25	-359.5	-340.5	-359.5	15.9	1
50	-359.5	-339.0	-359.0	17.3	2
75	-359.4	-337.4	-358.6	18.8	2
100	-359.4	-335.8	-358.1	20.1	2
150	-359.3	-332.6	-357.1	22.7	2
200	-359.2	-329.5	-356.0	25.0	2
250	-359.0	-326.4	-354.9	27.2	2
300	-358.8	-323.3	-353.8	29.2	2

ALF6--- (AQ)

(EXTRAPOLATED AS CRISS & COBBLE TYPE 3)

T(C)	DELTA H	DELTA F	ENTHALPY	ENTROPY	REF
25	-610.8	-539.6	-610.8	-25.0	5
50	-614.4	-533.5	-615.1	-38.9	3
75	-617.8	-527.1	-619.9	-53.2	3
100	-621.1	-520.5	-625.2	-67.9	3
150	-626.2	-506.9	-633.5	-88.6	3
200	-634.6	-491.9	-645.9	-117.5	3
250	-642.6	-476.4	-659.3	-144.5	3
300	-651.9	-460.1	-674.1	-171.5	3

ALI3 (C,L), ALUMINUM IODIDE

MELT PT OF ALI3 = 191C

T(C)	DELTA H	DELTA F	ENTHALPY	ENTROPY	REF
25	-75.0	-71.9	-75.0	38.0	1
50	-75.0	-71.6	-74.4	39.9	2
75	-75.1	-71.4	-73.8	41.8	2
100	-75.1	-71.1	-73.2	43.5	2
114	-75.2	-70.9	-72.8	44.4	2
114	-80.8	-70.9	-72.8	44.4	2
150	-81.1	-70.0	-71.9	46.7	2
185	-81.4	-69.1	-70.9	48.9	2
185	-96.4	-69.1	-70.9	48.9	2
191	-96.4	-68.7	-70.8	49.2	2
191	-92.6	-68.7	-67.0	57.4	2
200	-92.5	-68.3	-66.7	58.0	2
250	-92.0	-65.7	-65.3	60.9	2
300	-91.6	-63.3	-63.8	63.5	2

ALO2- (AQ)

T(C)	DELTA H	DELTA F	ENTHALPY	ENTROPY	REF
25	-219.6	-196.8	-219.6	0.0	1
50	-222.5	-194.8	-222.7	-10.0	3
75	-225.6	-192.5	-226.1	-20.4	3
100	-228.8	-190.0	-229.9	-31.0	3
150	-234.1	-184.6	-236.1	-46.4	3
200	-241.8	-177.9	-245.0	-67.0	3
250	-249.9	-170.7	-254.7	-86.5	3
300	-258.9	-162.8	-265.3	-106.0	3

AS (C), ARSENIC

T(C)	DELTA H	DELTA F	ENTHALPY	ENTROPY	REF
25	0.0	0.0	0.0	8.4	1
50	0.0	0.0	0.2	8.9	2
75	0.0	0.0	0.3	9.3	2
100	0.0	0.0	0.4	9.7	2
150	0.0	0.0	0.8	10.5	2
200	0.0	0.0	1.1	11.2	2
250	0.0	0.0	1.4	11.8	2
300	0.0	0.0	1.7	12.4	2

AS2O3 (C), ARSENIC TRIOXIDE (ARSENOLITE)
MELT PT OF AS2O3 = 278C

T(C)	DELTA H	DELTA F	ENTHALPY	ENTROPY	REF
25	-157.0	-137.8	-157.0	25.9	2
50	-157.0	-136.2	-156.4	27.8	2
75	-156.9	-134.6	-155.8	29.6	2
100	-156.9	-133.0	-155.1	31.4	2
150	-156.6	-129.8	-153.8	34.9	2
200	-156.3	-126.6	-152.3	38.3	2
250	-155.9	-123.5	-150.6	41.5	2
278	-155.6	-121.8	-149.7	43.3	2

AS2O3 (C), ARSENIC TRIOXIDE (CLAUDETITE)

T(C)	DELTA H	DELTA F	ENTHALPY	ENTROPY	REF
25	-156.2	-138.0	-156.2	29.3	2
50	-156.1	-136.5	-155.6	31.2	2
75	-156.1	-134.9	-155.0	33.1	2
100	-156.0	-133.4	-154.3	34.8	2
150	-155.8	-130.4	-152.9	38.3	2
200	-155.5	-127.4	-151.4	41.7	2
250	-155.0	-124.5	-149.8	45.0	2
300	-154.5	-121.6	-148.0	48.2	2

AS2O5 (C), ARSENIC PENTOXIDE

T(C)	DELTA H	DELTA F	ENTHALPY	ENTROPY	REF
25	-221.0	-187.0	-221.0	25.2	1
50	-221.1	-184.2	-220.4	27.4	2
75	-221.1	-181.3	-219.7	29.5	2
100	-221.2	-178.5	-219.0	31.5	2
150	-221.3	-172.7	-217.6	34.9	2
200	-221.5	-167.0	-216.2	38.1	2
250	-221.7	-161.2	-214.8	40.9	2
300	-221.9	-155.4	-213.4	43.4	2

ASCL3 (L,G), ARSENIC CHLORIDE
BOIL PT OF ASCL3 = 130C

T(C)	DELTA H	DELTA F	ENTHALPY	ENTROPY	REF
25	-72.9	-61.3	-72.9	49.6	1
50	-72.6	-60.4	-72.1	52.2	2
75	-72.2	-59.5	-71.3	54.6	2
100	-71.9	-58.6	-70.5	56.8	2
130	-71.5	-57.5	-69.6	59.2	2
130	-64.0	-57.5	-62.1	77.8	2
150	-63.9	-57.2	-61.6	78.9	2
200	-63.8	-56.4	-60.6	81.2	2
250	-63.7	-55.6	-59.5	83.3	2
300	-63.7	-54.8	-58.5	85.3	2

ASO2- (AQ)

T(C)	DELTA H	DELTA F	ENTHALPY	ENTROPY	REF
25	-102.5	-83.7	-102.5	14.9	1
50	-104.7	-82.0	-104.9	7.2	3
75	-107.0	-80.1	-107.5	-0.8	3
100	-109.4	-78.1	-110.5	-9.0	3
150	-113.4	-73.8	-115.4	-21.3	3
200	-118.9	-68.5	-122.1	-36.9	3
250	-124.8	-62.9	-129.6	-51.9	3
300	-131.4	-56.7	-137.8	-67.0	3

ASO4--- (AQ)

T(C)	DELTA H	DELTA F	ENTHALPY	ENTROPY	REF
25	-212.3	-155.0	-212.3	-23.9	1
50	-215.6	-150.1	-216.5	-37.6	3
75	-218.7	-144.9	-221.3	-51.8	3
100	-221.7	-139.5	-226.4	-66.3	3
150	-226.2	-128.4	-234.7	-86.7	3
200	-234.0	-116.0	-246.9	-115.3	3
250	-241.4	-103.1	-260.2	-142.0	3
300	-250.0	-89.5	-274.8	-168.6	3

AU (C), GOLD

T(C)	DELTA H	DELTA F	ENTHALPY	ENTROPY	REF
25	0.0	0.0	0.0	11.3	1
50	0.0	0.0	0.2	11.8	2
75	0.0	0.0	0.3	12.3	2
100	0.0	0.0	0.5	12.7	2
150	0.0	0.0	0.8	13.5	2
200	0.0	0.0	1.1	14.2	2
250	0.0	0.0	1.4	14.8	2
300	0.0	0.0	1.7	15.4	2

AU(CN)2- (AQ)
(EXTRAPOLATED AS CRISS & COBBLE TYPE 2)

T(C)	DELTA H	DELTA F	ENTHALPY	ENTROPY	REF
25	58.0	68.3	58.0	46.0	1
50	56.6	69.2	56.6	41.3	3
75	55.4	70.2	55.1	37.1	3
100	54.4	71.3	53.7	33.0	3
150	51.7	73.8	50.2	24.2	3
200	48.5	76.7	46.3	14.9	3
250	45.5	79.8	41.9	6.3	3
300	40.9	83.3	36.1	-4.5	3

AUBR(C), GOLD BROMIDE
THERMAL DECOMP TEMP OF AUBR = 115C

T(C)	DELTA H	DELTA F	ENTHALPY	ENTROPY	REF
25	-4.4	-3.7	-4.4	27.0	2
50	-4.5	-3.6	-4.1	28.0	2
58	-4.5	-3.6	-4.0	28.3	2
58	-8.0	-3.6	-4.0	28.3	2
75	-8.0	-3.3	-3.8	28.9	2
100	-8.0	-3.0	-3.5	29.7	2
115	-7.9	-2.8	-3.3	30.2	2

AUBR2- (AQ)
(EXTRAPOLATED AS CRISS & COBBLE TYPE 2)

T(C)	DELTA H	DELTA F	ENTHALPY	ENTROPY	REF
25	-30.7	-27.5	-30.7	57.5	1
50	-32.3	-27.2	-32.2	52.6	3
58	-32.9	-27.0	-32.8	51.0	2
58	-40.0	-27.0	-32.8	51.0	2
75	-40.6	-26.3	-33.7	48.3	3
100	-41.4	-25.3	-35.1	44.5	3
150	-44.0	-22.9	-38.6	35.6	3
200	-47.0	-20.1	-42.6	26.2	3
250	-49.8	-17.2	-46.9	17.5	3
300	-54.1	-13.9	-52.8	6.7	3

AUBR4- (AQ)
(EXTRAPOLATED AS CRISS & COBBLE TYPE 2)

T(C)	DELTA H	DELTA F	ENTHALPY	ENTROPY	REF
25	-45.8	-40.0	-45.8	85.3	1
50	-49.0	-39.4	-47.5	79.8	3
58	-48.8	-39.2	-48.1	77.9	2
58	-63.0	-39.2	-48.1	77.9	2
75	-63.6	-38.0	-48.9	75.6	3
100	-64.4	-36.1	-50.1	72.3	3
150	-67.6	-32.0	-53.7	63.1	3
200	-71.1	-27.5	-57.8	53.5	3
250	-74.3	-22.7	-62.2	44.7	3
300	-79.2	-17.6	-68.2	33.7	3

AUCL (C), GOLD MONOCHLORIDE
THERMAL DECOMP TEMP OF AUCL = 170C

T(C)	DELTA H	DELTA F	ENTHALPY	ENTROPY	REF
25	-8.3	-3.6	-8.3	22.2	2
50	-8.3	-3.2	-8.0	23.1	2
75	-8.2	-2.8	-7.7	24.0	2
100	-8.2	-2.4	-7.4	24.8	2
150	-8.1	-1.7	-6.8	26.3	2
170	-8.1	-1.4	-6.6	26.9	2

AUCL3 (C), GOLD TRICHLORIDE
THERMAL DECOMP TEMP OF AUCL3 = 254C

T(C)	DELTA H	DELTA F	ENTHALPY	ENTROPY	REF
25	-27.5	-10.8	-27.5	35.4	2
50	-27.4	-9.5	-26.9	37.2	2
75	-27.3	-8.1	-26.4	38.9	2
100	-27.2	-6.7	-25.8	40.5	2
150	-26.9	-4.0	-24.6	43.5	2
200	-26.7	-1.3	-23.4	46.1	2
250	-26.5	1.4	-22.3	48.5	2
254	-26.5	1.6	-22.2	48.6	2

AUCL4- (AQ)
(EXTRAPOLATED AS CRISS & COBBLE TYPE 2)

T(C)	DELTA H	DELTA F	ENTHALPY	ENTROPY	REF
25	-77.0	-56.2	-77.0	68.8	1
50	-78.7	-54.4	-78.6	63.6	3
75	-80.0	-52.5	-80.0	59.4	3
100	-81.0	-50.5	-81.4	55.8	3
150	-84.0	-46.2	-84.9	46.7	3
200	-87.4	-41.4	-88.9	37.3	3
250	-90.6	-36.4	-93.3	28.6	3
300	-95.4	-31.0	-99.3	17.7	3

AUF3 (C), GOLD TRIFLUORIDE

T(C)	DELTA H	DELTA F	ENTHALPY	ENTROPY	REF
25	-83.3	-66.4	-83.3	27.3	2
50	-83.2	-65.0	-82.8	29.1	2
75	-83.1	-63.6	-82.2	30.7	2
100	-83.0	-62.2	-81.6	32.3	2
150	-82.7	-59.4	-80.5	35.1	2
200	-82.5	-56.7	-79.4	37.7	2
250	-82.3	-53.9	-78.2	40.0	2
300	-82.1	-51.3	-77.1	42.1	2

AUI (C), GOLD MONOIODIDE
THERMAL DECOMP TEMP OF AUI = 120C

T(C)	DELTA H	DELTA F	ENTHALPY	ENTROPY	REF
25	0.2	-0.8	0.2	28.5	2
50	0.2	-0.9	0.5	29.5	2
75	0.2	-0.9	0.8	30.4	2
100	0.2	-1.0	1.1	31.3	2
114	0.2	-1.1	1.3	31.8	2
114	-1.7	-1.1	1.3	31.8	2
120	-1.7	-1.1	1.4	31.9	2

B (C), BORON

T(C)	DELTA H	DELTA F	ENTHALPY	ENTROPY	REF
25	0.0	0.0	0.0	1.4	1
50	0.0	0.0	0.1	1.6	2
75	0.0	0.0	0.2	1.9	2
100	0.0	0.0	0.2	2.1	2
150	0.0	0.0	0.4	2.6	2
200	0.0	0.0	0.6	3.1	2
250	0.0	0.0	0.9	3.5	2
300	0.0	0.0	1.1	4.0	2

B(OH)4- (AQ)
(EXTRAPOLATED AS CRISS & COBBLE TYPE 2)

T(C)	DELTA H	DELTA F	ENTHALPY	ENTROPY	REF
25	-321.2	-275.7	-321.2	29.5	1
50	-322.9	-271.8	-322.6	25.2	3
75	-324.4	-267.8	-324.0	20.9	3
100	-325.8	-263.6	-325.6	16.5	3
150	-329.2	-255.0	-329.0	7.9	3
200	-333.0	-246.0	-332.9	-1.3	3
250	-336.6	-236.6	-337.2	-9.9	3
300	-341.8	-226.8	-343.0	-20.5	3

B2O3 (C), BORON OXIDE

T(C)	DELTA H	DELTA F	ENTHALPY	ENTROPY	REF
25	-304.2	-285.3	-304.2	12.9	1
50	-304.2	-283.7	-303.8	14.1	2
75	-304.2	-282.1	-303.4	15.4	2
100	-304.3	-280.5	-303.0	16.6	2
150	-304.3	-277.4	-302.0	18.9	2
200	-304.3	-274.2	-301.0	21.1	2
250	-304.2	-271.0	-300.0	23.2	2
300	-304.2	-267.8	-298.9	25.2	2

BF3OH- (AQ)
(EXTRAPOLATED AS CRISS & COBBLE TYPE 2)

T(C)	DELTA H	DELTA F	ENTHALPY	ENTROPY	REF
25	-364.9	-338.1	-364.9	45.0	1
50	-366.4	-335.8	-366.3	40.4	3
75	-367.6	-333.4	-367.8	36.1	3
100	-368.8	-330.9	-369.2	32.0	3
150	-371.7	-325.6	-372.7	23.2	3
200	-375.1	-319.8	-376.6	13.9	3
250	-378.3	-313.8	-380.9	5.3	3
300	-383.0	-307.5	-386.8	-5.5	3

BF4- (AQ)
(EXTRAPOLATED AS CRISS & COBBLE TYPE 2)

T(C)	DELTA H	DELTA F	ENTHALPY	ENTROPY	REF
25	-376.5	-355.4	-376.5	48.0	1
50	-378.0	-353.6	-378.0	43.3	3
75	-379.2	-351.7	-379.4	39.0	3
100	-380.2	-349.7	-380.9	35.0	3
150	-383.0	-345.3	-384.3	26.2	3
200	-386.2	-340.6	-388.3	16.9	3
250	-389.3	-335.6	-392.6	8.2	3
300	-393.9	-330.3	-398.5	-2.5	3

BH4- (AQ)
(EXTRAPOLATED AS CRISS & COBBLE TYPE 2)

T(C)	DELTA H	DELTA F	ENTHALPY	ENTROPY	REF
25	11.5	27.3	11.5	31.4	1
50	10.2	28.7	10.2	27.1	3
75	9.0	30.2	8.7	22.7	3
100	8.0	31.7	7.2	18.4	3
150	5.3	35.1	3.8	9.8	3
200	2.3	38.9	-0.2	0.6	3
250	-0.6	42.9	-4.4	-8.0	3
300	-5.1	47.3	-10.3	-18.7	3

BO2- (AQ)

T(C)	DELTA H	DELTA F	ENTHALPY	ENTROPY	REF
25	-184.6	-162.3	-184.6	-3.9	1
50	-187.7	-160.3	-187.9	-14.5	3
75	-190.9	-158.0	-191.6	-25.5	3
100	-194.3	-155.5	-195.6	-36.8	3
150	-199.8	-150.1	-202.1	-53.0	3
200	-207.9	-143.4	-211.5	-74.9	3
250	-216.5	-136.1	-221.8	-95.6	3
300	-226.1	-128.0	-233.1	-116.2	3

BA (C), BARIUM

T(C)	DELTA H	DELTA F	ENTHALPY	ENTROPY	REF
25	0.0	0.0	0.0	15.0	1
50	0.0	0.0	0.2	15.5	2
75	0.0	0.0	0.3	16.0	2
100	0.0	0.0	0.5	16.4	2
150	0.0	0.0	0.8	17.3	2
200	0.0	0.0	1.2	18.0	2
250	0.0	0.0	1.5	18.7	2
300	0.0	0.0	1.9	19.4	2

BA++ (AQ)

T(C)	DELTA H	DELTA F	ENTHALPY	ENTROPY	REF
25	-128.5	-134.0	-128.5	-7.7	1
50	-128.7	-134.5	-127.6	-4.7	3
75	-129.0	-134.9	-126.3	-0.8	3
100	-129.4	-135.3	-124.7	3.6	3
150	-130.3	-136.0	-122.1	10.1	3
200	-131.0	-136.6	-118.8	17.8	3
250	-132.4	-137.2	-115.2	25.0	3
300	-133.4	-137.6	-111.2	32.4	3

BABR2 (C), BARIUM BROMIDE

T(C)	DELTA H	DELTA F	ENTHALPY	ENTROPY	REF
25	-181.0	-176.1	-181.0	35.0	1
50	-181.1	-175.7	-180.5	36.4	2
58	-181.2	-175.6	-180.4	36.9	2
58	-188.3	-175.6	-180.4	36.9	2
75	-188.2	-174.9	-180.1	37.8	2
100	-188.2	-174.0	-179.6	39.1	2
150	-188.0	-172.1	-178.7	41.4	2
200	-187.9	-170.2	-177.8	43.5	2
250	-187.7	-168.4	-176.8	45.4	2
300	-187.5	-166.5	-175.9	47.1	2

BACL2 (C), BARIUM CHLORIDE

T(C)	DELTA H	DELTA F	ENTHALPY	ENTROPY	REF
25	-205.2	-193.6	-205.2	29.6	1
50	-205.1	-192.7	-204.8	31.0	2
75	-205.0	-191.7	-204.3	32.4	2
100	-204.9	-190.8	-203.8	33.6	2
150	-204.8	-188.9	-202.9	35.9	2
200	-204.6	-187.0	-202.0	38.0	2
250	-204.5	-185.2	-201.1	39.9	2
300	-204.3	-183.3	-200.1	41.6	2

BACO3 (C), BARIUM CARBONATE

T(C)	DELTA H	DELTA F	ENTHALPY	ENTROPY	REF
25	-290.7	-271.9	-290.7	26.8	1
50	-290.6	-270.3	-290.2	28.5	2
75	-290.6	-268.9	-289.6	30.2	2
100	-290.5	-267.2	-289.0	31.8	2
150	-290.3	-264.1	-287.9	34.7	2
200	-290.2	-261.0	-286.6	37.5	2
250	-290.0	-257.9	-285.4	40.0	2
300	-289.8	-254.9	-284.0	42.4	2

BAF2 (C), BARIUM FLUORIDE

T(C)	DELTA H	DELTA F	ENTHALPY	ENTROPY	REF
25	-288.5	-276.5	-288.5	23.0	1
50	-288.4	-275.5	-288.0	24.5	2
75	-288.3	-274.5	-287.6	26.0	2
100	-288.2	-273.5	-287.1	27.3	2
150	-287.9	-271.5	-286.1	29.6	2
200	-287.7	-269.6	-285.2	31.8	2
250	-287.5	-267.7	-284.2	33.7	2
300	-287.3	-265.8	-283.3	35.5	2

BAI2 (C), BARIUM IODIDE

T(C)	DELTA H	DELTA F	ENTHALPY	ENTROPY	REF
25	-144.5	-143.7	-144.5	40.0	2
50	-144.5	-143.6	-144.0	41.6	2
75	-144.5	-143.5	-143.5	43.0	2
100	-144.5	-143.5	-143.0	44.4	2
114	-144.6	-143.4	-142.8	45.1	2
114	-148.3	-143.4	-142.8	45.1	2
150	-148.5	-143.0	-142.0	46.9	2
185	-148.7	-142.5	-141.3	48.5	2
185	-158.7	-142.5	-141.3	48.5	2
200	-158.7	-142.0	-141.0	49.2	2
250	-158.4	-140.2	-140.0	51.2	2
300	-158.2	-138.5	-139.9	53.2	2

BAO (C), BARIUM OXIDE

T(C)	DELTA H	DELTA F	ENTHALPY	ENTROPY	REF
25	-132.3	-125.5	-132.3	16.8	1
50	-132.3	-125.0	-132.0	17.7	2
75	-132.2	-124.4	-131.7	18.6	2
100	-132.2	-123.9	-131.5	19.4	2
150	-132.1	-122.7	-130.9	20.9	2
200	-132.0	-121.6	-130.3	22.2	2
250	-132.0	-120.5	-129.6	23.5	2
300	-131.9	-119.4	-129.0	24.6	2

BASO4 (C), BARIUM SULFATE

T(C)	DELTA H	DELTA F	ENTHALPY	ENTROPY	REF
25	-367.4	-335.0	-367.4	31.6	1
50	-367.4	-332.3	-366.8	33.6	2
75	-367.4	-329.6	-366.2	35.6	2
100	-367.3	-326.9	-365.5	37.5	2
150	-367.2	-321.5	-364.0	41.1	2
200	-367.0	-316.1	-362.6	44.3	2
250	-366.8	-310.8	-361.0	47.4	2
300	-366.5	-305.4	-359.5	50.2	2

BE (C), BERYLLIUM

T(C)	DELTA H	DELTA F	ENTHALPY	ENTROPY	REF
25	0.0	0.0	0.0	2.3	1
50	0.0	0.0	0.1	2.6	2
75	0.0	0.0	0.2	3.0	2
100	0.0	0.0	0.3	3.3	2
150	0.0	0.0	0.6	3.9	2
200	0.0	0.0	0.8	4.4	2
250	0.0	0.0	1.1	5.0	2
300	0.0	0.0	1.4	5.5	2

BE(OH)2 (C), BERYLLIUM HYDROXIDE (ALPHA)
DECOMP PT OF BE(OH)2 = 134C

T(C)	DELTA H	DELTA F	ENTHALPY	ENTROPY	REF
25	-215.7	-194.8	-215.7	12.4	1
50	-215.8	-193.0	-215.3	13.7	2
75	-215.8	-191.3	-214.9	15.0	2
100	-215.8	-189.5	-214.4	16.2	2
134	-215.8	-187.1	-213.8	17.8	2

BE(OH)2 (C), BERYLLIUM HYDROXIDE (BETA)
DECOMP PT OF BE(OH)2 = 144C

T(C)	DELTA H	DELTA F	ENTHALPY	ENTROPY	REF
25	-216.5	-195.5	-216.5	12.0	1
50	-216.5	-193.7	-216.1	13.3	2
75	-216.6	-192.0	-215.7	14.6	2
100	-216.6	-190.2	-215.2	15.8	2
144	-216.6	-187.1	-214.4	17.9	2

BE++ (AQ)

T(C)	DELTA H	DELTA F	ENTHALPY	ENTROPY	REF
25	-91.4	-90.8	-91.4	-41.0	1
50	-91.1	-90.7	-90.1	-36.9	3
75	-90.9	-90.7	-88.4	-31.6	3
100	-90.7	-90.7	-86.2	-25.6	3
150	-90.5	-90.7	-82.5	-16.3	3
200	-89.9	-90.8	-78.1	-5.9	3
250	-89.8	-90.9	-73.1	4.1	3
300	-89.3	-91.0	-67.6	14.1	3

BEBR2 (C), BERYLLIUM BROMIDE

T(C)	DELTA H	DELTA F	ENTHALPY	ENTROPY	REF
25	-88.4	-84.4	-88.4	25.4	2
50	-88.5	-84.1	-88.0	26.8	2
58	-88.6	-84.0	-87.8	27.2	2
58	-95.7	-84.0	-87.8	27.2	2
75	-95.6	-83.4	-87.6	28.0	2
100	-95.5	-82.6	-87.1	29.3	2
150	-95.3	-80.9	-86.2	31.5	2
200	-95.0	-79.2	-85.3	33.6	2
250	-94.8	-77.5	-84.3	35.5	2
300	-94.6	-75.9	-83.4	37.2	2

BECL2 (C), BERYLLIUM CHLORIDE

T(C)	DELTA H	DELTA F	ENTHALPY	ENTROPY	REF
25	-118.5	-107.3	-119.5	18.1	1
50	-118.4	-105.4	-118.1	19.3	2
75	-118.4	-105.5	-117.7	20.5	2
100	-118.3	-104.6	-117.3	21.6	2
150	-118.1	-102.7	-116.5	23.7	2
200	-118.0	-100.9	-115.7	25.6	2
250	-117.8	-99.1	-114.8	27.3	2
300	-117.6	-97.4	-113.9	28.9	2

BEF2 (C), BERYLLIUM FLUORIDE
PHASE TRANS OF BEF2 AT 228C

T(C)	DELTA H	DELTA F	ENTHALPY	ENTROPY	REF
25	-245.4	-234.1	-245.4	12.8	1
50	-245.4	-233.1	-245.1	13.8	2
75	-245.4	-232.2	-244.7	14.8	2
100	-245.3	-231.2	-244.4	15.8	2
150	-245.2	-229.4	-243.6	17.7	2
200	-245.1	-227.5	-242.9	19.4	2
228	-245.0	-226.5	-242.4	20.3	2
228	-244.9	-226.5	-242.3	20.5	2
250	-244.9	-225.7	-242.0	21.2	2
300	-244.8	-223.8	-241.2	22.6	2

BEI2 (C), BERYLLIUM IODIDE

T(C)	DELTA H	DELTA F	ENTHALPY	ENTROPY	REF
25	-50.6	-50.2	-50.6	28.8	2
50	-50.6	-50.2	-50.2	30.2	2
75	-50.6	-50.2	-49.7	31.5	2
100	-50.6	-50.1	-49.3	32.8	2
114	-50.7	-50.1	-49.0	33.4	2
114	-54.4	-50.1	-49.0	33.4	2
150	-54.6	-49.7	-48.3	35.1	2
185	-54.8	-49.3	-47.7	36.6	2
185	-64.8	-49.3	-47.7	36.6	2
200	-64.7	-48.8	-47.4	37.2	2
250	-64.5	-47.1	-46.4	39.1	2
300	-64.2	-45.5	-45.5	40.9	2

BEO (C), BERYLLIUM OXIDE

T(C)	DELTA H	DELTA F	ENTHALPY	ENTROPY	REF
25	-145.7	-138.7	-145.7	3.4	1
50	-145.7	-138.1	-145.5	3.9	2
75	-145.8	-137.5	-145.4	4.4	2
100	-145.8	-137.0	-145.2	4.9	2
150	-145.8	-135.8	-144.8	5.9	2
200	-145.8	-134.6	-144.4	6.9	2
250	-145.8	-133.4	-143.9	7.8	2
300	-145.8	-132.2	-143.4	8.7	2

BEO2-- (AQ)

T(C)	DELTA H	DELTA F	ENTHALPY	ENTROPY	REF
25	-188.9	-153.0	-188.9	-28.0	1
50	-192.7	-149.8	-193.4	-42.3	3
75	-196.5	-146.4	-198.3	-57.2	3
100	-200.5	-142.6	-203.7	-72.3	3
150	-206.5	-134.7	-212.4	-93.6	3
200	-216.2	-125.1	-225.1	-123.6	3
250	-226.1	-115.0	-239.0	-151.5	3
300	-237.3	-103.9	-254.3	-179.3	3

BESO4 (C), BERYLLIUM SULFATE

T(C)	DELTA H	DELTA F	ENTHALPY	ENTROPY	REF
25	-303.4	-270.9	-303.4	18.6	1
50	-303.4	-268.2	-302.9	20.3	2
75	-303.4	-265.5	-302.3	22.0	2
100	-303.4	-262.7	-301.7	23.6	2
150	-303.4	-257.3	-300.5	26.7	2
200	-303.3	-251.9	-299.2	29.6	2
250	-303.1	-246.4	-297.8	32.4	2
300	-302.9	-241.0	-296.3	35.0	2

BI (C,L), BISMUTH
MELT PT = 271C

T(C)	DELTA H	DELTA F	ENTHALPY	ENTROPY	REF
25	0.0	0.0	0.0	13.6	1
50	0.0	0.0	0.2	14.1	2
75	0.0	0.0	0.3	14.5	2
100	0.0	0.0	0.5	15.0	2
150	0.0	0.0	0.8	15.8	2
200	0.0	0.0	1.1	16.5	2
250	0.0	0.0	1.5	17.2	2
271	0.0	0.0	1.6	17.5	2
271	0.0	0.0	4.2	22.2	2
300	0.0	0.0	4.4	22.6	2

BI2O3 (C), BISMUTH TRIOXIDE

T(C)	DELTA H	DELTA F	ENTHALPY	ENTROPY	REF
25	-137.2	-117.9	-137.2	36.2	1
50	-137.1	-116.4	-136.5	38.4	2
75	-137.0	-114.8	-135.8	40.4	2
100	-136.9	-113.2	-135.1	42.3	2
150	-136.6	-110.0	-133.7	45.9	2
200	-136.4	-106.9	-132.3	49.0	2
250	-136.2	-103.7	-130.9	51.9	2
271	-136.1	-102.4	-130.2	53.1	2
271	-141.4	-102.4	-130.2	53.1	2
300	-141.3	-100.4	-129.4	54.6	2

BI2S3 (C), DIBISMUTH TRISULFIDE

T(C)	DELTA H	DELTA F	ENTHALPY	ENTROPY	REF
25	-80.2	-62.0	-80.2	47.9	1
50	-80.1	-60.5	-79.4	50.4	2
75	-79.9	-59.0	-78.7	52.7	2
100	-79.7	-57.5	-77.9	54.8	2
150	-79.4	-54.6	-76.3	58.8	2
200	-79.1	-51.6	-74.8	62.3	2
250	-78.8	-48.8	-73.1	65.5	2
271	-78.7	-47.5	-72.5	66.8	2
271	-83.9	-47.5	-72.5	66.8	2
300	-83.8	-45.6	-71.5	68.5	2

BR- (AQ)

T(C)	DELTA H	DELTA F	ENTHALPY	ENTROPY	REF
25	-29.0	-24.8	-29.0	24.7	1
50	-30.1	-24.5	-30.3	20.5	3
58	-30.5	-24.3	-30.8	19.1	2
58	-34.0	-24.3	-30.8	19.1	2
75	-34.5	-23.8	-31.8	16.2	3
100	-35.3	-23.0	-33.4	11.7	3
150	-37.3	-21.2	-36.8	3.1	3
200	-39.6	-19.1	-40.7	-6.0	3
250	-41.8	-16.8	-44.9	-14.5	3
300	-45.5	-14.3	-50.8	-25.2	3

BR2 (L,G), BROMINE
BOILING PT = 58C

T(C)	DELTA H	DELTA F	ENTHALPY	ENTROPY	REF
25	0.0	0.0	0.0	36.4	1
50	0.0	0.0	0.4	37.8	2
58	0.0	0.0	0.6	38.2	2
58	0.0	0.0	7.7	59.6	2
75	0.0	0.0	7.8	60.0	2
100	0.0	0.0	8.0	60.6	2
150	0.0	0.0	8.5	61.7	2
200	0.0	0.0	8.9	62.7	2
250	0.0	0.0	9.4	63.6	2
300	0.0	0.0	9.8	64.4	2

BR2CL- (AQ)
(EXTRAPOLATED AS CRISS & COBBLE TYPE 2)

T(C)	DELTA H	DELTA F	ENTHALPY	ENTROPY	REF
25	-40.7	-30.7	-40.7	50.1	1
50	-42.2	-29.8	-42.2	45.3	3
58	-42.8	-29.5	-42.7	43.8	2
58	-49.8	-29.5	-42.7	43.8	2
75	-50.4	-28.4	-43.6	41.1	3
100	-51.2	-26.8	-45.0	37.1	3
150	-53.7	-23.3	-48.5	28.3	3
200	-56.6	-19.5	-52.5	19.0	3
250	-59.2	-15.4	-56.8	10.3	3
300	-63.5	-11.0	-62.7	-0.5	3

BRI2- (AQ)
(EXTRAPOLATED AS CRISS & COBBLE TYPE 2)

T(C)	DELTA H	DELTA F	ENTHALPY	ENTROPY	REF
25	-30.6	-26.3	-30.6	52.2	1
50	-32.1	-25.9	-32.1	47.4	3
58	-32.7	-25.7	-32.6	45.8	2
58	-36.2	-25.7	-32.6	45.8	2
75	-36.9	-25.2	-33.5	43.1	3
100	-37.8	-24.3	-34.9	39.2	3
114	-38.6	-23.8	-35.8	36.7	2
114	-42.3	-23.8	-35.8	36.7	2
150	-44.6	-21.9	-38.4	30.3	3
185	-46.9	-20.0	-41.1	23.8	2
185	-57.0	-20.0	-41.1	23.8	2
200	-57.9	-18.7	-42.4	21.0	3
250	-60.6	-14.4	-46.7	12.4	3
300	-64.8	-9.8	-52.6	1.5	3

BRO- (AQ)

T(C)	DELTA H	DELTA F	ENTHALPY	ENTROPY	REF
25	-22.4	-8.0	-22.4	15.0	1
50	-24.6	-6.7	-24.8	7.3	3
58	-25.4	-6.2	-25.6	4.8	2
58	-28.9	-6.2	-25.6	4.8	2
75	-30.3	-5.1	-27.5	-0.7	3
100	-32.6	-3.2	-30.4	-8.9	3
150	-36.3	1.0	-35.3	-21.1	3
200	-41.6	5.9	-42.0	-36.7	3
250	-47.2	11.2	-49.5	-51.7	3
300	-53.5	17.0	-57.7	-66.7	3

BRO3- (AQ)

T(C)	DELTA H	DELTA F	ENTHALPY	ENTROPY	REF
25	-20.0	0.4	-20.0	44.0	1
50	-21.0	2.2	-21.0	40.8	3
58	-21.3	2.7	-21.3	39.8	2
58	-24.8	2.7	-21.3	39.8	2
75	-25.3	4.1	-22.1	37.4	3
100	-26.0	6.3	-23.3	33.9	3
150	-27.6	10.7	-25.7	27.8	3
200	-29.2	15.4	-28.3	21.9	3
250	-30.8	20.1	-31.5	15.6	3
300	-32.7	25.1	-35.0	9.2	3

C (C), CARBON (GRAPHITE)

T(C)	DELTA H	DELTA F	ENTHALPY	ENTROPY	REF
25	0.0	0.0	0.0	1.4	1
50	0.0	0.0	0.0	1.5	2
75	0.0	0.0	0.1	1.7	2
100	0.0	0.0	0.2	1.9	2
150	0.0	0.0	0.3	2.3	2
200	0.0	0.0	0.5	2.6	2
250	0.0	0.0	0.6	3.0	2
300	0.0	0.0	0.8	3.3	2

C2H2 (G), ACETYLENE

T(C)	DELTA H	DELTA F	ENTHALPY	ENTROPY	REF
25	54.2	50.0	54.2	48.0	1
50	54.2	49.7	54.5	48.9	2
75	54.2	49.3	54.7	49.7	2
100	54.2	48.9	55.0	50.5	2
150	54.1	48.3	55.6	52.0	2
200	54.1	47.6	56.3	53.4	2
250	54.1	46.9	56.9	54.7	2
300	54.0	46.2	57.6	56.0	2

C2H4 (G), ETHYLENE

T(C)	DELTA H	DELTA F	ENTHALPY	ENTROPY	REF
25	12.5	16.3	12.5	52.4	1
50	12.3	16.6	12.8	53.4	2
75	12.2	16.9	13.1	54.4	2
100	12.0	17.3	13.4	55.3	2
150	11.8	18.0	14.1	57.0	2
200	11.5	18.8	14.8	58.6	2
250	11.1	19.5	15.6	60.1	2
300	10.9	20.4	16.3	61.5	2

C2O4-- (AQ)

T(C)	DELTA H	DELTA F	ENTHALPY	ENTROPY	REF
25	-197.2	-161.1	-197.2	20.9	1
50	-198.8	-158.0	-199.3	14.1	3
75	-200.2	-154.8	-201.6	7.1	3
100	-201.5	-151.5	-204.2	-0.2	3
150	-203.7	-144.8	-208.6	-11.1	3
200	-207.0	-137.5	-214.5	-24.8	3
250	-210.0	-130.0	-221.1	-38.0	3
300	-213.8	-122.1	-228.4	-51.3	3

C6H6 (L,G), BENZENE
BOILING PT = 80C

T(C)	DELTA H	DELTA F	ENTHALPY	ENTROPY	REF
25	11.7	29.7	11.7	41.4	2
50	11.7	31.3	12.5	44.0	2
75	11.6	32.8	13.4	46.4	2
80	11.6	33.1	13.5	46.9	2
80	19.0	33.1	20.9	67.7	2
100	18.7	33.9	21.4	69.1	2
150	18.2	35.9	22.7	72.5	2
200	17.7	38.1	24.2	75.8	2
250	17.3	40.2	25.8	79.1	2
300	16.8	42.5	27.6	82.3	2

CH3COO- (AQ)
(EXTRAPOLATED AS CRISS & COBBLE TYPE 4)

T(C)	DELTA H	DELTA F	ENTHALPY	ENTROPY	REF
25	-116.2	-88.3	-116.2	25.7	1
50	-117.1	-85.9	-117.0	23.0	3
75	-117.7	-83.5	-117.8	20.6	3
100	-118.2	-81.0	-118.6	18.4	3
150	-120.5	-75.8	-121.3	11.2	3
200	-122.1	-70.4	-123.6	6.1	3
250	-123.7	-64.7	-126.3	0.7	3
300	-125.6	-59.2	-129.2	-4.6	3

CH3COOH (AQ), UNDISSOCIATED COMPLEX

T(C)	DELTA H	DELTA F	ENTHALPY	ENTROPY	REF
25	-116.1	-94.8	-116.1	42.7	1
50	-116.1	-93.0	-115.5	44.5	6
75	-115.8	-91.2	-114.5	47.6	6
100	-114.9	-89.5	-113.1	51.7	6
150	-113.6	-86.1	-110.6	58.1	6
200	-109.1	-83.0	-105.3	71.2	6

CH4(G), METHANE

T(C)	DELTA H	DELTA F	ENTHALPY	ENTROPY	REF
25	-17.9	-12.1	-17.9	44.5	1
50	-18.1	-11.6	-17.7	45.2	2
75	-18.2	-11.1	-17.4	45.8	2
100	-18.4	-10.6	-17.2	46.5	2
150	-18.8	-9.5	-16.7	47.7	2
200	-19.1	-8.4	-16.2	48.9	2
250	-19.4	-7.3	-15.6	50.0	2
300	-19.7	-6.1	-15.0	51.1	2

CH4 (AQ)

T(C)	DELTA H	DELTA F	ENTHALPY	ENTROPY	REF
25	-21.3	-8.2	-21.3	20.0	1
50	-20.1	-7.2	-19.6	25.3	14
75	-19.2	-6.2	-18.5	28.8	14
100	-18.4	-5.3	-17.2	32.2	14
150	-15.6	-3.0	-14.2	39.9	14

CN- (AQ)
(EXTRAPOLATED AS CRISS & COBBLE TYPE 2)

T(C)	DELTA H	DELTA F	ENTHALPY	ENTROPY	REF
25	36.0	41.2	36.0	27.5	1
50	35.0	41.7	34.7	23.3	3
75	34.1	42.2	33.3	18.9	3
100	33.3	42.8	31.7	14.5	3
150	31.2	44.3	28.3	5.9	3
200	28.8	46.0	24.4	-3.2	3
250	26.5	48.0	20.1	-11.8	3
300	22.6	50.2	14.3	-22.5	3

CNO- (AQ)
(EXTRAPOLATED AS CRISS & COBBLE TYPE 3)

T(C)	DELTA H	DELTA F	ENTHALPY	ENTROPY	REF
25	-34.9	-23.3	-34.9	30.5	1
50	-36.3	-22.3	-36.5	25.2	3
75	-37.7	-21.1	-38.4	19.7	3
100	-39.0	-19.9	-40.4	14.0	3
150	-41.5	-17.2	-44.0	5.1	3
200	-44.7	-14.0	-48.5	-5.4	3
250	-48.1	-10.6	-53.6	-15.7	3
300	-52.0	-6.9	-59.3	-26.2	3

CNS- (AQ)
(EXTRAPOLATED AS CRISS & COBBLE TYPE 3)

T(C)	DELTA H	DELTA F	ENTHALPY	ENTROPY	REF
25	2.9	12.7	2.9	39.5	1
50	2.0	13.5	1.7	35.6	3
75	1.1	14.5	0.4	31.5	3
100	0.2	15.4	-1.1	27.3	3
150	-1.5	17.6	-3.9	20.2	3
200	-3.5	20.0	-7.2	12.8	3
250	-5.5	22.6	-11.0	5.1	3
300	-8.0	25.4	-15.2	-2.6	3

CO (G), CARBON MONOXIDE

T(C)	DELTA H	DELTA F	ENTHALPY	ENTROPY	REF
25	-26.4	-32.8	-26.4	47.2	1
50	-26.4	-33.3	-26.2	47.8	2
75	-26.4	-33.8	-26.1	48.3	2
100	-26.3	-34.4	-25.9	48.8	2
150	-26.3	-35.5	-25.5	49.7	2
200	-26.3	-36.6	-25.2	50.5	2
250	-26.3	-37.6	-24.8	51.2	2
300	-26.3	-38.7	-24.5	51.9	2

CO2 (G), CARBON DIOXIDE

T(C)	DELTA H	DELTA F	ENTHALPY	ENTROPY	REF
25	-94.1	-94.3	-94.1	51.1	1
50	-94.1	-94.3	-93.8	51.8	2
75	-94.1	-94.3	-93.6	52.5	2
100	-94.1	-94.3	-93.3	53.2	2
150	-94.1	-94.3	-92.8	54.4	2
200	-94.1	-94.4	-92.3	55.6	2
250	-94.1	-94.4	-91.8	56.7	2
300	-94.1	-94.4	-91.2	57.7	2

CO3-- (AQ)

T(C)	DELTA H	DELTA F	ENTHALPY	ENTROPY	REF
25	-161.9	-126.2	-161.9	-3.6	1
50	-164.5	-123.1	-165.1	-14.2	3
75	-167.1	-119.8	-168.8	-25.1	3
100	-169.6	-116.3	-172.8	-36.3	3
150	-173.6	-109.0	-179.3	-52.5	3
200	-180.0	-100.7	-188.6	-74.3	3
250	-186.4	-92.0	-198.9	-94.9	3
300	-193.7	-82.6	-210.1	-11.0	8

(COOH)2 (AQ), UNDISSOCIATED COMPLEX

T(C)	DELTA H	DELTA F	ENTHALPY	ENTROPY	REF
25	-195.6	-166.8	-195.6	35.5	5
50	-195.2	-164.4	-194.6	38.6	6
75	-194.4	-162.1	-193.1	43.1	6
100	-193.2	-159.8	-191.2	48.4	6
150	-192.6	-155.3	-189.3	53.0	6
200	-190.6	-151.2	-185.9	61.0	6

COS (G), CARBON OXIDE SULFIDE (GAS)

T(C)	DELTA H	DELTA F	ENTHALPY	ENTROPY	REF
25	-49.3	-49.9	-49.3	55.3	1
50	-49.3	-50.0	-49.1	56.1	2
75	-49.3	-50.1	-48.8	56.9	2
100	-49.3	-50.1	-48.5	57.7	2
150	-49.2	-50.2	-48.0	59.0	2
200	-49.2	-50.4	-47.4	60.3	2
250	-49.2	-50.5	-46.8	61.5	2
300	-49.2	-50.6	-46.2	62.6	2

CS2 (L,G), CARBON DISULFIDE
BOILING PT = 46C

T(C)	DELTA H	DELTA F	ENTHALPY	ENTROPY	REF
25	-9.2	-3.4	-9.2	36.2	1
46	-9.1	-3.0	-8.9	37.4	2
46	-2.6	-3.0	-2.4	57.8	2
50	-2.6	-3.0	-2.3	57.9	2
75	-2.5	-3.0	-2.0	58.8	2
100	-2.5	-3.0	-1.7	59.6	2
150	-2.5	-3.1	-1.1	61.1	2
200	-2.4	-3.2	-0.5	62.4	2
250	-2.4	-3.3	0.1	63.7	2
300	-2.4	-3.3	0.7	64.9	2

CA (C), CALCIUM

T(C)	DELTA H	DELTA F	ENTHALPY	ENTROPY	REF
25	0.0	0.0	0.0	9.9	1
50	0.0	0.0	0.2	10.4	2
75	0.0	0.0	0.3	10.9	2
100	0.0	0.0	0.5	11.3	2
150	0.0	0.0	0.8	12.2	2
200	0.0	0.0	1.2	12.9	2
250	0.0	0.0	1.5	13.6	2
300	0.0	0.0	1.9	14.3	2

CA(NO3)2 (C), CALCIUM NITRATE

T(C)	DELTA H	DELTA F	ENTHALPY	ENTROPY	REF
25	-224.3	-177.6	-224.3	46.2	1
50	-224.2	-173.7	-223.4	49.1	2
75	-224.1	-169.8	-222.4	52.0	2
100	-224.0	-165.9	-221.4	54.7	2
150	-223.8	-158.1	-219.4	59.9	2
200	-223.4	-150.4	-217.2	64.8	2
250	-222.9	-142.7	-214.9	69.4	2
300	-222.4	-135.1	-212.5	73.8	2

36

CA(OH)2 (C), CALCIUM HYDROXIDE

T(C)	DELTA H	DELTA F	ENTHALPY	ENTROPY	REF
25	-235.7	-214.8	-235.7	19.9	1
50	-235.6	-213.0	-235.1	21.7	2
75	-235.6	-211.3	-234.6	23.3	2
100	-235.6	-209.5	-234.0	24.9	2
150	-235.4	-206.0	-232.9	27.8	2
200	-235.3	-202.5	-231.6	30.5	2
250	-235.1	-199.1	-230.4	33.0	2
300	-234.9	-195.7	-229.1	35.3	2

CA++ (AQ)

T(C)	DELTA H	DELTA F	ENTHALPY	ENTROPY	REF
25	-129.7	-132.3	-129.7	-22.7	1
50	-129.7	-132.5	-128.6	-19.2	3
75	-129.8	-132.7	-127.2	-14.7	3
100	-130.0	-132.9	-125.4	-9.6	3
150	-130.4	-133.3	-122.3	-1.8	3
200	-130.6	-133.6	-118.4	7.2	3
250	-131.3	-133.9	-114.2	15.6	3
300	-131.7	-134.1	-109.5	24.2	3

CA.167AL2.33SI3.67O10(OH)2 (C), CALCIUM MONTMORILLONITE

T(C)	DELTA H	DELTA F	ENTHALPY	ENTROPY	REF
25	-1368.0	-1279.1	-1368.0	61.2	10
50	-1368.2	-1271.7	-1366.1	67.1	10
75	-1368.3	-1264.2	-1364.2	72.9	10
100	-1368.4	-1256.7	-1362.2	78.6	10
150	-1368.4	-1241.8	-1357.9	89.2	10
200	-1368.3	-1226.8	-1353.5	99.2	10
250	-1368.1	-1211.9	-1348.8	108.5	10
300	-1367.8	-1196.9	-1344.0	117.3	10

CA2MG5SI8O22(OH)2 (C), TREMOLITE

T(C)	DELTA H	DELTA F	ENTHALPY	ENTROPY	REF
25	-2954.0	-2780.2	-2954.0	131.2	1
50	-2954.4	-2765.6	-2950.1	143.8	10
75	-2954.9	-2751.0	-2946.2	155.5	10
100	-2955.4	-2736.3	-2942.3	166.4	10
150	-2956.6	-2706.9	-2934.4	186.1	10
200	-2958.1	-2677.3	-2926.6	203.6	10
250	-2959.6	-2647.6	-2918.7	219.3	10
300	-2961.3	-2617.7	-2910.9	233.6	10

CA3(PO4)2 (C), CALCIUM ORTHOPHOSPHATE

T(C)	DELTA H	DELTA F	ENTHALPY	ENTROPY	REF
25	-984.9	-928.6	-984.9	56.4	1
44	-985.0	-924.9	-983.8	59.8	2
44	-985.3	-924.9	-983.8	59.8	2
50	-985.3	-923.8	-983.5	60.9	2
75	-985.4	-919.1	-982.1	65.1	2
100	-985.4	-914.3	-980.6	69.2	2
150	-985.5	-904.8	-977.6	76.8	2
200	-985.4	-895.2	-974.4	83.9	2
250	-985.3	-885.7	-971.1	90.6	2
274	-985.2	-881.2	-969.5	93.6	2
274	-991.1	-881.2	-969.5	93.6	2
300	-990.8	-876.0	-967.7	96.8	2

CAAL2SI2O8 (C), ANORTHITE

T(C)	DELTA H	DELTA F	ENTHALPY	ENTROPY	REF
25	-1009.3	-955.6	-1009.3	48.4	11
50	-1009.4	-951.1	-1008.0	52.6	12
75	-1009.5	-946.6	-1006.7	56.6	12
100	-1009.5	-942.1	-1005.3	60.5	12
150	-1009.5	-933.1	-1002.3	67.9	12
200	-1009.4	-924.0	-999.2	74.9	12
250	-1009.2	-915.0	-996.0	81.3	12
300	-1009.0	-906.0	-992.7	87.4	12

CABR2 (C), CALCIUM BROMIDE

T(C)	DELTA H	DELTA F	ENTHALPY	ENTROPY	REF
25	-163.2	-158.6	-163.2	31.0	1
50	-163.4	-158.3	-162.8	32.3	2
58	-163.4	-158.1	-162.6	32.7	2
58	-170.5	-158.1	-162.6	32.7	2
75	-170.5	-157.5	-162.4	33.6	2
100	-170.5	-156.6	-162.0	34.7	2
150	-170.4	-154.7	-161.1	36.9	2
200	-170.3	-152.9	-160.2	38.8	2
250	-170.2	-151.1	-159.3	40.6	2
300	-170.1	-149.2	-158.4	42.3	2

CACL2 (C), CALCIUM CHLORIDE

T(C)	DELTA H	DELTA F	ENTHALPY	ENTROPY	REF
25	-190.2	-178.8	-190.2	25.0	1
50	-190.1	-177.9	-189.8	26.4	2
75	-190.0	-176.9	-189.3	27.7	2
100	-190.0	-176.0	-188.9	29.0	2
150	-189.8	-174.1	-188.0	31.2	2
200	-189.7	-172.3	-187.1	33.3	2
250	-189.5	-170.4	-186.1	35.1	2
300	-189.4	-168.6	-185.2	36.8	2

CACO3 (C), CALCIUM CARBONATE
ARAGONITE TO CALCITE PHASE TRANS AT 50C

T(C)	DELTA H	DELTA F	ENTHALPY	ENTROPY	REF
25	-288.5	-269.5	-288.5	21.2	1
50	-288.5	-268.0	-288.0	22.8	2
50	-288.4	-268.0	-288.0	23.0	2
75	-288.4	-266.4	-287.4	24.5	2
100	-288.3	-264.8	-286.9	26.1	2
150	-288.2	-261.7	-285.7	29.0	2
200	-288.0	-258.5	-284.5	31.7	2
250	-287.9	-255.4	-283.3	34.2	2
300	-287.7	-252.3	-282.0	36.6	2

CACO3.MGCO3 (C), CALCIUM MAGNESIUM CARBONATE (DOLOMITE)

T(C)	DELTA H	DELTA F	ENTHALPY	ENTROPY	REF
25	-556.0	-517.1	-556.0	37.1	1
50	-556.0	-513.9	-555.1	40.2	2
75	-556.0	-510.6	-554.1	43.1	2
100	-556.0	-507.3	-553.1	45.9	2
150	-555.9	-500.8	-551.0	51.1	2
200	-555.8	-494.3	-548.8	55.9	2
250	-555.7	-487.8	-546.6	60.4	2
300	-555.6	-481.4	-544.3	64.7	2

CAF2 (C), CALCIUM FLUORIDE

T(C)	DELTA H	DELTA F	ENTHALPY	ENTROPY	REF
25	-291.5	-279.0	-291.5	16.5	1
50	-291.4	-278.0	-291.1	17.8	2
75	-291.3	-276.9	-290.6	19.1	2
100	-291.3	-275.9	-290.2	20.3	2
150	-291.1	-273.8	-289.3	22.5	2
200	-291.0	-271.8	-288.5	24.5	2
250	-290.8	-269.8	-287.5	26.3	2
300	-290.7	-267.8	-286.6	28.0	2

CAI2 (C), CALCIUM IODIDE

T(C)	DELTA H	DELTA F	ENTHALPY	ENTROPY	REF
25	-127.5	-126.4	-127.5	34.0	1
50	-127.5	-126.3	-127.1	35.4	2
75	-127.6	-126.2	-126.6	36.8	2
100	-127.6	-126.1	-126.1	38.1	2
114	-127.7	-126.1	-125.9	38.8	2
114	-131.4	-126.1	-125.9	38.8	2
150	-131.7	-125.6	-125.2	40.4	2
185	-131.9	-125.0	-124.6	41.9	2
185	-141.9	-125.0	-124.6	41.9	2
200	-141.9	-124.5	-124.3	42.5	2
250	-141.7	-122.6	-123.3	44.4	2
300	-141.6	-120.8	-122.3	46.2	2

CAMGSI2O6 (C), DIOPSIDE

T(C)	DELTA H	DELTA F	ENTHALPY	ENTROPY	REF
25	-766.3	-724.7	-766.3	34.2	1
50	-766.4	-721.2	-765.3	37.3	12
75	-766.5	-717.7	-764.3	40.4	12
100	-766.5	-714.2	-763.2	43.4	12
150	-766.5	-707.2	-760.9	49.2	12
200	-766.3	-700.2	-758.5	54.6	12
250	-766.2	-693.2	-755.9	59.7	12
300	-766.0	-686.2	-753.3	64.4	12

CAO (C), CALCIUM OXIDE

T(C)	DELTA H	DELTA F	ENTHALPY	ENTROPY	REF
25	-151.8	-144.4	-151.8	9.5	1
50	-151.8	-143.7	-151.5	10.3	2
75	-151.8	-143.1	-151.3	11.1	2
100	-151.7	-142.5	-151.0	11.9	2
150	-151.7	-141.3	-150.4	13.3	2
200	-151.6	-140.0	-149.9	14.6	2
250	-151.6	-138.8	-149.3	15.8	2
300	-151.5	-137.6	-148.7	16.9	2

CAS (C), CALCIUM SULFIDE

T(C)	DELTA H	DELTA F	ENTHALPY	ENTROPY	REF
25	-130.6	-123.6	-130.6	13.5	1
50	-130.6	-123.0	-130.4	14.4	2
75	-130.6	-122.4	-130.1	15.3	2
100	-130.6	-121.8	-129.8	16.1	2
150	-130.5	-120.7	-129.2	17.5	2
200	-130.5	-119.5	-128.6	18.9	2
250	-130.4	-118.3	-128.0	20.1	2
300	-130.4	-117.2	-127.4	21.2	2

CASIO3 (C), WOLLASTONITE

T(C)	DELTA H	DELTA F	ENTHALPY	ENTROPY	REF
25	-390.6	-370.3	-390.6	19.6	11
50	-390.7	-368.6	-390.1	21.3	12
75	-390.7	-366.9	-389.6	22.9	12
100	-390.7	-365.2	-389.0	24.5	12
150	-390.6	-361.7	-387.8	27.5	12
200	-390.5	-358.3	-386.5	30.3	12
250	-390.4	-354.9	-385.3	32.9	12
300	-390.3	-351.5	-383.9	35.3	12

CASO4 (C), CALCIUM SULFATE (ANHYDRITE)

T(C)	DELTA H	DELTA F	ENTHALPY	ENTROPY	REF
25	-358.1	-325.4	-358.1	25.5	1
50	-358.1	-322.7	-357.5	27.4	2
75	-358.1	-319.9	-356.9	29.3	2
100	-358.1	-317.2	-356.3	31.0	2
150	-358.0	-311.7	-354.9	34.3	2
200	-358.0	-306.2	-353.6	37.4	2
250	-357.8	-300.8	-352.1	40.2	2
300	-357.7	-295.3	-350.7	43.0	2

CASO4..5H2O (C), CALCIUM SULFATE 1/2-HYDRATE (ALPHA)

T(C)	DELTA H	DELTA F	ENTHALPY	ENTROPY	REF
25	-392.2	-352.9	-392.2	31.2	1
50	-392.1	-349.6	-391.4	33.9	2
75	-392.0	-346.3	-390.5	36.4	2
100	-391.9	-343.0	-389.7	38.8	2
150	-391.7	-336.5	-387.9	43.2	2
200	-391.4	-330.0	-386.1	47.2	2
250	-391.1	-323.5	-384.2	51.0	2
300	-390.8	-317.1	-382.3	54.6	2

CASO4.2H2O (C), CALCIUM SULFATE DIHYDRATE

T(C)	DELTA H	DELTA F	ENTHALPY	ENTROPY	REF
25	-498.8	-439.1	-498.8	46.4	1
50	-498.4	-434.1	-497.3	51.2	2
75	-498.0	-429.1	-495.7	55.8	2
100	-497.6	-424.2	-494.2	60.0	2
150	-496.8	-414.4	-491.1	67.9	2
200	-496.0	-404.7	-487.9	74.9	2
250	-495.2	-395.1	-484.7	81.4	2
300	-494.3	-385.6	-481.4	87.4	2

CD (C), CADMIUM

T(C)	DELTA H	DELTA F	ENTHALPY	ENTROPY	REF
25	0.0	0.0	0.0	12.4	1
50	0.0	0.0	0.2	12.9	2
75	0.0	0.0	0.3	13.3	2
100	0.0	0.0	0.5	13.8	2
150	0.0	0.0	0.8	14.6	2
200	0.0	0.0	1.1	15.3	2
250	0.0	0.0	1.5	16.0	2
300	0.0	0.0	1.8	16.7	2

CD(CN)4-- (AQ)
(EXTRAPOLATED AS CRISS & COBBLE TYPE 3)

T(C)	DELTA H	DELTA F	ENTHALPY	ENTROPY	REF
25	102.3	121.3	102.3	87.0	1
50	103.6	122.8	103.4	90.5	3
75	105.5	124.2	104.6	93.9	3
100	107.8	125.4	105.8	97.4	3
150	110.6	127.7	107.1	100.4	3
200	116.1	129.0	110.6	108.7	3
250	122.3	130.1	113.9	115.3	3
300	128.7	130.5	117.4	121.8	3

CD(NH3)2++ (AQ)

T(C)	DELTA H	DELTA F	ENTHALPY	ENTROPY	REF
25	-63.6	-38.0	-63.6	24.6	1
50	-64.8	-35.8	-63.0	26.6	3
75	-66.2	-33.5	-62.2	29.1	3
100	-67.9	-31.0	-61.2	31.8	3
150	-71.3	-25.8	-59.7	35.7	3
200	-74.5	-20.3	-57.5	40.8	3
250	-78.5	-14.3	-55.2	45.4	3
300	-82.4	-8.0	-52.6	50.1	3

CD(NH3)4++ (AQ)

T(C)	DELTA H	DELTA F	ENTHALPY	ENTROPY	REF
25	-107.6	-54.1	-107.6	70.4	1
50	-109.9	-49.5	-107.5	70.9	3
75	-112.7	-44.7	-107.3	71.4	3
100	-115.9	-39.7	-107.1	72.0	3
150	-122.1	-29.0	-107.1	72.0	3
200	-128.3	-17.5	-106.5	73.4	3
250	-135.7	-5.4	-106.1	74.3	3
300	-143.0	7.3	-105.6	75.2	3

CD++ (AQ)

T(C)	DELTA H	DELTA F	ENTHALPY	ENTROPY	REF
25	-18.1	-18.5	-18.1	-27.5	1
50	-18.1	-18.6	-17.0	-23.8	3
75	-18.1	-18.6	-15.5	-19.2	3
100	-18.2	-18.6	-13.6	-13.8	3
150	-18.5	-18.7	-10.3	-5.6	3
200	-18.4	-18.7	-6.3	3.8	3
250	-19.0	-18.7	-1.9	12.6	3
300	-19.1	-18.7	3.0	21.5	3

CDBR+ (AQ)

T(C)	DELTA H	DELTA F	ENTHALPY	ENTROPY	REF
25	-48.0	-46.3	-48.0	4.5	1
50	-48.0	-46.2	-47.2	7.1	3
58	-48.0	-46.2	-46.9	8.0	2
58	-51.5	-46.2	-46.9	8.0	2
75	-51.4	-45.9	-46.1	10.5	3
100	-51.3	-45.5	-44.7	14.2	3
150	-51.2	-44.7	-42.5	19.8	3
200	-50.7	-44.0	-39.6	26.5	3
250	-50.5	-43.3	-36.5	32.7	3
300	-49.9	-42.7	-33.1	39.1	3

CDCL+ (AQ)

T(C)	DELTA H	DELTA F	ENTHALPY	ENTROPY	REF
25	-57.5	-53.6	-57.5	5.4	1
50	-57.4	-53.3	-56.7	8.0	3
75	-57.3	-53.0	-55.6	11.3	3
100	-57.2	-52.7	-54.3	15.0	3
150	-57.1	-52.1	-52.1	20.5	3
200	-56.6	-51.6	-49.3	27.1	3
250	-56.4	-51.1	-46.2	33.3	3
300	-55.9	-50.6	-42.8	39.6	3

CDCL2 (C), CADMIUM CHLORIDE

T(C)	DELTA H	DELTA F	ENTHALPY	ENTROPY	REF
25	-93.6	-82.2	-93.6	27.5	1
50	-93.5	-81.3	-93.1	29.0	2
75	-93.4	-80.3	-92.6	30.4	2
100	-93.3	-79.4	-92.2	31.7	2
150	-93.1	-77.5	-91.2	34.1	2
200	-92.8	-75.7	-90.3	36.3	2
250	-92.6	-73.9	-89.3	38.3	2
300	-92.4	-72.1	-88.3	40.1	2

CDCL2 (AQ), UNDISSOCIATED COMPLEX

T(C)	DELTA H	DELTA F	ENTHALPY	ENTROPY	REF
25	-96.8	-85.9	-96.8	29.1	1
50	-97.4	-84.9	-97.0	28.3	6
75	-97.9	-84.0	-97.1	28.0	6
100	-98.3	-83.0	-97.2	27.9	6
150	-99.3	-80.8	-97.5	27.0	6
200	-99.6	-78.5	-97.3	27.8	6

CDCL3- (AQ)
(EXTRAPOLATED AS CRISS & COBBLE TYPE 2)

T(C)	DELTA H	DELTA F	ENTHALPY	ENTROPY	REF
25	-134.1	-116.4	-134.1	53.5	1
50	-135.6	-114.9	-135.6	48.7	3
75	-136.8	-113.2	-137.0	44.4	3
100	-137.8	-111.5	-138.5	40.5	3
150	-140.6	-107.8	-141.9	31.6	3
200	-143.8	-103.6	-145.9	22.5	3
250	-146.7	-99.2	-150.3	13.6	3
300	-151.3	-94.4	-156.2	2.9	3

CDI+ (AQ)

T(C)	DELTA H	DELTA F	ENTHALPY	ENTROPY	REF
25	-33.9	-33.8	-33.9	5.3	1
50	-33.9	-33.8	-33.1	7.9	3
75	-33.8	-33.8	-32.0	11.2	3
100	-33.8	-33.8	-30.7	14.9	3
114	-33.8	-33.8	-30.1	16.5	2
114	-35.6	-33.8	-30.1	16.5	2
150	-35.8	-33.6	-28.5	20.4	3
185	-35.6	-33.4	-26.5	25.0	2
185	-40.7	-33.4	-26.5	25.0	2
200	-40.5	-33.2	-25.7	27.1	3
250	-40.3	-32.4	-22.6	33.2	3
300	-39.8	-31.7	-19.2	39.5	3

CDO (C), CADMIUM OXIDE

T(C)	DELTA H	DELTA F	ENTHALPY	ENTROPY	REF
25	-61.7	-54.6	-61.7	13.1	1
50	-61.7	-54.0	-61.4	13.9	2
75	-61.7	-53.4	-61.2	14.7	2
100	-61.7	-52.8	-60.9	15.4	2
150	-61.7	-51.7	-60.4	16.7	2
200	-61.6	-50.5	-59.9	17.9	2
250	-61.6	-49.3	-59.3	19.0	2
300	-61.6	-48.1	-58.8	20.0	2

CDS (C), CADMIUM SULFIDE

T(C)	DELTA H	DELTA F	ENTHALPY	ENTROPY	REF
25	-54.0	-46.8	-54.0	15.5	1
50	-54.0	-46.3	-53.7	16.6	2
75	-53.9	-45.7	-53.4	17.5	2
100	-53.8	-45.1	-53.1	18.5	2
150	-53.7	-43.9	-52.4	20.1	2
200	-53.6	-42.8	-51.7	21.6	2
250	-53.4	-41.6	-51.1	23.0	2
300	-53.3	-40.5	-50.4	24.2	2

CDSO4 (C), CADMIUM SULFATE

T(C)	DELTA H	DELTA F	ENTHALPY	ENTROPY	REF
25	-238.4	-206.1	-238.4	29.4	1
50	-238.4	-203.4	-237.8	31.4	2
75	-238.4	-200.7	-237.2	33.2	2
100	-238.0	-198.0	-236.5	34.9	2
150	-238.4	-192.6	-235.3	38.2	2
200	-238.3	-187.2	-233.9	41.2	2
250	-238.2	-181.8	-232.5	44.0	2
300	-238.1	-176.4	-231.1	46.6	2

CE (C), CERIUM

T(C)	DELTA H	DELTA F	ENTHALPY	ENTROPY	REF
25	0.0	0.0	0.0	15.3	2
50	0.0	0.0	0.2	15.9	2
75	0.0	0.0	0.4	16.5	2
100	0.0	0.0	0.6	17.0	2
150	0.0	0.0	0.9	17.9	2
200	0.0	0.0	1.3	18.7	2
250	0.0	0.0	1.7	19.5	2
300	0.0	0.0	2.0	20.1	2

CEO2 (C), CERIUM OXIDE

T(C)	DELTA H	DELTA F	ENTHALPY	ENTROPY	REF
25	-260.2	-245.4	-260.2	14.7	2
50	-260.2	-244.2	-259.8	16.0	2
75	-260.1	-242.9	-259.4	17.2	2
100	-260.1	-241.7	-259.0	18.3	2
150	-260.0	-239.2	-258.2	20.3	2
200	-260.0	-236.8	-257.4	22.1	2
250	-259.9	-234.3	-256.6	23.7	2
300	-259.8	-231.9	-255.8	25.2	2

CL- (AQ)

T(C)	DELTA H	DELTA F	ENTHALPY	ENTROPY	REF
25	-39.9	-31.4	-39.9	18.5	1
50	-40.8	-30.6	-41.2	14.5	3
75	-41.7	-29.8	-42.7	10.1	3
100	-42.5	-28.9	-44.3	5.5	3
150	-44.5	-26.9	-47.6	-3.0	3
200	-46.8	-24.7	-51.5	-12.0	3
250	-48.9	-22.2	-55.8	-20.6	3
300	-52.6	-19.5	-61.6	-31.2	3

CL2 (G), CHLORINE

T(C)	DELTA H	DELTA F	ENTHALPY	ENTROPY	REF
25	0.0	0.0	0.0	53.3	1
50	0.0	0.0	0.2	53.9	2
75	0.0	0.0	0.4	54.6	2
100	0.0	0.0	0.6	55.1	2
150	0.0	0.0	1.0	56.2	2
200	0.0	0.0	1.5	57.1	2
250	0.0	0.0	1.9	58.0	2
300	0.0	0.0	2.3	58.8	2

CLO- (AQ)

T(C)	DELTA H	DELTA F	ENTHALPY	ENTROPY	REF
25	-25.7	-8.8	-25.7	15.0	1
50	-27.8	-7.3	-28.1	7.3	3
75	-30.0	-5.6	-30.8	-0.7	3
100	-32.2	-3.8	-33.7	-8.9	3
150	-35.9	0.2	-38.6	-21.1	3
200	-41.2	5.0	-45.3	-36.7	3
250	-46.7	10.1	-52.8	-51.7	3
300	-53.0	15.9	-61.0	-66.7	3

CLO2- (AQ)

T(C)	DELTA H	DELTA F	ENTHALPY	ENTROPY	REF
25	-15.9	4.1	-15.9	29.2	1
50	-17.4	5.8	-17.6	23.7	3
75	-18.9	7.7	-19.5	18.0	3
100	-20.4	9.6	-21.6	12.1	3
150	-23.0	13.8	-25.3	2.9	3
200	-26.5	18.5	-30.0	-8.0	3
250	-30.1	23.4	-35.3	-18.8	3
300	-34.3	28.7	-41.3	-29.5	3

CLO3- (AQ)

T(C)	DELTA H	DELTA F	ENTHALPY	ENTROPY	REF
25	-23.7	-0.8	-23.7	43.8	1
50	-24.6	1.2	-24.7	40.6	3
75	-25.4	3.2	-25.9	37.2	3
100	-26.2	5.3	-27.1	33.7	3
150	-27.8	9.6	-29.6	27.5	3
200	-29.3	14.1	-32.2	21.5	3
250	-31.0	18.7	-35.4	15.1	3
300	-32.9	23.6	-38.9	8.7	3

CLO4- (AQ)

T(C)	DELTA H	DELTA F	ENTHALPY	ENTROPY	REF
25	-30.9	-2.1	-30.9	48.5	1
50	-31.7	0.4	-31.7	46.0	3
75	-32.3	2.9	-32.6	43.4	3
100	-32.8	5.4	-33.6	40.6	3
150	-34.2	10.6	-35.6	35.4	3
200	-35.3	16.0	-37.6	31.0	3
250	-36.4	21.5	-40.0	26.0	3
300	-37.8	27.0	-42.8	21.0	3

CO (C), COBALT

T(C)	DELTA H	DELTA F	ENTHALPY	ENTROPY	REF
25	0.0	0.0	0.0	7.2	1
50	0.0	0.0	0.2	7.7	2
75	0.0	0.0	0.3	8.1	2
100	0.0	0.0	0.5	8.5	2
150	0.0	0.0	0.8	9.3	2
200	0.0	0.0	1.1	10.1	2
250	0.0	0.0	1.4	10.8	2
300	0.0	0.0	1.8	11.4	2

CO(NH3)++ (AQ)

T(C)	DELTA H	DELTA F	ENTHALPY	ENTROPY	REF
25	-34.8	-22.1	-34.8	-7.0	1
50	-35.3	-21.0	-33.9	-4.0	3
75	-36.0	-19.9	-32.6	-0.2	3
100	-36.7	-18.7	-31.1	4.2	3
150	-38.4	-16.1	-28.5	10.7	3
200	-39.7	-13.4	-25.2	18.3	3
250	-41.8	-10.5	-21.7	25.5	3
300	-43.6	-7.4	-17.7	32.8	3

CO(NH3)2++ (AQ)

T(C)	DELTA H	DELTA F	ENTHALPY	ENTROPY	REF
25	----	-30.5	----	----	15
50	----	-28.3	----	----	15
75	----	-25.9	----	----	15
100	----	-23.8	----	----	15

CO(NH3)3++ (AQ)

T(C)	DELTA H	DELTA F	ENTHALPY	ENTROPY	REF
25	----	-38.2	----	----	15
50	----	-34.8	----	----	15
75	----	-31.1	----	----	15
100	----	-27.9	----	----	15

CO(NH3)4++ (AQ)

T(C)	DELTA H	DELTA F	ENTHALPY	ENTROPY	REF
25	----	-45.5	----	----	15
50	----	-40.8	----	----	15
75	----	-35.9	----	----	15
100	----	-31.5	----	----	15

CO(NH3)5++

T(C)	DELTA H	DELTA F	ENTHALPY	ENTROPY	REF
25	----	-51.9	----	----	15
50	----	-45.9	----	----	15
75	----	-39.7	----	----	15
100	----	-34.1	----	----	15

CO(NH3)6++ (AQ)

T(C)	DELTA H	DELTA F	ENTHALPY	ENTROPY	REF
25	----	-57.3	----	----	15
50	----	-49.9	----	----	15
75	----	-42.2	----	----	15
100	----	-35.3	----	----	15

CO(NH3)6+++ (AQ)

T(C)	DELTA H	DELTA F	ENTHALPY	ENTROPY	REF
25	-139.8	-38.9	-139.8	25.0	1
50	-142.9	-30.3	-139.2	27.0	3
75	-146.4	-21.5	-138.4	29.4	3
100	-150.4	-12.3	-137.4	32.2	3
150	-158.1	6.9	-135.9	36.0	3
200	-165.9	26.8	-133.8	41.1	3
250	-175.1	47.7	-131.5	45.7	3
300	-184.2	69.4	-128.9	50.3	3

CO++ (AQ), COBALTOUS ION

T(C)	DELTA H	DELTA F	ENTHALPY	ENTROPY	REF
25	-13.9	-13.0	-13.9	-37.0	1
50	-13.7	-12.9	-12.6	-33.0	3
75	-13.6	-12.9	-11.0	-27.9	3
100	-13.5	-12.8	-8.9	-22.1	3
150	-13.5	-12.7	-5.4	-13.1	3
200	-13.1	-12.7	-1.0	-3.0	3
250	-13.3	-12.6	3.8	6.6	3
300	-13.0	-12.6	9.1	16.3	3

CO+++ (AQ), COBALTIC ION

T(C)	DELTA H	DELTA F	ENTHALPY	ENTROPY	REF
25	22.0	32.0	22.0	-88.0	1
50	22.2	32.8	23.8	-82.4	3
75	22.4	33.6	26.2	-75.1	3
100	22.4	34.5	29.1	-66.8	3
150	22.6	36.1	34.4	-53.5	3
200	22.9	37.6	40.5	-39.3	3
250	22.5	39.2	47.3	-25.5	3
300	22.7	40.8	55.0	-11.6	3

CO3O4 (C), COBALT(2,3) OXIDE

T(C)	DELTA H	DELTA F	ENTHALPY	ENTROPY	REF
25	-213.0	-184.7	-213.0	24.5	1
50	-213.0	-182.3	-212.2	26.9	2
75	-213.1	-179.9	-211.5	29.3	2
100	-213.1	-177.5	-210.6	31.5	2
150	-213.1	-172.8	-208.9	35.8	2
200	-213.0	-168.0	-207.2	39.8	2
250	-212.9	-163.3	-205.3	43.5	2
300	-212.8	-158.5	-203.4	47.0	2

COCL2 (C), COBALT(2) CHLORIDE

T(C)	DELTA H	DELTA F	ENTHALPY	ENTROPY	REF
25	-74.7	-64.4	-74.7	26.1	1
50	-74.6	-63.6	-74.2	27.6	2
75	-74.4	-62.8	-73.7	29.0	2
100	-74.3	-61.9	-73.3	30.4	2
150	-74.1	-60.3	-72.2	33.0	2
200	-73.8	-58.7	-71.2	35.3	2
250	-73.4	-57.1	-70.1	37.5	2
300	-73.1	-55.5	-69.0	39.5	2

COO (C), COBALT(2) OXIDE

T(C)	DELTA H	DELTA F	ENTHALPY	ENTROPY	REF
25	-56.9	-51.2	-56.9	12.7	1
50	-56.8	-50.7	-56.6	13.7	2
75	-56.7	-50.3	-56.2	14.6	2
100	-56.7	-49.8	-55.9	15.5	2
150	-56.5	-48.9	-55.3	17.1	2
200	-56.4	-48.0	-54.7	18.5	2
250	-56.3	-47.1	-54.0	19.8	2
300	-56.2	-46.2	-53.4	20.9	2

COS (C), COBALT SULFIDE

T(C)	DELTA H	DELTA F	ENTHALPY	ENTROPY	REF
25	-34.6	-29.2	-34.6	16.1	5
50	-34.6	-28.7	-34.4	17.0	12
75	-34.6	-28.3	-34.1	17.9	12
100	-34.5	-27.8	-33.8	18.7	12
150	-34.5	-26.9	-33.2	20.1	12
200	-34.4	-26.0	-32.6	21.4	12
250	-34.4	-25.1	-32.0	22.6	12
300	-34.3	-24.3	-31.4	23.7	12

COSO4 (C), COBALT(2) SULFATE

T(C)	DELTA H	DELTA F	ENTHALPY	ENTROPY	REF
25	-227.6	-196.6	-227.6	28.2	1
50	-227.4	-194.0	-226.8	30.9	2
75	-227.2	-191.4	-226.0	33.4	2
100	-227.0	-188.8	-225.1	35.7	2
150	-226.5	-183.7	-223.4	40.0	2
200	-226.0	-178.7	-221.7	43.8	2
250	-225.6	-173.7	-220.0	47.3	2
300	-225.1	-168.8	-218.2	50.6	2

CR (C), CHROMIUM

T(C)	DELTA H	DELTA F	ENTHALPY	ENTROPY	REF
25	0.0	0.0	0.0	5.7	1
50	0.0	0.0	0.1	6.1	2
75	0.0	0.0	0.3	6.6	2
100	0.0	0.0	0.4	7.0	2
150	0.0	0.0	0.7	7.7	2
200	0.0	0.0	1.0	8.4	2
250	0.0	0.0	1.3	9.0	2
300	0.0	0.0	1.7	9.6	2

CR(OH)++ (AQ)

T(C)	DELTA H	DELTA F	ENTHALPY	ENTROPY	REF
25	-112.2	-103.0	-112.2	-26.4	5
50	-112.4	-102.2	-111.1	-22.8	3
75	-112.6	-101.4	-109.6	-18.1	3
100	-112.8	-100.6	-107.7	-12.8	3
150	-113.4	-98.9	-104.5	-4.7	3
200	-113.8	-97.2	-100.5	4.5	3
250	-114.7	-95.4	-96.2	13.3	3
300	-115.2	-93.5	-91.3	22.1	3

CR++ (AQ)

T(C)	DELTA H	DELTA F	ENTHALPY	ENTROPY	REF
25	-33.2	-42.1	-33.2	-5.7	5
50	-33.4	-42.8	-32.3	-2.7	3
75	-33.7	-43.6	-31.0	1.0	3
100	-34.1	-44.2	-29.5	5.3	3
150	-35.1	-45.5	-27.0	11.7	3
200	-35.8	-46.7	-23.8	19.3	3
250	-37.2	-47.8	-20.2	26.3	3
300	-38.3	-48.7	-16.3	33.5	3

CR+++ (AQ)

T(C)	DELTA H	DELTA F	ENTHALPY	ENTROPY	REF
25	-61.2	-51.5	-61.2	-88.5	5
50	-60.9	-50.7	-59.4	-82.9	3
75	-60.8	-49.9	-57.0	-75.6	3
100	-60.7	-49.1	-54.0	-67.2	3
150	-60.5	-47.6	-48.8	-53.9	3
200	-60.1	-46.1	-42.6	-39.6	3
250	-60.6	-44.6	-35.8	-25.9	3
300	-60.3	-43.1	-28.1	-11.9	3

CR2(SO4)3 (C), CHROMIUM(3) SULFATE

T(C)	DELTA H	DELTA F	ENTHALPY	ENTROPY	REF
25	-191.7	-95.5	-191.7	64.5	2
50	-191.5	-87.5	-189.9	70.4	2
75	-191.4	-79.4	-188.1	75.8	2
100	-191.2	-71.4	-186.2	81.0	2
150	-190.8	-55.4	-182.4	90.5	2
200	-190.4	-39.4	-178.5	99.1	2
250	-189.9	-23.5	-174.6	107.1	2
300	-189.4	-7.6	-170.6	114.5	2

CR2O3 (C), CHROMIUM(3) OXIDE

T(C)	DELTA H	DELTA F	ENTHALPY	ENTROPY	REF
25	-272.4	-252.9	-272.4	19.4	1
50	-272.3	-251.3	-271.8	21.4	2
75	-272.2	-249.6	-271.1	23.4	2
100	-272.1	-248.0	-270.5	25.2	2
150	-271.9	-244.8	-269.1	28.6	2
200	-271.7	-241.6	-267.7	31.7	2
250	-271.5	-238.4	-266.3	34.5	2
300	-271.3	-235.3	-264.9	37.1	2

CR2O7-- (AQ)

T(C)	DELTA H	DELTA F	ENTHALPY	ENTROPY	REF
25	-356.2	-311.0	-356.2	72.6	1
50	-355.7	-307.2	-355.8	73.9	3
75	-354.8	-303.5	-355.4	75.0	3
100	-353.5	-299.9	-355.0	76.2	3
150	-352.2	-292.8	-354.9	76.1	3
200	-349.0	-286.3	-353.5	79.6	3
250	-345.2	-279.8	-352.4	81.9	3
300	-341.3	-273.8	-351.2	84.1	3

CRCL2 (C), CHROMIUM(2) CHLORIDE

T(C)	DELTA H	DELTA F	ENTHALPY	ENTROPY	REF
25	-94.5	-85.1	-94.5	27.6	1
50	-94.4	-84.4	-94.1	28.9	2
75	-94.4	-83.6	-93.6	30.2	2
100	-94.3	-82.8	-93.2	31.4	2
150	-94.1	-81.3	-92.4	33.6	2
200	-94.0	-79.8	-91.5	35.5	2
250	-93.8	-78.3	-90.6	37.3	2
300	-93.7	-76.8	-89.7	39.0	2

CRCL2+ (AQ)

T(C)	DELTA H	DELTA F	ENTHALPY	ENTROPY	REF
25	-130.0	-115.2	-130.0	-11.3	5
50	-129.8	-114.0	-129.0	-8.1	3
75	-129.6	-112.8	-127.7	-4.1	3
100	-129.2	-111.6	-126.1	0.4	3
150	-128.8	-109.2	-123.4	7.3	3
200	-127.9	-107.0	-119.9	15.3	3
250	-127.2	-104.9	-116.2	22.8	3
300	-126.1	-102.8	-112.0	30.4	3

CRCL3 (C), CHROMIUM(3) CHLORIDE

T(C)	DELTA H	DELTA F	ENTHALPY	ENTROPY	REF
25	-133.0	-116.2	-133.0	29.4	1
50	-132.9	-114.8	-132.5	31.1	2
75	-132.8	-113.4	-131.9	32.8	2
100	-132.7	-112.1	-131.4	34.3	2
150	-132.5	-109.3	-130.3	37.1	2
200	-132.4	-106.6	-129.1	39.6	2
250	-132.1	-103.9	-128.0	41.9	2
300	-132.0	-101.2	-126.8	44.0	2

CRO4-- (AQ)

T(C)	DELTA H	DELTA F	ENTHALPY	ENTROPY	REF
25	-210.6	-174.0	-210.6	22.0	1
50	-212.2	-170.8	-212.6	15.4	3
75	-213.6	-167.6	-214.9	8.5	3
100	-214.8	-164.3	-217.5	1.5	3
150	-216.9	-157.4	-221.8	-9.3	3
200	-220.0	-150.1	-227.5	-22.6	3
250	-222.9	-142.6	-233.9	-35.5	3
300	-226.4	-134.7	-241.0	-48.4	3

CS (C,L), CESIUM
MELT PT = 29C

T(C)	DELTA H	DELTA F	ENTHALPY	ENTROPY	REF
25	0.0	0.0	0.0	20.2	2
29	0.0	0.0	0.0	20.3	2
29	0.0	0.0	0.5	22.0	2
50	0.0	0.0	0.7	22.5	2
75	0.0	0.0	0.9	23.1	2
100	0.0	0.0	1.1	23.6	2
150	0.0	0.0	1.5	24.5	2
200	0.0	0.0	1.8	25.4	2
250	0.0	0.0	2.2	26.2	2
300	0.0	0.0	2.6	26.8	2

CS+ (AQ)

T(C)	DELTA H	DELTA F	ENTHALPY	ENTROPY	REF
25	-59.3	-67.4	-59.3	26.8	5
29	-59.3	-67.5	-59.2	27.1	2
29	-59.8	-67.5	-59.2	27.1	2
50	-59.9	-68.1	-58.7	28.7	3
75	-60.0	-69.7	-57.9	31.1	3
100	-60.1	-69.3	-57.0	33.8	3
150	-60.6	-70.5	-55.5	37.4	3
200	-60.7	-71.6	-53.4	42.3	3
250	-61.2	-72.8	-51.2	46.8	3
300	-61.5	-73.9	-48.7	51.3	3

CSBR (C), CESIUM BROMIDE

T(C)	DELTA H	DELTA F	ENTHALPY	ENTROPY	REF
25	-94.3	-90.9	-94.3	27.1	2
29	-94.3	-90.9	-94.3	27.3	2
29	-94.8	-90.9	-94.3	27.3	2
50	-94.9	-90.6	-94.0	28.1	2
58	-94.9	-90.5	-93.9	28.4	2
58	-98.5	-90.5	-93.9	28.4	2
75	-98.5	-90.1	-93.7	29.0	2
100	-98.5	-89.5	-93.4	29.9	2
150	-98.4	-88.3	-92.7	31.5	2
200	-98.4	-87.1	-92.1	32.9	2
250	-98.4	-85.9	-91.4	34.2	2
300	-98.3	-84.8	-90.8	35.4	2

CSBR (AQ), UNDISSOCIATED COMPLEX

T(C)	DELTA H	DELTA F	ENTHALPY	ENTROPY	REF
25	-88.2	-92.0	-88.2	51.1	5
29	-88.3	-92.0	-88.3	50.8	2
29	-88.9	-92.0	-88.3	50.8	2
50	-89.8	-92.2	-88.9	48.8	6
58	-90.2	-92.3	-89.1	48.0	2
58	-93.8	-92.3	-89.1	48.0	2
75	-94.4	-92.2	-89.6	46.8	6
100	-95.3	-92.0	-90.3	44.9	6
150	-97.9	-91.3	-92.2	39.8	6
200	-100.5	-90.4	-94.1	35.4	6

CSCL (C), CESIUM CHLORIDE

T(C)	DELTA H	DELTA F	ENTHALPY	ENTROPY	REF
25	-105.8	-99.1	-105.8	24.2	2
29	-105.8	-99.0	-105.8	24.3	2
29	-106.4	-99.0	-105.8	24.3	2
50	-106.3	-98.5	-105.5	25.2	2
75	-106.3	-97.9	-105.2	26.2	2
100	-106.3	-97.3	-104.9	27.0	2
150	-106.2	-96.1	-104.2	28.7	2
200	-106.1	-94.9	-103.6	30.2	2
250	-106.1	-93.7	-102.9	31.5	2
300	-106.0	-92.5	-102.2	32.8	2

CSCL (AQ), UNDISSOCIATED COMPLEX

T(C)	DELTA H	DELTA F	ENTHALPY	ENTROPY	REF
25	-89.3	-88.8	-89.3	45.0	5
29	-89.4	-88.8	-89.4	44.7	2
29	-90.0	-88.8	-89.4	44.7	2
50	-90.8	-89.6	-90.0	42.8	6
75	-91.8	-88.4	-90.6	40.8	6
100	-92.8	-88.2	-91.4	38.8	6
150	-95.3	-87.3	-93.3	33.9	6
200	-97.7	-86.2	-95.1	29.6	6

CSF (C), CESIUM FLUORIDE

T(C)	DELTA H	DELTA F	ENTHALPY	ENTROPY	REF
25	-132.6	-125.6	-132.6	21.1	2
29	-132.6	-125.5	-132.5	21.3	2
29	-133.1	-125.5	-132.5	21.3	2
50	-133.1	-125.0	-132.3	22.1	2
75	-133.0	-124.4	-131.9	23.0	2
100	-133.0	-123.8	-131.6	23.9	2
150	-132.9	-122.5	-131.0	25.5	2
200	-132.9	-121.3	-130.3	27.0	2
250	-132.8	-120.1	-129.7	28.3	2
300	-132.7	-118.9	-129.0	29.6	2

CSF (AQ), UNDISSOCIATED COMPLEX

T(C)	DELTA H	DELTA F	ENTHALPY	ENTROPY	REF
25	-137.9	-133.5	-137.9	29.5	5
29	-138.0	-133.4	-138.0	29.3	2
29	-138.6	-133.4	-138.0	29.3	2
50	-139.2	-133.0	-138.4	27.9	6
75	-140.2	-132.5	-139.1	26.0	6
100	-141.2	-131.9	-139.8	23.9	6
150	-143.4	-130.5	-141.5	19.6	6
200	-145.5	-128.8	-142.9	16.3	6

CSI (C), CESIUM IODIDE

T(C)	DELTA H	DELTA F	ENTHALPY	ENTROPY	REF
25	-80.5	-79.3	-80.5	30.0	2
29	-80.5	-79.3	-80.5	30.2	2
29	-81.0	-79.3	-80.5	30.2	2
50	-81.1	-79.2	-80.2	31.0	2
75	-81.1	-79.0	-79.9	31.9	2
100	-81.2	-78.9	-79.6	32.8	2
114	-81.2	-78.8	-79.4	33.2	2
114	-83.1	-78.8	-79.4	33.2	2
150	-83.3	-78.4	-79.0	34.3	2
185	-83.4	-77.9	-78.5	35.3	2
185	-88.4	-77.9	-78.5	35.3	2
200	-88.4	-77.6	-78.3	35.7	2
250	-88.4	-76.5	-77.7	37.0	2
300	-88.3	-75.3	-77.0	38.3	2

CSI (AQ), UNDISSOCIATED COMPLEX

T(C)	DELTA H	DELTA F	ENTHALPY	ENTROPY	REF
25	-72.6	-79.8	-72.6	57.9	5
29	-72.8	-79.9	-72.8	57.5	2
29	-73.3	-79.9	-72.8	57.5	2
50	-74.3	-80.3	-73.4	55.4	6
75	-75.3	-80.7	-74.1	53.4	6
100	-76.3	-81.1	-74.7	51.7	6
114	-77.1	-81.2	-75.3	50.2	2
114	-78.9	-81.2	-75.3	50.2	2
150	-81.1	-91.3	-76.7	46.5	6
185	-83.1	-81.3	-78.1	43.3	2
185	-88.1	-81.3	-78.1	43.3	2
200	-88.9	-91.0	-78.7	41.9	6

CSNO3 (AQ), UNDISSOCIATED COMPLEX

T(C)	DELTA H	DELTA F	ENTHALPY	ENTROPY	REF
25	-102.7	-93.8	-102.7	86.8	5
29	-102.8	-93.7	-102.7	86.9	2
29	-103.3	-93.7	-102.7	86.9	2
50	-103.6	-93.0	-102.5	87.4	6
75	-103.7	-92.2	-102.1	88.6	6
100	-103.6	-91.4	-101.5	90.3	6
150	-102.8	-89.8	-99.7	95.2	6
200	-99.4	-88.4	-95.7	105.3	6

CU (C), COPPER

T(C)	DELTA H	DELTA F	ENTHALPY	ENTROPY	REF
25	0.0	0.0	0.0	7.9	1
50	0.0	0.0	0.2	8.4	2
75	0.0	0.0	0.3	8.8	2
100	0.0	0.0	0.4	9.3	2
150	0.0	0.0	0.7	10.0	2
200	0.0	0.0	1.0	10.7	2
250	0.0	0.0	1.4	11.3	2
300	0.0	0.0	1.7	11.9	2

CU(NH3)2+ (AQ)

T(C)	DELTA H	DELTA F	ENTHALPY	ENTROPY	REF
25	-36.1	-15.6	-36.1	58.0	5
50	-37.1	-13.8	-35.8	58.9	3
75	-38.3	-12.0	-35.4	60.0	3
100	-39.7	-10.0	-35.1	61.1	3
150	-42.5	-5.8	-34.6	62.1	3
200	-45.0	-1.4	-33.6	64.5	3
250	-48.1	3.4	-32.7	66.4	3
300	-51.1	8.5	-31.6	68.4	3

CU(NH3)++ (AQ)

T(C)	DELTA H	DELTA F	ENTHALPY	ENTROPY	REF
25	-9.3	3.7	-9.3	-7.1	1
50	-9.8	4.8	-8.3	-4.1	3
75	-10.4	6.0	-7.1	-0.3	3
100	-11.2	7.2	-5.5	4.1	3
150	-12.8	9.8	-2.9	10.6	3
200	-14.1	12.5	0.4	18.3	3
250	-16.2	15.5	3.9	25.4	3
300	-17.9	18.6	7.9	32.7	3

CU(NH3)2++ (AQ)

T(C)	DELTA H	DELTA F	ENTHALPY	ENTROPY	REF
25	-34.0	-7.3	-34.0	16.6	1
50	-35.0	-5.0	-33.3	18.8	3
75	-36.4	-2.6	-32.3	21.7	3
100	-37.9	-0.1	-31.2	24.8	3
150	-41.0	5.2	-29.4	29.3	3
200	-43.8	10.8	-27.0	35.1	3
250	-47.6	16.8	-24.3	40.4	3
300	-51.1	23.1	-21.4	45.7	3

CU(NH3)3++ (AQ)

T(C)	DELTA H	DELTA F	ENTHALPY	ENTROPY	REF
25	-58.7	-17.5	-58.7	37.7	1
50	-60.3	-14.0	-58.2	39.3	3
75	-62.3	-10.3	-57.5	41.2	3
100	-64.5	-6.4	-56.8	43.3	3
150	-69.0	1.7	-55.7	46.1	3
200	-73.3	10.3	-54.0	50.1	3
250	-78.6	19.4	-52.2	53.7	3
300	-83.7	29.0	-50.2	57.3	3

CU(NH3)4++ (AQ)

T(C)	DELTA H	DELTA F	ENTHALPY	ENTROPY	REF
25	-83.3	-26.6	-83.3	55.4	1
50	-85.4	-21.8	-83.0	56.4	3
75	-88.0	-16.7	-82.6	57.6	3
100	-90.9	-11.5	-82.1	58.8	3
150	-96.6	-0.4	-81.6	60.1	3
200	-102.3	11.4	-80.5	62.7	3
250	-109.0	23.7	-79.5	64.8	3
300	-115.6	36.7	-78.3	67.0	3

CU(OH)2 (C), COPPER(2) HYDROXIDE
THERMAL DECOMP TEMP = 160C

T(C)	DELTA H	DELTA F	ENTHALPY	ENTROPY	REF
25	-107.6	-89.1	-107.6	25.9	2
50	-107.6	-87.5	-107.1	27.7	2
75	-107.5	-86.0	-106.5	29.5	2
100	-107.4	-84.4	-105.9	31.1	2
150	-107.2	-81.4	-104.7	34.1	2
160	-107.2	-80.8	-104.5	34.6	2

CU+ (AQ)

T(C)	DELTA H	DELTA F	ENTHALPY	ENTROPY	REF
25	17.1	12.0	17.1	4.7	1
50	17.3	11.5	18.0	7.3	3
75	17.6	11.0	19.0	10.6	3
100	17.9	10.6	20.4	14.4	3
150	18.2	9.6	22.6	19.9	3
200	18.9	8.5	25.5	26.6	3
250	19.4	7.3	28.5	32.9	3
300	20.2	6.1	32.0	39.2	3

CU++ (AQ)

T(C)	DELTA H	DELTA F	ENTHALPY	ENTROPY	REF
25	15.5	15.7	15.5	-33.8	1
50	15.6	15.7	16.7	-29.9	3
75	15.7	15.7	18.3	-25.0	3
100	15.7	15.7	20.3	-19.3	3
150	15.7	15.7	23.8	-10.6	3
200	16.0	15.6	28.0	-0.7	3
250	15.7	15.6	32.7	8.6	3
300	15.9	15.6	37.9	18.1	3

CU2CO3(OH)2 (C), MALACHITE

T(C)	DELTA H	DELTA F	ENTHALPY	ENTROPY	REF
25	-251.3	-213.6	-251.3	44.5	1
50	-251.3	-210.4	-250.3	47.7	13
75	-251.2	-207.3	-249.3	50.7	13
100	-251.2	-204.1	-248.3	53.5	13
150	-251.1	-197.8	-246.2	58.7	13
200	-251.1	-191.5	-244.1	63.4	13
250	-251.0	-185.3	-241.9	67.8	13
300	-250.9	-179.0	-239.8	71.8	13

CU2O (C), COPPER(1) OXIDE

T(C)	DELTA H	DELTA F	ENTHALPY	ENTROPY	REF
25	-40.3	-34.9	-40.3	22.3	1
50	-40.3	-34.5	-39.9	23.5	2
75	-40.3	-34.0	-39.5	24.7	2
100	-40.3	-33.6	-39.1	25.8	2
150	-40.2	-32.7	-38.3	27.9	2
200	-40.2	-31.8	-37.5	29.7	2
250	-40.2	-30.9	-36.6	31.4	2
300	-40.1	-30.0	-35.7	33.0	2

CU2S (C), COPPER(1) SULFIDE
PHASE TRANS OF CU2S AT 103C

T(C)	DELTA H	DELTA F	ENTHALPY	ENTROPY	REF
25	-34.3	-30.1	-34.3	28.9	1
50	-34.2	-29.8	-33.8	30.5	2
75	-34.2	-29.4	-33.4	31.9	2
100	-34.1	-29.1	-32.9	33.3	2
103	-34.1	-29.0	-32.8	33.4	2
103	-33.1	-29.0	-31.9	35.9	2
150	-32.8	-28.5	-30.8	38.6	2
200	-32.4	-28.1	-29.6	41.2	2
250	-32.1	-27.6	-28.5	43.6	2
300	-31.8	-27.2	-27.3	45.7	2

CU2SE (C), COPPER SELENIDE
SOL PHASE TRANSITION AT 110C

T(C)	DELTA H	DELTA F	ENTHALPY	ENTROPY	REF
25	-14.2	-17.7	-14.2	37.6	2
50	-14.1	-18.0	-13.7	39.3	2
75	-14.0	-18.3	-13.1	40.9	2
100	-14.0	-18.6	-12.6	42.4	2
110	-13.9	-18.7	-12.4	42.9	2
110	-12.8	-18.7	-11.2	45.9	2
150	-12.7	-19.3	-10.4	48.1	2
200	-12.5	-20.1	-9.3	50.5	2
221	-12.5	-20.4	-8.8	51.5	2
221	-13.8	-20.4	-8.8	51.5	2
250	-13.8	-20.8	-8.2	52.7	2
300	-13.7	-21.5	-7.1	54.7	2

CU3(OH)2(CO3)2 (C), AZURITE

T(C)	DELTA H	DELTA F	ENTHALPY	ENTROPY	REF
25	-399.3	-343.9	-398.3	71.1	13
50	-398.0	-339.3	-396.6	76.6	13
75	-397.8	-334.8	-394.9	81.7	13
100	-397.5	-330.3	-393.1	86.6	13
150	-396.9	-321.3	-389.5	95.6	13
200	-396.3	-312.4	-385.9	103.8	13
250	-395.7	-303.6	-382.2	111.3	13
300	-395.1	-294.8	-378.4	118.2	13

CU3SO4(OH)4 (C), ANTLERITE

T(C)	DELTA H	DELTA F	ENTHALPY	ENTROPY	REF
25	-428.5	-355.3	-428.5	63.9	13
50	-428.5	-349.1	-426.9	69.0	13
75	-428.5	-343.0	-425.3	73.7	13
100	-428.5	-336.9	-423.7	78.2	13
150	-428.5	-324.6	-420.5	86.3	13
200	-428.5	-312.3	-417.2	93.6	13
250	-428.6	-300.0	-413.9	100.3	13
300	-428.6	-287.7	-410.5	106.4	13

CU4SO4(OH)6 (C), BROCHANTITE

T(C)	DELTA H	DELTA F	ENTHALPY	ENTROPY	REF
25	-538.4	-445.0	-538.4	84.1	13
50	-538.4	-437.2	-536.4	90.8	13
75	-538.4	-429.3	-534.3	97.1	13
100	-538.5	-421.5	-532.2	102.9	13
150	-538.5	-405.8	-527.9	113.6	13
200	-538.5	-390.1	-523.6	123.2	13
250	-538.5	-374.4	-519.3	131.9	13
300	-538.5	-358.8	-514.9	139.9	13

CU5FES4 (C), BORNITE
PHASE TRANSITIONS FOR CU5FES4 AT 212C, 267C

T(C)	DELTA H	DELTA F	ENTHALPY	ENTROPY	REF
25	-152.3	-131.8	-152.3	86.6	4
50	-152.0	-130.1	-150.8	91.4	8
75	-151.8	-128.4	-149.3	95.9	8
100	-151.6	-126.8	-147.7	100.1	8
150	-151.1	-123.4	-144.6	108.0	8
200	-150.5	-120.2	-141.4	115.3	8
212	-150.4	-119.5	-140.6	116.9	8
212	-149.0	-119.5	-139.1	119.8	8
250	-147.6	-117.2	-135.7	126.7	8
267	-146.9	-116.2	-134.1	129.7	8
267	-146.9	-116.2	-134.1	129.7	8
300	-146.1	-114.4	-131.4	134.5	8

CUBR (C), COPPER(1) BROMIDE

T(C)	DELTA H	DELTA F	ENTHALPY	ENTROPY	REF
25	-25.0	-24.1	-25.0	23.0	1
50	-25.0	-24.0	-24.7	24.0	2
58	-25.0	-24.0	-24.6	24.4	2
58	-28.6	-24.0	-24.6	24.4	2
75	-28.5	-23.7	-24.3	25.0	2
100	-28.5	-23.4	-24.0	25.9	2
150	-28.3	-22.7	-23.3	27.6	2
200	-28.2	-22.1	-22.7	29.2	2
250	-28.0	-21.4	-22.0	30.6	2
300	-27.8	-20.8	-21.3	31.8	2

CUCL (C), COPPER(1) CHLORIDE

T(C)	DELTA H	DELTA F	ENTHALPY	ENTROPY	REF
25	-32.8	-28.6	-32.8	20.6	1
50	-32.8	-28.3	-32.5	21.5	2
75	-32.7	-27.9	-32.2	22.4	2
100	-32.7	-27.6	-31.9	23.3	2
150	-32.5	-26.9	-31.3	24.8	2
200	-32.4	-26.3	-30.6	26.3	2
250	-32.3	-25.6	-30.0	27.6	2
300	-32.1	-25.0	-29.3	28.9	2

CUCL2 (C), COPPER(2) CHLORIDE

T(C)	DELTA H	DELTA F	ENTHALPY	ENTROPY	REF
25	-52.6	-42.1	-52.6	25.8	1
50	-52.5	-41.2	-52.2	27.2	2
75	-52.4	-40.3	-51.7	28.5	2
100	-52.3	-39.4	-51.3	29.8	2
150	-52.2	-37.7	-50.4	32.0	2
200	-52.0	-36.0	-49.5	34.0	2
250	-51.8	-34.3	-48.6	35.8	2
300	-51.7	-32.7	-47.7	37.5	2

CUCL2- (AQ)
(EXTRAPOLATED AS CRISS & COBBLE TYPE 2)

T(C)	DELTA H	DELTA F	ENTHALPY	ENTROPY	REF
25	-66.1	-57.9	-66.1	54.4	5
50	-67.5	-57.2	-67.6	49.6	3
75	-68.5	-56.3	-69.0	45.3	3
100	-69.4	-55.5	-70.4	41.4	3
150	-72.0	-53.4	-73.9	32.5	3
200	-74.9	-50.9	-77.9	23.2	3
250	-77.6	-48.2	-82.2	14.5	3
300	-82.0	-45.2	-88.1	3.7	3

CUF+ (AQ)

T(C)	DELTA H	DELTA F	ENTHALPY	ENTROPY	REF
25	-62.4	-52.7	-62.4	-21.0	1
50	-62.0	-51.9	-61.3	-17.5	3
75	-61.5	-51.1	-59.9	-13.1	3
100	-60.9	-50.4	-58.1	-8.1	3
150	-60.0	-49.1	-55.1	-0.4	3
200	-58.5	-48.0	-51.3	8.4	3
250	-57.2	-46.9	-47.1	16.7	3
300	-55.4	-46.0	-42.5	25.1	3

CUF2 (C), COPPER(2) FLUORIDE

T(C)	DELTA H	DELTA F	ENTHALPY	ENTROPY	REF
25	-131.2	-119.3	-131.2	16.4	2
50	-131.1	-119.3	-130.8	17.8	2
75	-131.0	-117.3	-130.4	19.0	2
100	-130.9	-116.3	-129.9	20.2	2
150	-130.8	-114.4	-129.0	22.4	2
200	-130.6	-112.4	-128.2	24.4	2
250	-130.4	-110.5	-127.3	26.2	2
300	-130.2	-108.6	-126.3	27.9	2

CUFE2O4 (C), CUPRIC FERRITE

T(C)	DELTA H	DELTA F	ENTHALPY	ENTROPY	REF
25	-231.0	-206.0	-231.0	35.1	4
50	-230.9	-203.9	-230.1	38.0	9
75	-230.8	-201.8	-229.1	40.8	9
100	-230.6	-199.7	-228.2	43.5	9
150	-230.2	-195.6	-226.1	48.7	9
200	-229.8	-191.6	-224.0	53.5	9
250	-229.3	-187.5	-221.7	58.1	9
300	-228.8	-183.6	-219.4	62.3	9

CUFEO2 (C), CUPROUS FERRITE

T(C)	DELTA H	DELTA F	ENTHALPY	ENTROPY	REF
25	-122.5	-109.9	-122.5	21.2	4
50	-122.5	-108.9	-122.0	22.8	9
75	-122.5	-107.8	-121.5	24.3	9
100	-122.4	-106.8	-121.0	25.8	9
150	-122.4	-104.7	-119.9	28.4	9
200	-122.3	-102.6	-118.8	30.9	9
250	-122.2	-100.5	-117.7	33.2	9
300	-122.1	-98.4	-116.6	35.2	9

CUFES2 (C), CHALCOPYRITE

T(C)	DELTA H	DELTA F	ENTHALPY	ENTROPY	REF
25	-76.2	-64.5	-76.2	29.9	4
50	-76.1	-63.6	-75.6	31.8	8
75	-76.0	-62.6	-75.0	33.5	8
100	-75.9	-61.6	-74.4	35.2	8
150	-75.7	-59.7	-73.1	38.4	8
200	-75.4	-57.9	-71.8	41.3	8
250	-75.2	-56.0	-70.5	43.9	8
300	-74.9	-54.2	-69.1	46.4	8

CUI (C), COPPER(1) IODIDE

T(C)	DELTA H	DELTA F	ENTHALPY	ENTROPY	REF
25	-16.2	-16.6	-16.2	23.1	1
50	-16.2	-16.6	-15.9	24.2	2
75	-16.2	-16.7	-15.5	25.1	2
100	-16.2	-16.7	-15.2	26.0	2
114	-16.2	-16.7	-15.0	26.5	2
114	-18.0	-16.7	-15.0	26.5	2
150	-18.1	-16.6	-14.6	27.7	2
185	-18.2	-16.5	-14.1	28.8	2
185	-23.2	-16.5	-14.1	28.8	2
200	-23.2	-16.2	-13.9	29.2	2
250	-23.0	-15.5	-13.2	30.5	2
300	-22.9	-14.8	-12.5	31.8	2

CUO (C), COPPER(2) OXIDE

T(C)	DELTA H	DELTA F	ENTHALPY	ENTROPY	REF
25	-37.6	-31.0	-37.6	10.2	1
50	-37.6	-30.4	-37.3	11.0	2
75	-37.6	-29.9	-37.1	11.8	2
100	-37.5	-29.3	-36.8	12.6	2
150	-37.4	-28.2	-36.3	14.0	2
200	-37.3	-27.1	-35.7	15.3	2
250	-37.3	-26.1	-35.1	16.5	2
300	-37.1	-25.0	-34.5	17.6	2

CUO.CUSO4 (C), COPPER(2) OXIDE SULFATE

T(C)	DELTA H	DELTA F	ENTHALPY	ENTROPY	REF
25	-237.0	-198.9	-237.0	37.6	2
50	-237.0	-195.7	-236.2	40.4	2
75	-237.0	-192.5	-235.3	43.1	2
100	-236.9	-189.3	-234.3	45.7	2
150	-236.6	-182.9	-232.4	50.6	2
200	-236.3	-176.6	-230.3	55.2	2
250	-236.0	-170.3	-228.2	59.4	2
300	-235.5	-164.1	-226.0	63.4	2

CUS (C), COPPER(2) SULFIDE

T(C)	DELTA H	DELTA F	ENTHALPY	ENTROPY	REF
25	-28.0	-22.3	-28.0	15.9	1
50	-28.0	-21.8	-27.8	16.8	2
75	-28.0	-21.3	-27.5	17.7	2
100	-27.9	-20.9	-27.2	18.5	2
150	-27.8	-19.9	-26.6	19.9	2
200	-27.8	-19.0	-26.0	21.3	2
250	-27.7	-18.1	-25.4	22.5	2
300	-27.6	-17.2	-24.8	23.5	2

CUSIO3.2H2O (C), CHRYSOCOLLA

T(C)	DELTA H	DELTA F	ENTHALPY	ENTROPY	REF
25	-320.7	-271.3	-320.7	31.6	13
50	-321.0	-267.1	-320.0	34.0	13
75	-321.3	-263.0	-319.2	36.4	13
100	-321.6	-258.8	-318.4	38.6	13
150	-322.1	-250.3	-316.7	42.7	13
200	-322.6	-241.8	-315.0	46.5	13
250	-323.1	-233.2	-313.3	49.9	13
300	-323.6	-224.6	-311.6	53.2	13

CUSO4 (C), COPPER(2) SULFATE

T(C)	DELTA H	DELTA F	ENTHALPY	ENTROPY	REF
25	-199.7	-167.7	-199.7	26.0	1
50	-199.7	-165.1	-199.1	28.0	2
75	-199.7	-162.4	-198.5	29.8	2
100	-199.6	-159.7	-197.8	31.7	2
150	-199.5	-154.4	-196.4	35.1	2
200	-199.3	-149.0	-195.0	38.3	2
250	-199.0	-143.8	-193.5	41.4	2
300	-198.7	-138.5	-191.9	44.3	2

CUSO4.H2O (C), COPPER(2) SULFATE MONOHYDRATE

T(C)	DELTA H	DELTA F	ENTHALPY	ENTROPY	REF
25	-274.9	-229.0	-274.9	34.9	1
50	-274.9	-225.1	-274.1	37.5	4
75	-275.0	-221.2	-273.2	39.9	4
100	-275.0	-217.4	-272.4	42.3	4
150	-275.0	-209.7	-270.6	46.7	4
200	-274.9	-201.9	-268.8	50.8	4
250	-274.8	-194.2	-266.9	54.6	4

CUSO4.3H2O (C), COPPER(2) SULFATE 3-HYDRATE

T(C)	DELTA H	DELTA F	ENTHALPY	ENTROPY	REF
25	-417.9	-344.1	-417.9	52.9	1
50	-418.0	-337.9	-416.7	56.9	4
75	-418.2	-331.7	-415.4	60.7	4
100	-418.3	-325.5	-414.1	64.3	4
150	-418.4	-313.1	-411.5	70.9	4
200	-418.6	-300.6	-408.7	77.0	4
250	-418.7	-288.2	-406.0	82.6	4

CUSO4.5H2O (C), COPPER(2) SULFATE 5-HYDRATE

T(C)	DELTA H	DELTA F	ENTHALPY	ENTROPY	REF
25	-560.2	-458.8	-560.2	71.8	1
50	-560.4	-450.3	-558.5	77.3	4
75	-560.6	-441.8	-556.8	82.4	4
100	-560.8	-433.3	-555.0	87.2	4
150	-561.1	-416.2	-551.5	96.2	4
200	-561.4	-399.0	-547.8	104.3	4
250	-561.6	-381.9	-544.2	111.7	4

F- (AQ)

T(C)	DELTA H	DELTA F	ENTHALPY	ENTROPY	REF
25	-79.5	-66.6	-79.5	1.7	1
50	-80.3	-65.5	-80.6	-2.0	3
75	-81.1	-64.4	-82.1	-6.4	3
100	-82.1	-63.1	-83.8	-11.3	3
150	-83.9	-60.4	-87.1	-19.6	3
200	-86.1	-57.4	-90.9	-28.5	3
250	-89.3	-54.3	-95.2	-37.0	3
300	-91.9	-50.9	-100.9	-47.6	3

F2 (G), FLUORINE

T(C)	DELTA H	DELTA F	ENTHALPY	ENTROPY	REF
25	0.0	0.0	0.0	48.4	1
50	0.0	0.0	0.2	49.1	2
75	0.0	0.0	0.4	49.6	2
100	0.0	0.0	0.6	50.2	2
150	0.0	0.0	1.0	51.2	2
200	0.0	0.0	1.4	52.1	2
250	0.0	0.0	1.8	52.9	2
300	0.0	0.0	2.2	53.7	2

FE (C), IRON

T(C)	DELTA H	DELTA F	ENTHALPY	ENTROPY	REF
25	0.0	0.0	0.0	6.5	1
50	0.0	0.0	0.2	7.0	2
75	0.0	0.0	0.3	7.5	2
100	0.0	0.0	0.5	7.9	2
150	0.0	0.0	0.8	8.7	2
200	0.0	0.0	1.1	9.5	2
250	0.0	0.0	1.5	10.2	2
300	0.0	0.0	1.9	10.9	2

FE(CN)6--- (AQ)
(EXTRAPOLATED AS CRISS & COBBLE TYPE 2)

T(C)	DELTA H	DELTA F	ENTHALPY	ENTROPY	REF
25	134.3	174.3	134.3	79.6	1
50	133.0	177.7	132.6	74.2	3
75	132.7	181.1	131.2	70.0	3
100	133.1	184.5	129.9	66.6	3
150	132.1	191.5	126.4	57.4	3
200	131.1	198.6	122.3	47.9	3
250	131.2	205.7	117.9	39.2	3
300	129.7	212.9	111.9	28.2	3

FE(CN)6---- (AQ)
(EXTRAPOLATED AS CRISS & COBBLE TYPE 2)

T(C)	DELTA H	DELTA F	ENTHALPY	ENTROPY	REF
25	108.9	166.1	108.9	42.7	1
50	108.4	170.9	107.5	38.1	3
75	108.7	175.7	106.0	33.8	3
100	109.8	180.4	104.6	29.7	3
150	110.5	189.8	101.1	20.9	3
200	111.5	199.0	97.2	11.7	3
250	114.0	208.1	92.9	3.1	3
300	114.9	217.1	87.0	-7.7	3

FE(OH)2 (C), IRON(2) HYDROXIDE

T(C)	DELTA H	DELTA F	ENTHALPY	ENTROPY	REF
25	-136.0	-116.4	-136.0	21.0	1
50	-135.9	-114.8	-135.4	22.9	2
75	-135.8	-113.1	-134.8	24.6	2
100	-135.7	-111.5	-134.2	26.3	2
150	-135.6	-108.3	-133.0	29.4	2
200	-135.4	-105.1	-131.8	32.2	2
250	-135.2	-101.9	-130.5	34.7	2
300	-135.0	-98.7	-129.2	37.1	2

FE(OH)3 (C), IRON(3) HYDROXIDE

T(C)	DELTA H	DELTA F	ENTHALPY	ENTROPY	REF
25	-196.7	-166.5	-196.7	25.5	1
50	-196.8	-164.0	-196.1	27.5	2
75	-196.8	-161.4	-195.4	29.4	2
100	-196.8	-158.9	-194.8	31.3	2
150	-196.8	-153.8	-193.3	34.9	2
200	-196.7	-148.7	-191.9	38.2	2
250	-196.6	-143.6	-190.3	41.3	2
300	-196.4	-138.6	-188.7	44.3	2

FE++ (AQ)

T(C)	DELTA H	DELTA F	ENTHALPY	ENTROPY	REF
25	-21.3	-18.8	-21.3	-42.9	1
50	-21.1	-18.7	-20.0	-38.7	3
75	-20.9	-18.5	-18.2	-33.4	3
100	-20.7	-18.3	-16.1	-27.3	3
150	-20.5	-18.0	-12.3	-17.8	3
200	-19.9	-17.8	-7.8	-7.2	3
250	-19.9	-17.5	-2.8	2.9	3
300	-19.3	-17.3	2.8	13.1	3

46

FE+++ (AQ)

T(C)	DELTA H	DELTA F	ENTHALPY	ENTROPY	REF
25	-11.6	-1.1	-11.6	-90.5	1
50	-11.4	-0.2	-9.8	-84.8	3
75	-11.2	0.6	-7.4	-77.4	3
100	-11.1	1.5	-4.4	-69.0	3
150	-10.9	3.2	0.9	-55.5	3
200	-10.5	4.8	7.1	-41.1	3
250	-10.8	6.4	14.1	-27.1	3
300	-10.5	8.0	21.8	-13.0	3

FE.9470 (C), WUESTITE

T(C)	DELTA H	DELTA F	ENTHALPY	ENTROPY	REF
25	-63.6	-59.6	-63.6	13.7	1
50	-63.6	-58.2	-63.3	14.7	2
75	-63.5	-57.8	-63.1	15.5	2
100	-63.5	-57.3	-62.8	16.4	2
150	-63.4	-56.5	-62.2	17.9	2
200	-63.3	-55.7	-61.6	19.3	2
250	-63.2	-54.9	-60.9	20.5	2
300	-63.1	-54.2	-60.3	21.6	2

FE2(OH)2++++ (AQ)

T(C)	DELTA H	DELTA F	ENTHALPY	ENTROPY	REF
25	-146.2	-111.7	-146.2	-105.0	1
50	-146.8	-108.8	-144.3	-98.8	3
75	-147.7	-105.8	-141.7	-90.9	3
100	-148.7	-102.7	-138.4	-81.7	3
150	-150.7	-96.4	-132.6	-67.0	3
200	-152.6	-89.8	-125.9	-51.4	3
250	-155.8	-83.0	-118.4	-36.3	3
300	-158.2	-75.9	-110.0	-20.9	3

FE2(SO4)3 (C), IRON(3) SULFATE

T(C)	DELTA H	DELTA F	ENTHALPY	ENTROPY	REF
25	-663.4	-569.4	-663.4	73.5	2
50	-663.4	-561.5	-661.8	78.7	2
75	-663.4	-553.6	-660.1	83.8	2
100	-663.3	-545.7	-658.3	88.7	2
150	-663.1	-530.0	-654.6	98.0	2
200	-662.7	-514.3	-650.7	106.7	2
250	-662.2	-498.6	-646.6	114.8	2
300	-661.7	-483.0	-642.4	122.4	2

FE2O3 (C), IRON(3) OXIDE (HEMATITE)

T(C)	DELTA H	DELTA F	ENTHALPY	ENTROPY	REF
25	-197.0	-177.4	-197.0	20.9	1
50	-196.9	-175.8	-196.4	23.0	2
75	-196.8	-174.1	-195.7	24.9	2
100	-196.7	-172.5	-195.0	26.8	2
150	-196.5	-169.3	-193.6	30.4	2
200	-196.2	-166.1	-192.1	33.8	2
250	-196.0	-162.9	-190.5	36.9	2
300	-195.6	-159.8	-188.9	39.9	2

FE2SIO4 (C), IRON ORTHOSILICATE (FAYALITE)

T(C)	DELTA H	DELTA F	ENTHALPY	ENTROPY	REF
25	-353.7	-329.6	-353.7	34.7	1
50	-353.7	-327.6	-352.9	37.3	10
75	-353.6	-325.6	-352.0	39.8	10
100	-353.5	-323.5	-351.2	42.2	10
150	-353.4	-319.5	-349.4	46.8	10
200	-353.2	-315.5	-347.5	50.9	10
250	-353.0	-311.6	-345.6	54.8	10
300	-352.9	-307.6	-343.6	58.4	10

FE3O4 (C), IRON(2,3) OXIDE (MAGNETITE)

T(C)	DELTA H	DELTA F	ENTHALPY	ENTROPY	REF
25	-267.3	-242.7	-267.3	35.0	1
50	-267.2	-240.6	-266.4	37.9	2
75	-267.1	-238.6	-265.5	40.7	2
100	-267.0	-236.5	-264.5	43.4	2
150	-266.6	-232.5	-262.5	48.5	2
200	-266.3	-228.5	-260.3	53.3	2
250	-265.8	-224.5	-258.0	57.8	2
300	-265.3	-220.6	-255.6	62.2	2

FEBR++ (AQ)

T(C)	DELTA H	DELTA F	ENTHALPY	ENTROPY	REF
25	-34.7	-26.8	-34.7	-43.0	1
50	-34.7	-26.1	-33.4	-38.8	3
58	-34.7	-25.9	-33.0	-37.5	2
58	-38.3	-25.9	-33.0	-37.5	2
75	-38.2	-25.3	-31.6	-33.5	3
100	-38.1	-24.4	-29.5	-27.4	3
150	-38.1	-22.5	-25.7	-17.9	3
200	-37.7	-20.7	-21.2	-7.3	3
250	-37.9	-18.9	-16.2	2.8	3
300	-37.6	-17.1	-10.5	13.0	3

FEBR2 (C), IRON(2) BROMIDE

T(C)	DELTA H	DELTA F	ENTHALPY	ENTROPY	REF
25	-59.5	-56.7	-59.5	33.6	2
50	-59.6	-56.5	-59.0	35.2	2
58	-59.6	-56.4	-58.9	35.7	2
58	-66.7	-56.4	-58.9	35.7	2
75	-66.7	-55.9	-58.5	36.6	2
100	-66.6	-55.2	-58.1	38.0	2
150	-66.3	-53.6	-57.1	40.4	2
200	-66.1	-52.2	-56.1	42.7	2
250	-65.9	-50.7	-55.1	44.7	2
300	-65.7	-49.3	-54.0	46.6	2

FECL++ (AQ)

T(C)	DELTA H	DELTA F	ENTHALPY	ENTROPY	REF
25	-43.0	-34.4	-43.0	-37.0	1
50	-43.0	-33.7	-41.8	-33.0	3
75	-43.0	-33.0	-40.1	-27.9	3
100	-43.0	-32.2	-38.1	-22.1	3
150	-43.2	-30.8	-34.5	-13.1	3
200	-43.0	-29.3	-30.2	-3.0	3
250	-43.4	-27.9	-25.4	6.6	3
300	-43.4	-26.4	-20.0	16.3	3

FECL2 (C), IRON(2) CHLORIDE

T(C)	DELTA H	DELTA F	ENTHALPY	ENTROPY	REF
25	-81.7	-72.3	-81.7	28.2	1
50	-81.6	-71.5	-81.2	29.7	2
75	-81.5	-70.7	-80.8	31.0	2
100	-81.4	-69.9	-80.3	32.4	2
150	-81.2	-68.4	-79.3	34.8	2
200	-81.0	-66.9	-78.4	36.9	2
250	-80.8	-65.4	-77.4	38.9	2
300	-80.6	-64.0	-76.4	40.7	2

FECL3 (C), IRON(3) CHLORIDE

T(C)	DELTA H	DELTA F	ENTHALPY	ENTROPY	REF
25	-95.5	-79.8	-95.5	34.0	1
50	-95.4	-78.5	-94.9	35.9	2
75	-95.2	-77.2	-94.3	37.7	2
100	-95.1	-75.9	-93.7	39.4	2
150	-94.7	-73.4	-92.4	42.7	2
200	-94.3	-70.9	-91.0	45.7	2
250	-93.9	-69.5	-89.6	48.6	2
300	-93.4	-66.1	-88.1	51.3	2

FECO3 (C), IRON(2) CARBONATE

T(C)	DELTA H	DELTA F	ENTHALPY	ENTROPY	REF
25	-177.0	-159.4	-177.0	22.2	1
50	-177.0	-157.9	-176.5	23.8	2
75	-176.9	-156.4	-176.0	25.3	2
100	-176.9	-154.9	-175.5	26.8	2
150	-176.8	-152.0	-174.3	29.6	2
200	-176.7	-149.1	-173.2	32.3	2
250	-176.5	-146.1	-171.9	34.8	2
300	-176.3	-143.3	-170.6	37.2	2

FEF++ (AQ)

T(C)	DELTA H	DELTA F	ENTHALPY	ENTROPY	REF
25	-88.6	-77.1	-88.6	-49.0	1
50	-88.4	-76.1	-87.2	-44.6	3
75	-88.2	-75.2	-85.4	-39.0	3
100	-88.0	-74.3	-83.1	-32.6	3
150	-87.8	-72.4	-79.2	-22.6	3
200	-87.2	-70.7	-74.4	-11.5	3
250	-87.1	-69.0	-69.1	-1.0	3
300	-86.5	-67.3	-63.3	9.8	3

FEF2 (C), IRON(2) FLUORIDE

T(C)	DELTA H	DELTA F	ENTHALPY	ENTROPY	REF
25	-168.7	-158.5	-168.7	20.8	2
50	-168.6	-157.7	-168.3	22.1	2
75	-168.6	-156.8	-167.9	23.4	2
100	-168.5	-156.0	-167.5	24.5	2
150	-168.3	-154.3	-166.6	26.7	2
200	-168.2	-152.7	-165.7	28.7	2
250	-168.1	-151.0	-164.8	30.5	2
300	-167.9	-149.4	-163.9	32.2	2

FEF2+ (AQ)

T(C)	DELTA H	DELTA F	ENTHALPY	ENTROPY	REF
25	-166.4	-150.2	-166.4	-20.0	1
50	-166.1	-148.9	-165.3	-16.6	3
75	-165.8	-147.5	-163.9	-12.2	3
100	-165.3	-146.3	-162.1	-7.2	3
150	-164.6	-143.7	-159.1	0.4	3
200	-163.4	-141.4	-155.4	9.1	3
250	-162.4	-139.2	-151.3	17.3	3
300	-160.9	-137.0	-146.7	25.6	3

FEF3 (C), IRON(3) FLUORIDE

T(C)	DELTA H	DELTA F	ENTHALPY	ENTROPY	REF
25	-249.0	-232.4	-249.0	23.5	2
50	-248.9	-231.0	-248.5	25.3	2
75	-248.8	-229.6	-247.9	26.9	2
100	-248.7	-228.3	-247.3	28.5	2
150	-248.5	-225.5	-246.2	31.3	2
200	-248.3	-222.9	-245.0	33.9	2
250	-248.1	-220.2	-243.9	36.2	2
300	-247.9	-217.5	-242.7	38.3	2

FEI2 (C), IRON(2) IODIDE

T(C)	DELTA H	DELTA F	ENTHALPY	ENTROPY	REF
25	-25.0	-26.7	-25.0	40.0	2
50	-25.0	-26.8	-24.5	41.6	2
75	-25.0	-27.0	-24.0	43.1	2
100	-25.0	-27.1	-23.5	44.5	2
114	-25.0	-27.2	-23.2	45.2	2
114	-28.7	-27.2	-23.2	45.2	2
150	-28.9	-27.1	-22.5	47.0	2
185	-29.1	-26.9	-21.8	48.6	2
185	-39.2	-26.9	-21.8	48.6	2
200	-39.1	-26.5	-21.5	49.3	2
250	-38.9	-25.2	-20.5	51.3	2
300	-38.7	-23.9	-19.5	53.1	2

FEOH+ (AQ)

T(C)	DELTA H	DELTA F	ENTHALPY	ENTROPY	REF
25	-77.6	-66.3	-77.6	-12.0	1
50	-77.4	-65.4	-76.6	-8.8	3
75	-77.1	-64.4	-75.3	-4.8	3
100	-76.8	-63.6	-73.7	-0.2	3
150	-76.3	-61.8	-71.0	6.7	3
200	-75.4	-60.2	-67.5	14.8	3
250	-74.6	-58.7	-63.7	22.3	3
300	-73.5	-57.2	-59.5	30.0	3

FEOH++ (AQ)

T(C)	DELTA H	DELTA F	ENTHALPY	ENTROPY	REF
25	-69.6	-54.8	-69.6	-44.0	1
50	-69.5	-53.6	-68.3	-39.8	3
75	-69.5	-52.4	-66.5	-34.4	3
100	-69.4	-51.1	-64.3	-28.2	3
150	-69.5	-48.7	-60.5	-18.7	3
200	-69.3	-46.3	-55.9	-8.0	3
250	-69.6	-43.8	-50.9	2.2	3
300	-69.3	-41.3	-45.2	12.5	3

FES (C), IRON(2) SULFIDE
PHASE TRANS OF FES AT 138C

T(C)	DELTA H	DELTA F	ENTHALPY	ENTROPY	REF
25	-39.2	-33.5	-39.2	14.4	1
50	-39.2	-33.0	-38.9	15.5	2
75	-39.1	-32.5	-38.6	16.5	2
100	-38.9	-32.1	-38.2	17.5	2
138	-38.8	-31.4	-37.6	19.1	2
138	-38.2	-31.4	-37.0	20.4	2
150	-38.1	-31.2	-36.8	21.0	2
200	-37.8	-30.3	-35.9	22.9	2
250	-37.5	-29.6	-35.1	24.6	2
300	-37.2	-28.8	-34.2	26.2	2

FES2 (C), IRON DISULFIDE (PYRITE)

T(C)	DELTA H	DELTA F	ENTHALPY	ENTROPY	REF
25	-73.3	-58.9	-73.3	12.6	1
50	-73.3	-57.7	-72.9	13.9	2
75	-73.2	-56.4	-72.5	15.0	2
100	-73.2	-55.2	-72.1	16.1	2
150	-73.1	-52.8	-71.3	18.2	2
200	-73.0	-50.5	-70.4	20.1	2
250	-72.9	-48.1	-69.6	21.8	2
300	-72.8	-45.7	-68.7	23.5	2

FESO4 (C), IRON(2) SULFATE

T(C)	DELTA H	DELTA F	ENTHALPY	ENTROPY	REF
25	-237.2	-205.6	-237.2	25.7	1
50	-237.2	-203.0	-236.6	27.7	2
75	-237.2	-200.3	-236.0	29.6	2
100	-237.2	-197.7	-235.3	31.4	2
150	-237.0	-192.4	-233.9	34.9	2
200	-236.8	-187.1	-232.5	38.2	2
250	-236.6	-181.9	-230.9	41.3	2
300	-236.4	-176.6	-229.3	44.2	2

FESO4+ (AQ)

T(C)	DELTA H	DELTA F	ENTHALPY	ENTROPY	REF
25	-238.1	-194.2	-238.1	-36.0	1
50	-237.9	-190.5	-236.8	-32.1	3
75	-237.5	-186.9	-235.2	-27.0	3
100	-237.0	-183.3	-233.1	-21.2	3
150	-236.4	-176.1	-229.6	-12.3	3
200	-235.1	-169.1	-225.3	-2.3	3
250	-234.0	-162.2	-220.5	7.2	3
300	-232.4	-155.4	-215.3	16.9	3

GA (C,L), GALLIUM
MELT PT = 30C

T(C)	DELTA H	DELTA F	ENTHALPY	ENTROPY	REF
25	0.0	0.0	0.0	9.8	1
30	0.0	0.0	0.0	9.9	2
30	0.0	0.0	1.4	14.3	2
50	0.0	0.0	1.5	14.7	2
75	0.0	0.0	1.7	15.2	2
100	0.0	0.0	1.8	15.7	2
150	0.0	0.0	2.2	16.5	2
200	0.0	0.0	2.5	17.2	2
250	0.0	0.0	2.8	17.9	2
300	0.0	0.0	3.2	18.5	2

GA+++ (AQ)

T(C)	DELTA H	DELTA F	ENTHALPY	ENTROPY	REF
25	-50.5	-38.0	-50.5	-94.0	1
30	-50.5	-37.8	-50.2	-92.9	2
30	-51.8	-37.8	-50.2	-92.9	2
50	-51.6	-36.9	-48.7	-88.2	3
75	-51.4	-35.7	-46.2	-80.7	3
100	-51.3	-34.6	-43.2	-72.0	3
150	-50.9	-32.4	-37.8	-58.3	3
200	-50.4	-30.3	-31.4	-43.5	3
250	-50.6	-28.1	-24.3	-29.3	3
300	-50.1	-26.0	-16.4	-14.9	3

GA2O3 (C), GALLIUM SESQUIOXIDE

T(C)	DELTA H	DELTA F	ENTHALPY	ENTROPY	REF
25	-260.3	-238.6	-260.3	20.3	1
30	-260.3	-238.3	-260.2	20.6	2
30	-263.0	-238.3	-260.2	20.6	2
50	-263.1	-236.6	-259.8	21.9	2
75	-263.2	-234.6	-259.3	23.4	2
100	-263.3	-232.5	-258.8	24.8	2
150	-263.4	-228.4	-257.7	27.6	2
200	-263.4	-224.2	-256.5	30.2	2
250	-263.5	-220.1	-255.3	32.6	2
300	-263.4	-216.0	-254.0	34.9	2

GABR4- (AQ)
(EXTRAPOLATED AS CRISS & COBBLE TYPE 2)

T(C)	DELTA H	DELTA F	ENTHALPY	ENTROPY	REF
25	-158.2	-131.5	-158.2	13.6	1
30	-158.5	-131.1	-158.4	12.9	2
30	-159.9	-131.1	-158.4	12.9	2
50	-161.3	-129.1	-159.4	9.7	3
58	-161.9	-128.3	-159.9	8.4	2
58	-176.1	-129.3	-159.9	8.4	2
75	-177.0	-125.8	-160.9	5.3	3
100	-178.3	-122.1	-162.5	0.6	3
150	-181.3	-114.4	-165.9	-7.9	3
200	-184.6	-106.2	-169.7	-16.9	3
250	-187.7	-97.8	-174.0	-25.4	3
300	-192.4	-89.0	-179.8	-36.0	3

GAF++ (AQ)

T(C)	DELTA H	DELTA F	ENTHALPY	ENTROPY	REF
25	-128.0	-111.1	-128.0	-64.0	1
30	-128.0	-110.8	-127.8	-63.1	2
30	-129.3	-110.8	-127.8	-63.1	2
50	-129.1	-109.6	-126.5	-59.2	3
75	-128.7	-108.1	-124.5	-52.9	3
100	-128.2	-106.7	-121.9	-45.8	3
150	-127.5	-103.8	-117.5	-34.5	3
200	-126.4	-101.1	-112.2	-22.2	3
250	-125.7	-98.5	-106.3	-10.4	3
300	-124.4	-96.0	-99.8	1.5	3

GAF2+ (AQ)

T(C)	DELTA H	DELTA F	ENTHALPY	ENTROPY	REF
25	-205.9	-182.8	-205.9	-40.0	1
30	-205.9	-182.4	-205.7	-39.2	2
30	-207.2	-182.4	-205.7	-39.2	2
50	-206.8	-180.8	-204.7	-35.9	3
75	-206.2	-178.8	-202.9	-30.7	3
100	-205.3	-176.9	-200.8	-24.7	3
150	-204.0	-173.2	-197.2	-15.5	3
200	-202.1	-169.7	-192.7	-5.1	3
250	-200.2	-166.4	-187.8	4.7	3
300	-197.9	-163.3	-182.4	14.7	3

GD (C), GADOLINIUM

T(C)	DELTA H	DELTA F	ENTHALPY	ENTROPY	REF
25	0.0	0.0	0.0	16.3	1
50	0.0	0.0	0.2	17.0	2
75	0.0	0.0	0.4	17.6	2
100	0.0	0.0	0.7	18.3	2
150	0.0	0.0	1.1	19.3	2
200	0.0	0.0	1.5	20.3	2
250	0.0	0.0	2.0	21.2	2
300	0.0	0.0	2.4	22.0	2

GD+++ (AQ)

T(C)	DELTA H	DELTA F	ENTHALPY	ENTROPY	REF
25	-163.6	-158.0	-163.6	-64.2	1
50	-163.7	-157.5	-162.1	-59.3	3
75	-164.0	-157.0	-160.0	-53.1	3
100	-164.4	-156.5	-157.5	-45.9	3
150	-165.1	-155.4	-153.0	-34.7	3
200	-165.7	-154.2	-147.7	-22.3	3
250	-167.2	-152.9	-141.8	-10.5	3
300	-168.1	-151.5	-135.3	1.4	3

GE (C), GERMANIUM

T(C)	DELTA H	DELTA F	ENTHALPY	ENTROPY	REF
25	0.0	0.0	0.0	7.4	1
50	0.0	0.0	0.1	7.9	2
75	0.0	0.0	0.3	8.3	2
100	0.0	0.0	0.4	8.7	2
150	0.0	0.0	0.7	9.5	2
200	0.0	0.0	1.0	10.1	2
250	0.0	0.0	1.3	10.8	2
300	0.0	0.0	1.7	11.3	2

GEO2 (C), GERMANIUM DIOXIDE

T(C)	DELTA H	DELTA F	ENTHALPY	ENTROPY	REF
25	-131.7	-118.8	-131.7	13.2	1
50	-131.7	-117.7	-131.4	14.3	2
75	-131.7	-116.6	-131.0	15.3	2
100	-131.6	-115.6	-130.7	16.3	2
150	-131.6	-113.4	-130.0	18.0	2
200	-131.6	-111.3	-129.3	19.6	2
250	-131.5	-109.1	-128.5	21.1	2
300	-131.4	-107.0	-127.8	22.5	2

H+ (AQ)

T(C)	DELTA H	DELTA F	ENTHALPY	ENTROPY	REF
25	0.0	0.0	0.0	-5.0	1
50	0.0	0.0	0.5	-3.2	3
75	0.0	0.0	1.3	-0.8	3
100	0.0	0.0	2.3	2.0	3
150	0.0	0.0	4.1	6.5	3
200	0.0	0.0	6.1	11.1	3
250	0.0	0.0	8.6	16.1	3
300	0.0	0.0	11.3	20.7	3

H2 (G), HYDROGEN

T(C)	DELTA H	DELTA F	ENTHALPY	ENTROPY	REF
25	0.0	0.0	0.0	31.2	1
50	0.0	0.0	0.2	31.8	2
75	0.0	0.0	0.3	32.3	2
100	0.0	0.0	0.5	32.8	2
150	0.0	0.0	0.9	33.6	2
200	0.0	0.0	1.2	34.4	2
250	0.0	0.0	1.6	35.1	2
300	0.0	0.0	1.9	35.7	2

H2 (AQ)

T(C)	DELTA H	DELTA F	ENTHALPY	ENTROPY	REF
25	-1.0	4.2	-1.0	13.8	1
50	0.1	4.7	0.2	17.5	14
75	0.8	5.0	1.0	20.1	14
100	1.4	5.2	1.9	22.6	14
150	2.6	5.7	3.5	26.3	14
200	3.8	5.9	5.1	29.9	14
250	5.6	6.5	6.9	33.4	14
300	7.5	6.9	9.7	36.8	14

H2ASO3- (AQ)

T(C)	DELTA H	DELTA F	ENTHALPY	ENTROPY	REF
25	-170.9	-140.4	-170.9	31.4	1
50	-171.3	-137.8	-171.2	30.3	3
75	-171.4	-135.2	-171.4	29.6	3
100	-171.3	-132.6	-171.6	29.2	3
150	-172.5	-127.3	-173.2	24.8	3
200	-172.7	-122.0	-174.1	22.9	3
250	-172.7	-116.6	-175.1	20.8	3
300	-172.7	-111.3	-176.3	18.7	3

H2ASO4- (AQ)

T(C)	DELTA H	DELTA F	ENTHALPY	ENTROPY	REF
25	-217.4	-180.0	-217.4	33.0	1
50	-217.8	-176.9	-217.6	32.3	3
75	-217.8	-173.7	-217.7	32.1	3
100	-217.6	-170.6	-217.8	32.2	3
150	-218.7	-164.1	-219.0	28.6	3
200	-218.7	-157.8	-219.4	27.7	3
250	-218.4	-151.4	-220.0	26.5	3
300	-218.1	-145.0	-220.6	25.3	3

H2BO3- (AQ)

T(C)	DELTA H	DELTA F	ENTHALPY	ENTROPY	REF
25	-251.7	-217.6	-251.7	12.3	5
50	-253.7	-214.7	-253.7	6.0	3
75	-255.7	-211.6	-255.8	-0.5	3
100	-257.6	-208.3	-258.2	-7.0	3
150	-262.5	-201.3	-263.5	-20.7	3
200	-267.3	-193.7	-269.1	-33.6	3
250	-272.6	-185.6	-275.6	-46.6	3
300	-278.6	-177.0	-282.7	-59.6	3

H2CO3 (AQ), UNDISSOCIATED COMPLEX

T(C)	DELTA H	DELTA F	ENTHALPY	ENTROPY	REF
25	-167.2	-148.9	-167.2	44.7	6
50	-167.0	-147.4	-166.5	46.9	6
75	-166.3	-145.9	-165.4	50.5	6
100	-165.2	-144.5	-163.7	55.1	6
150	-163.2	-141.7	-160.9	62.4	6
200	-157.9	-139.5	-155.1	76.6	6

H2FE(CN)6-- (AQ)
(EXTRAPOLATED AS CRISS & COBBLE TYPE 2)

T(C)	DELTA H	DELTA F	ENTHALPY	ENTROPY	REF
25	108.9	157.4	108.9	62.0	1
50	107.1	161.5	107.4	57.0	3
75	105.9	165.7	105.9	52.7	3
100	105.1	170.0	104.6	49.0	3
150	102.3	178.9	101.1	40.0	3
200	99.2	188.2	97.0	30.6	3
250	96.7	197.8	92.7	21.9	3
300	92.5	207.6	86.8	11.1	3

H2GEO3 (AQ), UNDISSOCIATED COMPLEX

T(C)	DELTA H	DELTA F	ENTHALPY	ENTROPY	REF
25	-202.6	-182.0	-202.6	43.0	5
50	-202.6	-180.3	-202.0	44.9	6
75	-202.1	-178.6	-201.0	48.1	6
100	-201.2	-176.9	-199.5	52.3	6
150	-199.7	-173.6	-197.0	58.8	6
200	-195.0	-170.8	-191.6	72.1	6

H2O (L), WATER

T(C)	DELTA H	DELTA F	ENTHALPY	ENTROPY	REF
25	-68.3	-56.7	-68.3	16.7	1
50	-68.1	-55.7	-67.9	18.2	7
75	-67.9	-54.8	-67.4	19.5	7
100	-67.8	-53.8	-67.0	20.8	7
150	-67.4	-52.0	-66.1	23.1	7
200	-66.9	-50.2	-65.1	25.2	7
250	-66.5	-48.4	-64.1	27.2	7
300	-65.9	-46.8	-63.0	29.3	7

H2O (G), WATER (GAS)

T(C)	DELTA H	DELTA F	ENTHALPY	ENTROPY	REF
25	-57.8	-54.6	-57.8	45.1	1
50	-57.9	-54.4	-57.6	45.8	2
75	-57.9	-54.1	-57.4	46.4	2
100	-58.0	-53.8	-57.2	46.9	2
150	-58.1	-53.3	-56.8	48.0	2
200	-58.2	-52.7	-56.4	48.9	2
250	-58.3	-52.1	-55.9	49.7	2
300	-58.4	-51.5	-55.5	50.5	2

H2O2 (L,G), HYDROGEN PEROXIDE
BOIL PT = 158C

T(C)	DELTA H	DELTA F	ENTHALPY	ENTROPY	REF
25	-44.9	-28.8	-44.9	26.2	1
50	-44.7	-27.4	-44.3	27.9	2
75	-44.5	-26.1	-43.8	29.5	2
100	-44.3	-24.8	-43.3	31.0	2
150	-44.0	-22.2	-42.2	33.7	2
158	-43.9	-21.8	-42.0	34.1	2
158	-33.6	-21.8	-31.7	58.0	2
200	-33.7	-20.6	-31.2	59.1	2
250	-33.8	-19.3	-30.6	60.4	2
300	-33.9	-17.8	-29.9	61.6	2

H2O2 (AQ), UNDISSOCIATED COMPLEX

T(C)	DELTA H	DELTA F	ENTHALPY	ENTROPY	REF
25	-45.7	-32.1	-45.7	34.4	1
50	-47.0	-30.8	-46.6	31.4	6
75	-48.1	-29.6	-47.4	29.2	6
100	-48.9	-28.2	-47.9	27.7	6
150	-48.5	-25.5	-47.0	30.9	6
200	-46.5	-22.6	-45.0	36.2	6

H2P2O7-- (AQ)

T(C)	DELTA H	DELTA F	ENTHALPY	ENTROPY	REF
25	-544.5	-480.5	-544.5	49.0	1
44	-543.7	-476.4	-543.6	51.8	2
44	-544.0	-476.4	-543.6	51.8	2
50	-543.8	-475.2	-543.3	52.7	3
75	-541.9	-469.9	-541.8	57.3	3
100	-539.3	-464.9	-540.0	62.5	3
150	-536.6	-455.0	-538.2	66.7	3
200	-531.6	-445.8	-534.6	75.0	3
250	-525.3	-437.1	-530.7	83.0	3
274	-522.2	-433.2	-528.6	86.8	2
274	-527.9	-433.2	-528.6	86.8	2
300	-524.5	-429.0	-526.3	90.9	3

50

H2PO3- (AQ)

T(C)	DELTA H	DELTA F	ENTHALPY	ENTROPY	REF
25	-235.5	-202.4	-235.5	24.0	5
44	-236.3	-200.2	-236.2	21.6	2
44	-236.5	-200.2	-236.2	21.6	2
50	-236.7	-199.5	-236.5	20.9	3
75	-237.6	-196.6	-237.4	18.0	3
100	-238.3	-193.7	-238.4	15.2	3
150	-241.0	-187.4	-241.5	7.1	3
200	-243.0	-181.0	-244.2	1.0	3
250	-245.1	-174.3	-247.3	-5.3	3
274	-246.2	-171.1	-248.9	-8.3	2
274	-249.1	-171.1	-248.9	-8.3	2
300	-250.4	-167.4	-250.8	-11.6	3

H2PO4- (AQ)

T(C)	DELTA H	DELTA F	ENTHALPY	ENTROPY	REF
25	-299.8	-260.2	-299.8	26.6	1
44	-300.6	-257.6	-300.4	24.7	2
44	-300.7	-257.6	-300.4	24.7	2
50	-301.0	-256.8	-300.6	24.2	3
75	-301.6	-253.4	-301.3	22.0	3
100	-302.1	-249.9	-302.0	20.1	3
150	-304.5	-242.6	-304.6	13.3	3
200	-306.1	-235.3	-306.6	8.7	3
250	-307.6	-227.7	-309.0	3.9	3
274	-308.4	-224.1	-310.3	1.6	2
274	-311.3	-224.1	-310.3	1.6	2
300	-312.3	-220.0	-311.7	-0.9	3

H2S (AQ), UNDISSOCIATED COMPLEX

T(C)	DELTA H	DELTA F	ENTHALPY	ENTROPY	REF
25	-24.9	-16.1	-24.9	29.0	1
50	-25.4	-15.4	-25.1	28.4	6
75	-25.6	-14.6	-25.1	28.4	6
100	-25.7	-13.8	-24.9	29.0	6
150	-25.6	-12.2	-24.3	30.7	6
200	-24.1	-10.6	-22.7	34.9	6

H2S (G), HYDROGEN SULFIDE (GAS)

T(C)	DELTA H	DELTA F	ENTHALPY	ENTROPY	REF
25	-20.3	-17.5	-20.3	49.2	1
50	-20.3	-17.3	-20.1	49.8	2
75	-20.4	-17.0	-19.9	50.4	2
100	-20.5	-16.8	-19.7	51.0	2
150	-20.6	-16.3	-19.2	52.1	2
200	-20.7	-15.8	-18.8	53.1	2
250	-20.8	-15.2	-18.3	53.9	2
300	-20.9	-14.7	-17.9	54.8	2

H2SE (AQ), UNDISSOCIATED COMPLEX

T(C)	DELTA H	DELTA F	ENTHALPY	ENTROPY	REF
25	18.0	18.4	18.0	39.9	5
50	17.7	18.5	18.0	40.1	6
75	17.7	18.5	18.3	41.1	6
100	18.0	18.6	18.9	42.8	6
150	19.2	18.6	20.6	47.3	6
200	22.4	18.5	24.0	55.8	6

H2SEO3 (AQ), UNDISSOCIATED COMPLEX

T(C)	DELTA H	DELTA F	ENTHALPY	ENTROPY	REF
25	-121.3	-101.9	-121.3	49.7	1
50	-120.5	-100.3	-119.9	54.1	6
75	-119.1	-98.8	-118.0	60.1	6
100	-117.2	-97.4	-115.5	67.1	6
150	-114.3	-94.8	-111.4	77.3	6
200	-108.1	-93.0	-104.4	94.2	6

H2SEO4 (AQ), UNDISSOCIATED COMPLEX

T(C)	DELTA H	DELTA F	ENTHALPY	ENTROPY	REF
25	-145.3	-105.4	-145.3	5.7	5
50	-146.1	-102.1	-145.4	5.2	6
75	-146.7	-98.6	-145.3	5.5	6
100	-147.3	-95.2	-145.2	6.0	6
150	-151.6	-87.9	-147.9	-1.7	6
200	-158.0	-80.2	-152.0	-12.1	6

H2SO3 (AQ), UNDISSOCIATED COMPLEX

T(C)	DELTA H	DELTA F	ENTHALPY	ENTROPY	REF
25	-160.8	-139.0	-160.8	55.5	1
50	-159.7	-136.2	-159.2	60.8	6
75	-158.0	-134.4	-156.9	67.7	6
100	-155.6	-132.9	-154.0	75.9	6
150	-151.5	-130.0	-149.0	88.5	6
200	-143.6	-128.0	-140.6	109.0	6

H2SO4 (L), SULFURIC ACID
BOIL PT = 280C

T(C)	DELTA H	DELTA F	ENTHALPY	ENTROPY	REF
25	-209.9	-174.4	-209.9	37.5	1
50	-209.7	-171.5	-209.0	40.2	2
75	-209.4	-168.5	-208.2	42.8	2
100	-209.2	-165.6	-207.3	45.3	2
150	-208.6	-159.8	-205.5	49.9	2
200	-208.0	-154.0	-203.6	54.1	2
250	-207.4	-148.4	-201.6	58.0	2
280	-207.0	-145.0	-200.5	60.2	2
280	-192.9	-145.0	-186.3	83.5	2
300	-192.9	-142.0	-185.8	84.4	2

H2VO4- (AQ)

T(C)	DELTA H	DELTA F	ENTHALPY	ENTROPY	REF
25	-280.6	-244.0	-280.6	34.0	1
50	-280.9	-240.9	-280.7	33.6	3
75	-280.9	-237.8	-280.7	33.7	3
100	-280.5	-234.8	-280.6	34.1	3
150	-281.4	-228.5	-281.7	31.0	3
200	-281.2	-222.4	-281.9	30.6	3
250	-280.6	-216.2	-282.2	30.0	3
300	-280.0	-210.1	-282.5	29.4	3

H3ASO3 (AQ), UNDISSOCIATED COMPLEX

T(C)	DELTA H	DELTA F	ENTHALPY	ENTROPY	REF
25	-177.4	-152.9	-177.4	46.6	1
50	-177.1	-150.9	-176.4	49.8	6
75	-176.3	-148.9	-174.9	54.3	6
100	-174.9	-147.0	-173.0	60.0	6
150	-172.9	-143.3	-169.6	68.4	6
200	-167.5	-140.2	-163.4	83.6	6

H3ASO4 (AQ), UNDISSOCIATED COMPLEX

T(C)	DELTA H	DELTA F	ENTHALPY	ENTROPY	REF
25	-215.7	-183.1	-215.7	44.0	1
50	-215.5	-180.4	-214.7	47.1	6
75	-214.8	-177.7	-213.2	51.5	6
100	-213.6	-175.1	-211.3	57.0	6
150	-212.0	-169.9	-208.4	64.5	6
200	-207.6	-165.3	-202.8	78.1	6

H3P2O7- (AQ)

T(C)	DELTA H	DELTA F	ENTHALPY	ENTROPY	REF
25	-544.0	-483.6	-544.0	56.0	1
44	-543.2	-479.8	-542.7	60.3	2
44	-543.5	-479.8	-542.7	60.3	2
50	-543.3	-478.6	-542.3	61.6	3
75	-541.5	-473.6	-540.0	68.3	3
100	-539.1	-468.9	-537.4	75.8	3
150	-536.8	-459.5	-534.2	83.3	3
200	-532.0	-450.9	-528.9	95.8	3
250	-526.2	-442.6	-523.0	107.7	3
274	-523.3	-438.8	-519.9	113.4	2
274	-529.1	-438.8	-519.9	113.4	2
300	-525.8	-434.9	-516.5	119.6	3

H3PO3 (AQ), UNDISSOCIATED COMPLEX

T(C)	DELTA H	DELTA F	ENTHALPY	ENTROPY	REF
25	-231.7	-204.8	-231.7	40.0	5
44	-231.9	-203.1	-231.4	40.9	2
44	-232.0	-203.1	-231.4	40.9	2
50	-232.1	-202.5	-231.3	41.3	6
75	-232.0	-200.3	-230.5	43.7	6
100	-231.4	-198.0	-229.3	47.1	6
150	-230.7	-193.5	-227.3	52.2	6
200	-226.9	-189.2	-222.7	63.7	6

H3PO4 (C,L), ORTHOPHOSPHORIC ACID
        MELT PT = 42C

T(C)	DELTA H	DELTA F	ENTHALPY	ENTROPY	REF
25	-305.7	-267.5	-305.7	26.4	1
42	-305.8	-265.2	-305.3	27.9	2
42	-302.7	-265.2	-302.2	37.7	2
44	-302.6	-265.0	-302.1	38.0	2
44	-302.8	-265.0	-302.1	38.0	2
50	-302.7	-264.3	-301.8	38.8	2
75	-302.3	-261.4	-300.6	42.4	2
100	-301.8	-258.5	-299.4	45.8	2
150	-301.0	-252.7	-297.0	51.8	2
200	-300.2	-247.0	-294.6	57.1	2
250	-299.4	-241.5	-292.2	62.0	2
274	-299.0	-238.8	-291.0	64.1	2
274	-301.9	-238.8	-291.0	64.1	2
300	-301.4	-235.9	-289.8	66.4	2

H3PO4 (AQ), UNDISSOCIATED COMPLEX

T(C)	DELTA H	DELTA F	ENTHALPY	ENTROPY	REF
25	-307.9	-273.1	-307.9	37.8	1
44	-308.2	-270.8	-307.6	38.8	2
44	-308.3	-270.8	-307.6	38.8	2
50	-308.4	-270.2	-307.5	39.2	6
75	-308.3	-267.2	-306.7	41.7	6
100	-307.9	-264.3	-305.5	45.2	6
150	-307.6	-258.3	-303.7	49.6	6
200	-304.7	-252.7	-299.7	59.6	6

H4P2O7 (AQ), UNDISSOCIATED COMPLEX

T(C)	DELTA H	DELTA F	ENTHALPY	ENTROPY	REF
25	-542.2	-485.7	-542.2	64.0	1
44	-541.1	-482.1	-540.1	70.8	2
44	-541.4	-482.1	-540.1	70.8	2
50	-541.0	-481.0	-539.4	72.9	6
75	-538.6	-476.5	-535.8	83.9	6
100	-535.5	-472.2	-531.4	96.3	6
150	-531.0	-463.8	-524.5	113.7	6
200	-522.6	-456.6	-513.9	138.8	6

HASO2 (AQ), UNDISSOCIATED COMPLEX

T(C)	DELTA H	DELTA F	ENTHALPY	ENTROPY	REF
25	-109.0	-96.3	-109.0	30.1	1
50	-110.5	-95.1	-110.1	26.7	6
75	-111.9	-93.9	-111.0	23.9	6
100	-113.0	-92.5	-111.8	21.8	6
150	-113.7	-89.8	-111.8	22.3	6
200	-113.7	-86.7	-111.4	23.8	6

HASO4-- (AQ)

T(C)	DELTA H	DELTA F	ENTHALPY	ENTROPY	REF
25	-216.6	-170.8	-216.6	9.6	1
50	-218.5	-166.9	-218.8	2.6	3
75	-220.1	-162.9	-221.2	-4.7	3
100	-221.5	-158.7	-223.9	-12.1	3
150	-225.4	-150.0	-229.7	-27.1	3
200	-229.2	-140.8	-236.0	-41.6	3
250	-233.1	-131.3	-243.3	-56.1	3
300	-237.6	-121.3	-251.2	-70.6	3

HBR (G), HYDROGEN BROMIDE (GAS)

T(C)	DELTA H	DELTA F	ENTHALPY	ENTROPY	REF
25	-8.7	-12.8	-8.7	47.5	1
50	-8.8	-13.1	-8.5	48.0	2
58	-8.9	-13.2	-8.5	48.2	2
58	-12.4	-13.2	-8.5	48.2	2
75	-12.4	-13.3	-8.4	48.5	2
100	-12.5	-13.3	-8.2	49.0	2
150	-12.5	-13.5	-7.8	49.9	2
200	-12.5	-13.6	-7.5	50.7	2
250	-12.6	-13.7	-7.1	51.4	2
300	-12.6	-13.8	-6.8	52.0	2

HBRO (AQ), UNDISSOCIATED COMPLEX

T(C)	DELTA H	DELTA F	ENTHALPY	ENTROPY	REF
25	-26.9	-19.7	-26.9	34.0	1
50	-28.2	-19.0	-27.8	31.1	6
58	-28.7	-18.8	-28.1	30.2	2
58	-32.2	-18.8	-28.1	30.2	2
75	-32.8	-18.1	-28.6	28.9	6
100	-33.7	-17.0	-29.2	27.4	6
150	-33.4	-14.8	-28.6	29.5	6
200	-32.1	-12.5	-27.2	33.3	6

HBRO3 (AQ), UNDISSOCIATED COMPLEX

T(C)	DELTA H	DELTA F	ENTHALPY	ENTROPY	REF
25	-11.6	5.0	-11.6	51.6	5
50	-12.1	6.4	-11.5	51.8	6
58	-12.3	6.9	-11.5	51.9	2
58	-15.8	6.9	-11.5	51.9	2
75	-15.9	8.0	-11.3	52.6	6
100	-15.9	9.8	-10.8	54.0	6
150	-15.3	13.2	-9.5	57.6	6
200	-13.4	16.5	-6.9	64.0	6

HC2O4- (AQ)

T(C)	DELTA H	DELTA F	ENTHALPY	ENTROPY	REF
25	-195.6	-166.9	-195.6	40.7	1
50	-195.3	-164.5	-195.2	42.1	3
75	-194.4	-162.2	-194.5	44.2	3
100	-193.2	-160.0	-193.6	46.8	3
150	-192.6	-155.5	-193.4	46.9	3
200	-190.5	-151.4	-191.9	50.5	3
250	-187.9	-147.4	-190.3	53.7	3
300	-185.1	-143.6	-188.6	56.9	3

HCL (G), HYDROGEN CHLORIDE (GAS)

T(C)	DELTA H	DELTA F	ENTHALPY	ENTROPY	REF
25	-22.1	-22.8	-22.1	44.7	1
50	-22.1	-22.8	-21.9	45.2	2
75	-22.1	-22.9	-21.7	45.7	2
100	-22.1	-23.0	-21.5	46.2	2
150	-22.1	-23.1	-21.2	47.1	2
200	-22.2	-23.2	-20.8	47.8	2
250	-22.2	-23.3	-20.5	48.6	2
300	-22.3	-23.4	-20.1	49.2	2

HCL (AQ), UNDISSOCIATED COMPLEX

T(C)	DELTA H	DELTA F	ENTHALPY	ENTROPY	REF
25	-40.0	-31.3	-40.0	13.2	5
50	-40.9	-30.6	-40.7	10.9	6
75	-41.8	-29.8	-41.4	8.9	6
100	-42.6	-28.9	-42.1	7.1	6
150	-44.6	-26.9	-43.7	2.9	6
200	-47.0	-24.5	-45.7	-1.7	6

HCLO (AQ), UNDISSOCIATED COMPLEX

T(C)	DELTA H	DELTA F	ENTHALPY	ENTROPY	REF
25	-28.9	-19.1	-28.9	34.0	1
50	-30.0	-18.2	-29.8	31.1	6
75	-31.1	-17.3	-30.5	28.9	6
100	-31.9	-16.3	-31.1	27.4	6
150	-31.7	-14.2	-30.5	29.5	6
200	-30.2	-12.0	-29.1	33.3	6

HCLO2 (AQ), UNDISSOCIATED COMPLEX

T(C)	DELTA H	DELTA F	ENTHALPY	ENTROPY	REF
25	-12.4	1.4	-12.4	45.0	1
50	-13.1	2.6	-12.7	43.9	6
75	-13.6	3.8	-12.9	43.5	6
100	-13.9	5.1	-12.8	43.8	6
150	-13.1	7.6	-11.5	47.6	6
200	-10.8	10.0	-8.9	54.2	6

HCLO3 (AQ), UNDISSOCIATED COMPLEX

T(C)	DELTA H	DELTA F	ENTHALPY	ENTROPY	RE
25	-23.5	-0.6	-23.5	39.0	5
50	-24.4	1.3	-23.9	37.6	6
75	-25.2	3.3	-24.3	36.6	6
100	-25.9	5.4	-24.5	35.9	6
150	-27.5	9.7	-25.2	34.4	6
200	-29.0	14.2	-25.8	33.1	6

HCLO4 (AQ), UNDISSOCIATED COMPLEX

T(C)	DELTA H	DELTA F	ENTHALPY	ENTROPY	REF
25	-31.4	-2.5	-31.4	43.2	5
50	-32.2	-0.0	-31.6	42.5	6
75	-32.8	2.5	-31.7	42.2	6
100	-33.4	5.0	-31.8	42.2	6
150	-34.8	10.3	-32.1	41.4	6
200	-36.0	15.6	-32.1	41.3	6

HCN (AQ), UNDISSOCIATED COMPLEX

T(C)	DELTA H	DELTA F	ENTHALPY	ENTROPY	REF
25	25.6	28.6	25.6	29.8	1
50	24.9	28.9	25.1	28.2	6
75	24.3	29.2	24.8	27.3	6
100	24.0	29.6	24.6	26.9	6
150	23.1	30.4	24.2	25.8	6
200	22.7	31.4	24.1	25.8	6

HCNO (AQ), UNDISSOCIATED COMPLEX

T(C)	DELTA H	DELTA F	ENTHALPY	ENTROPY	REF
25	-35.1	-28.9	-35.1	43.6	5
50	-35.8	-28.3	-35.5	42.3	6
75	-36.3	-27.5	-35.7	41.8	6
100	-36.6	-27.1	-35.7	41.8	6
150	-36.1	-25.9	-34.7	44.8	6
200	-34.3	-24.7	-32.6	50.2	6

HCO3- (AQ)

T(C)	DELTA H	DELTA F	ENTHALPY	ENTROPY	REF
25	-165.4	-140.3	-165.4	26.8	1
50	-166.1	-138.1	-166.1	24.4	3
75	-166.5	-136.0	-166.8	22.3	3
100	-166.7	-133.8	-167.5	20.5	3
150	-168.5	-129.2	-170.0	13.8	3
200	-169.5	-124.5	-172.0	9.3	3
250	-170.4	-119.7	-174.4	4.6	3
300	-171.6	-114.8	-177.0	-0.1	3

HCOO- (AQ)

T(C)	DELTA H	DELTA F	ENTHALPY	ENTROPY	REF
25	-101.7	-83.9	-101.7	27.0	1
50	-102.2	-82.4	-102.4	24.7	3
75	-102.5	-80.9	-103.1	22.7	3
100	-102.6	-79.3	-103.7	20.8	3
150	-104.2	-76.0	-106.2	14.3	3
200	-105.0	-72.6	-108.1	9.9	3
250	-105.7	-69.2	-110.4	5.3	3
300	-106.6	-65.7	-113.0	0.7	3

HCOOH (L,G), FORMIC ACID
     BOILING PT = 101C

T(C)	DELTA H	DELTA F	ENTHALPY	ENTROPY	REF
25	-101.5	-86.4	-101.5	30.8	1
50	-101.3	-85.1	-100.9	32.7	2
75	-101.1	-83.9	-100.3	34.5	2
100	-101.0	-82.6	-99.7	36.1	2
101	-101.0	-82.6	-99.7	36.2	2
101	-95.6	-82.6	-94.4	50.4	2
150	-95.8	-80.9	-93.7	52.1	2
200	-96.0	-79.1	-93.0	53.7	2
250	-96.1	-77.3	-92.2	55.3	2
300	-96.2	-75.5	-91.4	56.7	2

HCOOH (AQ), UNDISSOCIATED COMPLEX

T(C)	DELTA H	DELTA F	ENTHALPY	ENTROPY	REF
25	-101.7	-89.0	-101.7	39.0	1
50	-101.6	-87.9	-101.2	40.6	6
75	-101.1	-86.9	-100.3	43.3	6
100	-100.2	-85.9	-99.0	47.1	6
150	-99.0	-84.0	-97.1	52.0	6
200	-95.0	-82.4	-92.7	62.8	6

HCRO4- (AQ)

T(C)	DELTA H	DELTA F	ENTHALPY	ENTROPY	REF
25	-209.9	-182.8	-209.9	49.0	1
50	-208.9	-180.6	-208.8	52.7	3
75	-207.2	-178.5	-207.2	57.3	3
100	-205.1	-176.5	-205.4	62.5	3
150	-202.9	-172.5	-203.6	66.7	3
200	-198.7	-169.7	-200.0	75.0	3
250	-193.7	-166.9	-196.1	83.0	3
300	-188.2	-164.6	-191.7	90.9	3

HF (G), HYDROGEN FLUORIDE (GAS)

T(C)	DELTA H	DELTA F	ENTHALPY	ENTROPY	REF
25	-64.8	-65.3	-64.8	41.5	1
50	-64.8	-65.3	-64.6	42.1	2
75	-64.8	-65.4	-64.4	42.6	2
100	-64.8	-65.4	-64.3	43.1	2
150	-64.9	-65.5	-63.9	43.9	2
200	-64.9	-65.6	-63.6	44.7	2
250	-64.9	-65.6	-63.2	45.4	2
300	-64.9	-65.7	-62.9	46.0	2

HF (AQ), UNDISSOCIATED COMPLEX

T(C)	DELTA H	DELTA F	ENTHALPY	ENTROPY	REF
25	-76.5	-70.9	-76.5	21.2	1
50	-76.3	-70.5	-76.1	22.3	6
75	-76.0	-70.0	-75.7	23.8	6
100	-75.5	-69.6	-75.0	25.6	6
150	-73.4	-68.9	-72.7	31.9	6
200	-68.8	-68.5	-68.4	42.7	6

HF2- (AQ)
  (EXTRAPOLATED AS CRISS & COBBLE TYPE 2)

T(C)	DELTA H	DELTA F	ENTHALPY	ENTROPY	REF
25	-155.3	-138.2	-155.3	27.1	1
50	-156.5	-136.7	-156.7	22.9	3
75	-157.5	-135.1	-158.1	18.5	3
100	-158.4	-133.5	-159.7	14.1	3
150	-160.8	-130.0	-163.0	5.5	3
200	-163.5	-126.1	-167.0	-3.6	3
250	-166.0	-122.0	-171.3	-12.2	3
300	-170.1	-117.6	-177.1	-22.9	3

HFE(CN)6--- (AQ)
(EXTRAPOLATED AS CRISS & COBBLE TYPE 2)

T(C)	DELTA H	DELTA F	ENTHALPY	ENTROPY	REF
25	109.0	160.4	109.0	57.0	1
50	107.8	164.8	107.4	52.1	3
75	107.4	169.1	106.0	47.8	3
100	107.5	173.5	104.6	44.0	3
150	106.4	182.4	101.1	35.1	3
200	105.4	191.5	97.1	25.7	3
250	105.4	200.6	92.8	17.0	3
300	103.7	209.8	86.9	6.2	3

HGEO3- (AQ)

T(C)	DELTA H	DELTA F	ENTHALPY	ENTROPY	REF
25	-197.2	-170.0	-197.2	26.0	5
50	-198.0	-167.7	-198.0	23.4	3
75	-198.6	-165.3	-198.8	21.1	3
100	-198.9	-163.0	-199.5	18.9	3
150	-201.0	-157.9	-202.2	11.9	3
200	-202.4	-152.7	-204.4	7.0	3
250	-203.7	-147.4	-207.0	1.8	3
300	-205.3	-142.0	-209.8	-3.4	3

HIO (AQ), UNDISSOCIATED COMPLEX

T(C)	DELTA H	DELTA F	ENTHALPY	ENTROPY	REF
25	-33.0	-23.7	-33.0	22.8	1
50	-34.8	-22.8	-34.4	18.1	6
75	-36.5	-21.8	-35.8	14.1	6
100	-38.0	-20.7	-37.0	10.8	6
114	-38.4	-20.1	-37.2	10.4	2
114	-40.2	-20.1	-37.2	10.4	2
150	-40.8	-19.2	-37.3	10.6	6
185	-41.6	-16.2	-37.8	9.9	2
185	-46.6	-16.2	-37.8	9.9	2
200	-46.4	-15.2	-37.6	10.7	6

HIO3 (AQ), UNDISSOCIATED COMPLEX

T(C)	DELTA H	DELTA F	ENTHALPY	ENTROPY	REF
25	-50.5	-31.7	-50.5	39.9	1
50	-51.5	-30.1	-51.0	38.3	6
75	-52.4	-28.4	-51.4	37.2	6
100	-53.1	-26.6	-51.6	36.7	6
114	-53.4	-25.7	-51.6	36.8	2
114	-55.3	-25.7	-51.6	36.8	2
150	-55.8	-22.9	-51.3	37.6	6
185	-56.1	-20.1	-50.9	38.3	2
185	-61.1	-20.1	-50.9	38.8	2
200	-60.9	-18.8	-50.5	39.8	6

HN3 (AQ), UNDISSOCIATED COMPLEX

T(C)	DELTA H	DELTA F	ENTHALPY	ENTROPY	REF
25	62.2	76.9	62.2	34.9	1
50	61.4	78.2	61.7	33.5	6
75	60.8	79.5	61.5	32.8	6
100	60.4	80.8	61.5	32.8	6
150	59.6	83.6	61.3	32.4	6
200	59.5	86.6	61.6	33.5	6

HNO2 (AQ), UNDISSOCIATED COMPLEX

T(C)	DELTA H	DELTA F	ENTHALPY	ENTROPY	REF
25	-28.5	-13.3	-28.5	36.5	1
50	-29.4	-12.0	-29.1	34.6	6
75	-30.3	-10.6	-29.6	33.2	6
100	-31.0	-9.2	-30.0	32.1	6
150	-32.4	-6.2	-30.6	30.6	6
200	-33.6	-3.0	-31.3	29.2	6

HNO3 (G), NITRIC ACID (GAS)

T(C)	DELTA H	DELTA F	ENTHALPY	ENTROPY	REF
25	-32.3	-17.9	-32.3	63.6	1
50	-32.4	-16.7	-32.0	64.7	2
75	-32.5	-15.4	-31.6	65.7	2
100	-32.6	-14.2	-31.3	66.7	2
150	-32.7	-11.7	-30.5	68.5	2
200	-32.8	-9.2	-29.7	70.3	2
250	-32.9	-6.7	-28.9	72.0	2
300	-32.9	-4.2	-28.0	73.6	2

HNO3 (AQ), UNDISSOCIATED COMPLEX

T(C)	DELTA H	DELTA F	ENTHALPY	ENTROPY	REF
25	-45.5	-24.7	-45.5	42.2	6
50	-46.3	-22.9	-45.8	41.1	6
75	-46.9	-21.0	-46.0	40.5	6
100	-47.4	-19.2	-46.1	40.3	6
150	-48.1	-15.4	-46.0	40.8	6
200	-48.2	-11.5	-45.3	42.6	6

HO2- (AQ)

T(C)	DELTA H	DELTA F	ENTHALPY	ENTROPY	REF
25	-38.3	-16.1	-38.3	10.7	1
50	-40.7	-14.1	-40.9	2.4	3
75	-43.2	-12.0	-43.8	-6.3	3
100	-45.7	-9.6	-47.0	-15.2	3
150	-49.9	-4.6	-52.3	-28.3	3
200	-55.9	1.3	-59.6	-45.4	3
250	-62.3	7.7	-67.7	-61.7	3
300	-69.4	14.7	-76.6	-78.0	3

HP2O7--- (AQ)

T(C)	DELTA H	DELTA F	ENTHALPY	ENTROPY	REF
25	-543.7	-471.4	-543.7	26.0	1
44	-544.0	-466.8	-544.3	24.0	2
44	-544.3	-466.8	-544.3	24.0	2
50	-544.4	-465.3	-544.5	23.4	3
75	-544.1	-459.2	-545.3	21.1	3
100	-543.1	-453.2	-546.1	18.9	3
150	-543.1	-441.0	-548.8	11.9	3
200	-541.8	-429.2	-550.9	7.0	3
250	-539.6	-417.4	-553.5	1.8	3
274	-538.7	-411.8	-554.8	-0.7	2
274	-544.4	-411.8	-554.8	-0.7	2
300	-543.5	-405.8	-556.3	-3.4	3

HPO3-- (AQ)

T(C)	DELTA H	DELTA F	ENTHALPY	ENTROPY	REF
25	-233.8	-194.0	-233.8	6.6	5
44	-235.3	-191.4	-235.6	0.6	2
44	-235.5	-191.4	-235.6	0.6	2
50	-236.0	-190.6	-236.2	-1.2	3
75	-237.8	-187.0	-239.0	-9.3	3
100	-239.4	-183.3	-241.9	-17.7	3
150	-243.7	-175.5	-248.4	-34.2	3
200	-248.1	-167.0	-255.4	-50.3	3
250	-252.6	-158.2	-263.5	-66.5	3
274	-255.1	-153.9	-267.6	-74.3	2
274	-258.0	-153.9	-267.6	-74.3	2
300	-260.9	-149.0	-272.3	-82.7	3

HPO4-- (AQ)

T(C)	DELTA H	DELTA F	ENTHALPY	ENTROPY	REF
25	-308.8	-260.3	-308.8	2.0	1
44	-310.7	-257.2	-311.0	-5.0	2
44	-310.9	-257.2	-311.0	-5.0	2
50	-311.5	-256.2	-311.7	-7.1	3
75	-313.8	-251.8	-314.8	-16.7	3
100	-316.1	-247.3	-318.4	-26.5	3
150	-321.5	-237.6	-325.7	-45.2	3
200	-327.3	-227.3	-333.9	-64.1	3
250	-333.3	-216.4	-343.3	-82.9	3
274	-336.5	-211.0	-348.1	-91.9	2
274	-339.4	-211.0	-348.1	-91.9	2
300	-343.2	-204.9	-353.6	-101.8	3

HS- (AQ)
(EXTRAPOLATED AS CRISS & COBBLE TYPE 2)

T(C)	DELTA H	DELTA F	ENTHALPY	ENTROPY	REF
25	-19.6	-6.6	-19.6	20.0	1
50	-20.5	-5.5	-20.8	15.9	3
75	-21.5	-4.3	-22.3	11.5	3
100	-22.4	-3.0	-23.9	7.0	3
150	-24.5	-0.3	-27.2	-1.5	3
200	-27.0	2.8	-31.2	-10.6	3
250	-29.3	6.1	-35.4	-19.1	3
300	-33.2	9.7	-41.2	-29.8	3

HSE- (AQ)
(EXTRAPOLATED AS CRISS & COBBLE TYPE 2)

T(C)	DELTA H	DELTA F	ENTHALPY	ENTROPY	REF
25	3.8	10.5	3.8	24.0	1
50	2.8	11.1	2.5	19.8	3
75	1.8	11.8	1.1	15.5	3
100	0.9	12.5	-0.5	11.0	3
150	-1.4	14.3	-3.9	2.4	3
200	-4.0	16.4	-7.8	-6.7	3
221	-5.1	17.3	-9.5	-10.2	2
221	-6.4	17.3	-9.5	-10.2	2
250	-7.9	18.7	-12.0	-15.2	3
300	-12.0	21.4	-17.9	-25.9	3

HSEO3- (AQ)

T(C)	DELTA H	DELTA F	ENTHALPY	ENTROPY	REF
25	-123.0	-98.4	-123.0	37.3	1
50	-122.9	-96.3	-122.8	37.8	3
75	-122.3	-94.3	-122.5	38.9	3
100	-121.4	-92.4	-122.0	40.3	3
150	-121.3	-88.3	-122.4	38.8	3
200	-119.9	-84.7	-121.8	40.4	3
221	-119.2	-83.1	-121.5	40.9	2
221	-120.5	-83.1	-121.5	40.9	2
250	-119.5	-81.0	-121.2	41.7	3
300	-117.6	-77.4	-120.5	42.9	3

HSEO4- (AQ)

T(C)	DELTA H	DELTA F	ENTHALPY	ENTROPY	REF
25	-139.0	-108.1	-139.0	40.7	1
50	-138.7	-105.5	-138.6	42.1	3
75	-137.9	-103.0	-137.9	44.2	3
100	-136.7	-100.6	-137.0	46.8	3
150	-136.1	-95.7	-136.8	46.9	3
200	-134.1	-91.2	-135.3	50.5	3
221	-133.0	-89.3	-134.7	51.8	2
221	-134.4	-89.3	-134.7	51.8	2
250	-132.9	-86.8	-133.7	53.7	3
300	-130.1	-82.5	-132.0	56.9	3

HSO3- (AQ)

T(C)	DELTA H	DELTA F	ENTHALPY	ENTROPY	REF
25	-165.0	-135.6	-165.0	38.4	1
50	-164.8	-133.2	-164.8	39.2	3
75	-164.0	-130.8	-164.3	40.6	3
100	-163.0	-128.5	-163.7	42.4	3
150	-162.5	-123.8	-163.9	41.4	3
200	-160.7	-119.4	-163.0	43.7	3
250	-158.4	-115.2	-162.1	45.6	3
300	-156.0	-111.2	-161.0	47.4	3

HSO4- (AQ)

T(C)	DELTA H	DELTA F	ENTHALPY	ENTROPY	REF
25	-227.4	-190.2	-227.4	36.5	1
50	-227.4	-187.0	-227.4	36.8	3
75	-227.0	-184.0	-227.1	37.6	3
100	-226.2	-180.9	-226.6	38.8	3
150	-226.3	-174.8	-227.3	36.9	3
200	-225.2	-168.9	-226.9	38.0	3
250	-223.6	-163.0	-226.5	38.8	3
300	-222.0	-157.3	-226.0	39.7	3

HVO4-- (AQ)

T(C)	DELTA H	DELTA F	ENTHALPY	ENTROPY	REF
25	-277.0	-233.0	-277.0	14.0	1
50	-278.5	-229.3	-278.9	8.2	3
75	-279.7	-225.4	-280.8	2.2	3
100	-280.6	-221.5	-283.0	-3.8	3
150	-283.6	-213.3	-288.0	-16.7	3
200	-286.4	-204.8	-293.2	-28.6	3
250	-289.0	-196.1	-299.2	-40.6	3
300	-292.1	-187.0	-305.7	-52.6	3

HF (C), HAFNIUM

T(C)	DELTA H	DELTA F	ENTHALPY	ENTROPY	REF
25	0.0	0.0	0.0	10.4	1
50	0.0	0.0	0.2	10.9	2
75	0.0	0.0	0.3	11.4	2
100	0.0	0.0	0.5	11.8	2
150	0.0	0.0	0.8	12.6	2
200	0.0	0.0	1.1	13.3	2
250	0.0	0.0	1.4	14.0	2
300	0.0	0.0	1.8	14.6	2

HG (L), MERCURY

T(C)	DELTA H	DELTA F	ENTHALPY	ENTROPY	REF
25	0.0	0.0	0.0	18.2	1
50	0.0	0.0	0.2	18.7	2
75	0.0	0.0	0.3	19.2	2
100	0.0	0.0	0.5	19.7	2
150	0.0	0.0	0.8	20.5	2
200	0.0	0.0	1.2	21.2	2
250	0.0	0.0	1.5	21.9	2
300	0.0	0.0	1.8	22.5	2

HG(CN)2 (AQ), UNDISSOCIATED COMPLEX

T(C)	DELTA H	DELTA F	ENTHALPY	ENTROPY	REF
25	65.8	74.6	65.8	37.3	1
50	64.6	75.4	65.0	34.7	6
75	63.4	76.3	64.3	32.4	6
100	62.2	77.2	63.5	30.4	6
150	58.8	79.5	61.2	24.2	6
200	54.9	82.2	58.4	17.6	6

HG(CN)3- (AQ)
(EXTRAPOLATED AS CRISS & COBBLE TYPE 2)

T(C)	DELTA H	DELTA F	ENTHALPY	ENTROPY	REF
25	94.3	110.7	94.3	56.4	1
50	92.6	112.1	92.7	51.5	3
75	91.3	113.7	91.3	47.2	3
100	90.2	115.3	89.9	43.4	3
150	87.0	118.9	86.4	34.5	3
200	83.5	123.0	82.4	25.1	3
250	80.1	127.4	78.1	16.5	3
300	75.1	132.1	72.2	5.6	3

HG(CN)4-- (AQ)
(EXTRAPOLATED AS CRISS & COBBLE TYPE 2)

T(C)	DELTA H	DELTA F	ENTHALPY	ENTROPY	REF
25	125.3	147.8	125.3	81.0	1
50	123.8	149.8	123.6	75.6	3
75	123.1	151.8	122.2	71.4	3
100	122.9	153.7	121.0	68.0	3
150	120.9	158.0	117.4	58.8	3
200	118.8	162.6	113.3	49.3	3
250	117.3	167.3	108.9	40.5	3
300	114.2	172.3	102.9	29.5	3

HG(NH3)2++ (AQ)

T(C)	DELTA H	DELTA F	ENTHALPY	ENTROPY	REF
25	-22.3	2.5	-22.3	33.0	1
50	-23.6	4.6	-21.8	34.7	3
75	-25.2	6.9	-21.1	36.8	3
100	-27.0	9.3	-20.3	39.2	3
150	-30.7	14.5	-19.0	42.3	3
200	-34.1	20.0	-17.1	46.8	3
250	-38.5	26.0	-15.2	50.7	3
300	-42.8	32.3	-13.0	54.7	3

HG(NH3)3++ (AQ)

T(C)	DELTA H	DELTA F	ENTHALPY	ENTROPY	REF
25	-44.9	-5.2	-44.9	53.0	1
50	-46.7	-1.8	-44.5	54.1	3
75	-48.9	1.8	-44.1	55.3	3
100	-51.4	5.5	-43.6	56.7	3
150	-56.4	13.6	-43.0	58.2	3
200	-61.3	22.2	-41.8	61.0	3
250	-67.2	31.3	-40.7	63.3	3
300	-73.0	41.0	-39.4	65.6	3

HG(NH3)4++ (AQ)

T(C)	DELTA H	DELTA F	ENTHALPY	ENTROPY	REF
25	-67.8	-12.4	-67.8	70.0	1
50	-70.1	-7.7	-67.6	70.5	3
75	-72.9	-2.7	-67.4	71.1	3
100	-76.1	2.5	-67.2	71.6	3
150	-82.3	13.5	-67.2	71.6	3
200	-88.5	25.3	-66.6	73.1	3
250	-95.8	37.6	-66.2	74.0	3
300	-103.1	50.7	-65.6	75.0	3

HG(OH)+ (AQ)

T(C)	DELTA H	DELTA F	ENTHALPY	ENTROPY	REF
25	-20.2	-12.5	-20.2	12.0	1
50	-20.2	-11.9	-19.4	14.4	3
75	-20.3	-11.2	-18.4	17.4	3
100	-20.3	-10.6	-17.2	20.8	3
150	-20.6	-9.2	-15.3	25.7	3
200	-20.5	-7.9	-12.6	31.8	3
250	-20.7	-6.6	-9.8	37.5	3
300	-20.6	-5.2	-6.7	43.2	3

HG(OH)2 (AQ), UNDISSOCIATED COMPLEX

T(C)	DELTA H	DELTA F	ENTHALPY	ENTROPY	REF
25	-84.9	-65.7	-84.9	34.0	1
50	-85.1	-64.1	-84.6	35.1	6
75	-85.1	-62.4	-84.1	36.5	6
100	-85.0	-60.8	-83.5	38.1	6
150	-84.2	-57.6	-81.8	42.8	6
200	-81.0	-54.5	-78.1	52.1	6

HG++ (AQ)

T(C)	DELTA H	DELTA F	ENTHALPY	ENTROPY	REF
25	40.9	39.3	40.9	-17.7	1
50	40.8	39.2	41.9	-14.4	3
75	40.7	39.1	43.3	-10.1	3
100	40.4	39.0	45.1	-5.2	3
150	39.8	38.8	48.0	2.2	3
200	39.5	38.7	51.7	10.7	3
250	38.6	38.7	55.7	18.8	3
300	38.0	38.7	60.1	26.9	3

HG2++ (AQ)

T(C)	DELTA H	DELTA F	ENTHALPY	ENTROPY	REF
25	41.2	36.7	41.2	10.2	1
50	40.7	36.3	42.0	12.7	3
75	40.0	36.0	43.0	15.7	3
100	39.1	35.8	44.2	19.2	3
150	37.2	35.6	46.2	24.3	3
200	35.6	35.4	48.9	30.5	3
250	33.2	35.6	51.8	36.3	3
300	31.1	35.9	55.0	42.2	3

HG2CL2 (C), MERCURY(1) CHLORIDE

T(C)	DELTA H	DELTA F	ENTHALPY	ENTROPY	REF
25	-63.4	-50.4	-63.4	46.0	1
50	-63.3	-49.3	-62.8	48.0	2
75	-63.2	-48.2	-62.2	49.8	2
100	-63.1	-47.1	-61.5	51.6	2
150	-62.9	-45.0	-60.3	54.7	2
200	-62.7	-42.9	-59.0	57.6	2
250	-62.5	-40.8	-57.7	60.2	2
300	-62.3	-38.8	-56.3	62.7	2

HGBR+ (AQ)

T(C)	DELTA H	DELTA F	ENTHALPY	ENTROPY	REF
25	1.6	2.1	1.6	14.0	1
50	1.5	2.2	2.3	16.3	3
58	1.4	2.2	2.6	17.1	2
58	-2.1	2.2	2.6	17.1	2
75	-2.1	2.4	3.3	19.3	3
100	-2.2	2.7	4.4	22.6	3
150	-2.4	3.4	6.3	27.3	3
200	-2.2	4.0	8.9	33.3	3
250	-2.4	4.7	11.6	38.7	3
300	-2.2	5.4	14.6	44.3	3

HGBR2 (C,L), MERCURY(2) BROMIDE
MELTING PT = 241C

T(C)	DELTA H	DELTA F	ENTHALPY	ENTROPY	REF
25	-40.8	-36.8	-40.8	41.0	1
50	-40.9	-36.4	-40.3	42.5	2
58	-41.0	-36.3	-40.2	42.9	2
58	-48.1	-36.3	-40.2	42.9	2
75	-49.0	-35.7	-39.9	43.9	2
100	-48.0	-34.8	-39.4	45.1	2
150	-47.8	-33.1	-38.5	47.4	2
200	-47.6	-31.4	-37.5	49.6	2
241	-47.4	-30.0	-36.8	51.2	2
241	-43.2	-30.0	-32.5	59.5	2
250	-43.1	-29.8	-32.3	59.9	2
300	-42.6	-28.5	-31.0	62.2	2

HGBR2 (AQ), UNDISSOCIATED COMPLEX

T(C)	DELTA H	DELTA F	ENTHALPY	ENTROPY	REF
25	-38.2	-34.2	-38.2	41.0	1
50	-39.3	-33.8	-38.7	39.4	6
58	-39.7	-33.7	-38.9	38.9	2
58	-46.8	-33.7	-38.9	38.9	2
75	-47.2	-33.0	-39.1	38.3	6
100	-47.9	-32.0	-39.4	37.5	6
150	-49.8	-29.7	-40.5	34.6	6
200	-51.3	-27.2	-41.3	32.9	6

HGBR3- (AQ)
(EXTRAPOLATED AS CRISS & COBBLE TYPE 2)

T(C)	DELTA H	DELTA F	ENTHALPY	ENTROPY	REF
25	-69.9	-62.0	-69.9	67.0	1
50	-71.8	-61.3	-71.4	61.9	3
58	-72.4	-61.0	-72.0	60.2	2
58	-83.1	-61.0	-72.0	60.2	2
75	-83.7	-59.8	-72.9	57.6	3
100	-84.6	-58.1	-74.2	54.0	3
150	-87.5	-54.3	-77.7	45.0	3
200	-90.7	-50.1	-81.8	35.5	3
250	-93.7	-45.7	-86.1	26.8	3
300	-98.3	-40.8	-92.1	15.9	3

HGBR4-- (AQ)
(EXTRAPOLATED AS CRISS & COBBLE TYPE 2)

T(C)	DELTA H	DELTA F	ENTHALPY	ENTROPY	REF
25	-103.1	-88.7	-103.1	84.0	1
50	-104.9	-87.4	-104.8	78.5	3
58	-105.4	-87.0	-105.4	76.7	2
58	-119.6	-87.0	-105.4	76.7	2
75	-119.7	-85.3	-106.2	74.3	3
100	-119.6	-82.9	-107.4	71.0	3
150	-121.2	-77.7	-111.0	61.8	3
200	-122.8	-72.4	-115.1	52.2	3
250	-123.8	-67.0	-119.5	43.4	3
300	-126.3	-61.5	-125.5	32.4	3

HGBRI (AQ), UNDISSOCIATED COMPLEX

T(C)	DELTA H	DELTA F	ENTHALPY	ENTROPY	REF
25	-28.0	-26.7	-28.0	46.0	1
50	-29.1	-26.5	-28.6	44.0	6
58	-29.5	-26.5	-28.8	43.4	2
58	-33.1	-26.5	-28.8	43.4	2
75	-33.6	-26.1	-29.0	42.7	6
100	-34.4	-25.6	-29.4	41.7	6
114	-35.0	-25.2	-29.8	40.7	2
114	-36.9	-25.2	-29.8	40.7	2
150	-39.7	-24.0	-30.8	38.0	6
185	-40.3	-22.8	-31.7	35.9	2
185	-45.4	-22.8	-31.7	35.9	2
200	-46.0	-22.0	-32.1	35.0	6

HGCL+ (AQ)

T(C)	DELTA H	DELTA F	ENTHALPY	ENTROPY	REF
25	-4.6	-1.3	-4.6	13.0	1
50	-4.6	-1.0	-3.9	15.4	3
75	-4.6	-0.7	-2.9	18.3	3
100	-4.6	-0.5	-1.7	21.7	3
150	-4.8	0.1	0.2	26.5	3
200	-4.6	0.6	2.8	32.5	3
250	-4.7	1.2	5.5	38.1	3
300	-4.5	1.8	8.6	43.7	3

HGCL2 (C,L), MERCURY(2) CHLORIDE
MELTING PT = 277C

T(C)	DELTA H	DELTA F	ENTHALPY	ENTROPY	REF
25	-53.6	-42.7	-53.6	34.9	1
50	-53.5	-41.8	-53.2	36.3	2
75	-53.4	-40.9	-52.7	37.7	2
100	-53.4	-40.0	-52.3	38.9	2
150	-53.2	-38.2	-51.3	41.2	2
200	-53.0	-36.4	-50.4	43.3	2
250	-52.8	-34.7	-49.4	45.2	2
277	-52.7	-33.8	-48.9	46.2	2
277	-48.1	-33.8	-44.3	54.6	2
300	-47.9	-33.2	-43.7	55.6	2

HGCL2 (AQ), UNDISSOCIATED COMPLEX

T(C)	DELTA H	DELTA F	ENTHALPY	ENTROPY	REF
25	-51.7	-41.4	-51.7	37.0	1
50	-52.3	-40.5	-52.0	36.0	6
75	-52.9	-39.6	-52.2	35.4	6
100	-53.4	-38.6	-52.3	35.0	6
150	-54.7	-36.5	-52.9	33.6	6
200	-55.2	-34.2	-52.8	34.0	6

HGCL3- (AQ)
(EXTRAPOLATED AS CRISS & COBBLE TYPE 2)

T(C)	DELTA H	DELTA F	ENTHALPY	ENTROPY	REF
25	-92.9	-73.9	-92.9	55.0	1
50	-94.4	-72.3	-94.4	50.1	3
75	-95.6	-70.5	-95.8	45.9	3
100	-96.6	-68.7	-97.2	42.0	3
150	-99.4	-64.7	-100.7	33.1	3
200	-102.6	-60.3	-104.7	23.8	3
250	-105.5	-55.7	-109.1	15.1	3
300	-110.1	-50.8	-115.0	4.3	3

HGCL4-- (AQ)
(EXTRAPOLATED AS CRISS & COBBLE TYPE 2)

T(C)	DELTA H	DELTA F	ENTHALPY	ENTROPY	REF
25	-132.4	-106.8	-132.4	80.0	1
50	-133.7	-104.6	-134.1	74.6	3
75	-134.3	-102.4	-135.5	70.4	3
100	-134.3	-100.1	-136.8	67.0	3
150	-135.9	-95.4	-140.4	57.8	3
200	-137.5	-90.5	-144.5	48.3	3
250	-138.4	-85.5	-148.8	39.5	3
300	-140.9	-80.3	-154.8	28.6	3

HGCN+ (AQ)

T(C)	DELTA H	DELTA F	ENTHALPY	ENTROPY	REF
25	54.9	56.9	54.9	15.2	1
50	54.9	57.1	55.6	17.5	3
75	54.8	57.2	56.6	20.4	3
100	54.7	57.4	57.7	23.6	3
150	54.3	57.8	59.6	28.2	3
200	54.4	58.2	62.1	34.1	3
250	54.0	58.6	64.8	39.5	3
300	54.0	59.1	67.8	44.9	3

HGF+ (AQ)

T(C)	DELTA H	DELTA F	ENTHALPY	ENTROPY	REF
25	-37.8	-29.5	-37.8	-6.0	1
50	-37.6	-28.8	-36.9	-3.0	3
75	-37.3	-28.2	-35.6	0.8	3
100	-37.0	-27.5	-34.1	5.0	3
150	-36.5	-26.3	-31.6	11.5	3
200	-35.6	-25.2	-28.3	19.0	3
250	-34.9	-24.1	-24.8	26.1	3
300	-33.9	-23.1	-20.8	33.3	3

HGF2 (C), MERCURY(2) FLUORIDE

T(C)	DELTA H	DELTA F	ENTHALPY	ENTROPY	REF
25	-101.0	-89.4	-101.0	27.8	2
50	-100.9	-88.5	-100.6	29.3	2
75	-100.8	-87.5	-100.1	30.6	2
100	-100.7	-86.6	-99.6	31.8	2
150	-100.5	-84.7	-98.7	34.2	2
200	-100.3	-82.8	-97.8	36.3	2
250	-100.1	-81.0	-96.9	38.2	2
300	-99.9	-79.1	-95.9	39.9	2

HGI+ (AQ)

T(C)	DELTA H	DELTA F	ENTHALPY	ENTROPY	REF
25	10.3	9.5	10.3	14.0	1
50	10.2	9.4	11.0	16.3	3
75	10.1	9.4	12.0	19.3	3
100	10.0	9.3	13.1	22.6	3
114	9.9	9.3	13.6	23.9	2
114	8.1	9.3	13.6	23.9	2
150	7.7	9.5	15.0	27.3	3
185	7.6	9.6	16.8	31.5	2
185	2.6	9.6	16.8	31.5	2
200	2.7	9.8	17.6	33.3	3
250	2.6	10.6	20.3	38.7	3
300	2.7	11.3	23.3	44.3	3

HGI2 (C,L), MERCURY(2) IODIDE (PHASE TRANSITION OF HGI2 AT 130C; MELT PT = 250C)

T(C)	DELTA H	DELTA F	ENTHALPY	ENTROPY	REF
25	-25.2	-24.3	-25.2	43.0	1
50	-25.2	-24.3	-24.7	44.5	2
75	-25.3	-24.2	-24.3	45.9	2
100	-25.3	-24.1	-23.8	47.2	2
114	-25.4	-24.0	-23.6	47.8	2
114	-29.1	-24.0	-23.6	47.8	2
130	-29.2	-23.8	-23.3	48.6	2
130	-28.6	-23.8	-22.6	50.2	2
150	-28.7	-23.6	-22.2	51.2	2
185	-28.9	-23.2	-21.5	52.8	2
185	-38.9	-23.2	-21.5	52.8	2
200	-38.8	-22.7	-21.2	53.4	2
250	-38.6	-21.0	-20.2	55.4	2
250	-34.1	-21.0	-15.7	64.1	2
250	-34.1	-21.0	-15.7	64.1	2
300	-33.6	-19.7	-14.4	66.4	2

HGI2 (AQ), UNDISSOCIATED COMPLEX

T(C)	DELTA H	DELTA F	ENTHALPY	ENTROPY	REF
25	-19.2	-18.0	-19.2	42.0	1
50	-20.4	-17.8	-19.9	39.6	6
75	-21.6	-17.6	-20.6	37.6	6
100	-22.7	-17.3	-21.2	36.0	6
114	-23.5	-17.1	-21.7	34.6	2
114	-27.2	-17.1	-21.7	34.6	2
150	-29.8	-16.0	-23.2	30.6	6
185	-32.3	-14.8	-24.8	27.0	2
185	-42.3	-14.8	-24.8	27.0	2
200	-43.4	-13.9	-25.5	25.2	6

HGI3- (AQ)
(EXTRAPOLATED AS CRISS & COBBLE TYPE 2)

T(C)	DELTA H	DELTA F	ENTHALPY	ENTROPY	REF
25	-36.5	-35.5	-36.5	77.0	1
50	-38.4	-35.3	-38.2	71.6	3
75	-39.8	-35.1	-39.6	67.4	3
100	-40.8	-34.7	-40.9	64.0	3
114	-41.7	-34.5	-41.8	61.5	2
114	-47.3	-34.5	-41.8	61.5	2
150	-50.1	-33.2	-44.4	54.8	3
185	-52.9	-31.7	-47.2	48.2	2
185	-68.0	-31.7	-47.2	48.2	2
200	-69.0	-30.4	-48.5	45.3	3
250	-72.0	-26.2	-52.9	36.6	3
300	-76.7	-21.6	-58.9	25.6	3

HGI4-- (AQ)
(EXTRAPOLATED AS CRISS & COBBLE TYPE 2)

T(C)	DELTA H	DELTA F	ENTHALPY	ENTROPY	REF
25	-56.2	-50.6	-56.2	96.0	1
50	-57.9	-50.1	-58.0	90.2	3
75	-58.8	-49.5	-59.4	86.1	3
100	-59.0	-48.9	-60.6	83.0	3
114	-59.6	-48.5	-61.5	80.4	2
114	-67.0	-48.5	-61.5	80.4	2
150	-69.0	-46.6	-64.2	73.6	3
185	-70.9	-44.7	-67.1	66.9	2
185	-91.0	-44.7	-67.1	66.9	2
200	-91.5	-43.2	-68.4	64.0	3
250	-92.5	-38.0	-72.7	55.2	3
300	-95.1	-32.7	-78.8	44.1	3

HGICL (AQ), UNDISSOCIATED COMPLEX

T(C)	DELTA H	DELTA F	ENTHALPY	ENTROPY	REF
25	-35.8	-30.8	-35.8	42.0	1
50	-36.9	-30.3	-36.5	39.7	6
75	-37.9	-29.8	-37.1	38.0	6
100	-38.9	-29.2	-37.6	36.6	6
114	-39.7	-28.8	-38.1	35.3	2
114	-41.5	-28.8	-38.1	35.3	2
150	-43.7	-27.5	-39.5	31.7	6
185	-45.8	-26.1	-40.8	28.5	2
185	-50.8	-26.1	-40.8	28.5	2
200	-51.8	-25.3	-41.5	27.0	6

HGO (C), MERCURY(2) OXIDE (RED)

T(C)	DELTA H	DELTA F	ENTHALPY	ENTROPY	REF
25	-21.7	-14.0	-21.7	16.8	1
50	-21.7	-13.4	-21.4	17.7	2
75	-21.7	-12.7	-21.2	18.5	2
100	-21.7	-12.1	-20.9	19.3	2
150	-21.6	-10.8	-20.3	20.7	2
200	-21.5	-9.5	-19.7	22.0	2
250	-21.4	-8.3	-19.1	23.2	2
300	-21.3	-7.0	-18.5	24.4	2

HGOH+ (AQ)

T(C)	DELTA H	DELTA F	ENTHALPY	ENTROPY	REF
25	-20.2	-12.5	-20.2	12.0	1
50	-20.2	-11.9	-19.4	14.4	3
75	-20.3	-11.2	-18.4	17.4	3
100	-20.3	-10.6	-17.2	20.8	3
150	-20.6	-9.2	-15.3	25.7	3
200	-20.5	-7.9	-12.6	31.8	3
250	-20.7	-6.6	-9.8	37.5	3
300	-20.6	-5.2	-6.7	43.2	3

HGS (C), MERCURY(2) SULFIDE (RED)

T(C)	DELTA H	DELTA F	ENTHALPY	ENTROPY	REF
25	-29.2	-21.6	-29.2	19.7	1
50	-29.2	-20.9	-29.0	20.6	2
75	-29.2	-20.3	-28.7	21.4	2
100	-29.2	-19.7	-28.4	22.2	2
150	-29.2	-18.4	-27.8	23.7	2
200	-29.1	-17.1	-27.2	25.0	2
250	-29.0	-15.8	-26.7	26.1	2
300	-29.0	-14.6	-26.0	27.2	2

I- (AQ)

T(C)	DELTA H	DELTA F	ENTHALPY	ENTROPY	REF
25	-13.2	-13.2	-13.2	31.6	1
50	-14.2	-12.2	-14.5	27.3	3
75	-15.1	-12.0	-16.0	22.9	3
100	-16.0	-11.8	-17.5	18.6	3
114	-16.5	-11.6	-18.4	16.2	2
114	-18.4	-11.6	-18.4	16.2	2
150	-20.1	-10.9	-20.9	10.0	3
185	-21.9	-10.1	-23.6	3.6	2
185	-27.0	-10.1	-23.6	3.6	2
200	-27.7	-9.5	-24.9	0.8	3
250	-29.9	-7.5	-29.1	-7.8	3
300	-33.6	-5.1	-35.0	-18.5	3

I2 (C,L,G), IODINE
MELT PT = 114C;   BOIL PT = 185C

T(C)	DELTA H	DELTA F	ENTHALPY	ENTROPY	REF
25	0.0	0.0	0.0	27.8	1
50	0.0	0.0	0.3	28.8	2
75	0.0	0.0	0.7	29.8	2
100	0.0	0.0	1.0	30.8	2
114	0.0	0.0	1.2	31.4	2
114	0.0	0.0	4.9	41.0	2
150	0.0	0.0	5.6	42.7	2
185	0.0	0.0	6.3	44.2	2
185	0.0	0.0	16.3	66.1	2
200	0.0	0.0	16.5	66.4	2
250	0.0	0.0	16.9	67.3	2
300	0.0	0.0	17.4	68.1	2

IO- (AQ)

T(C)	DELTA H	DELTA F	ENTHALPY	ENTROPY	REF
25	-25.7	-9.2	-25.7	3.7	1
50	-28.4	-7.7	-28.6	-5.7	3
75	-31.2	-6.0	-31.9	-15.5	3
100	-34.2	-4.1	-35.5	-25.5	3
114	-35.4	-3.0	-37.0	-29.6	2
114	-37.3	-3.0	-37.0	-29.6	2
150	-40.9	0.4	-41.3	-40.2	3
185	-45.9	4.1	-47.0	-53.7	2
185	-50.9	4.1	-47.0	-53.7	2
200	-53.2	5.9	-49.6	-59.5	3
250	-60.4	12.6	-58.8	-77.9	3
300	-68.6	19.9	-68.9	-96.3	3

IO3- (AQ)

T(C)	DELTA H	DELTA F	ENTHALPY	ENTROPY	REF
25	-52.9	-30.6	-52.9	33.3	1
50	-54.3	-28.7	-54.4	28.5	3
75	-55.8	-26.6	-56.1	23.4	3
100	-57.2	-24.5	-57.9	18.2	3
114	-57.9	-23.3	-58.8	15.8	2
114	-59.7	-23.3	-58.8	15.8	2
150	-61.7	-19.8	-61.3	9.8	3
185	-64.1	-16.2	-64.1	3.2	2
185	-69.1	-16.2	-64.1	3.2	2
200	-70.1	-14.5	-65.4	0.3	3
250	-73.3	-8.4	-70.1	-9.2	3
300	-77.0	-2.0	-75.4	-18.8	3

IN (C,L), INDIUM
  MELTING PT = 156C

T(C)	DELTA H	DELTA F	ENTHALPY	ENTROPY	REF
25	0.0	0.0	0.0	13.8	1
50	0.0	0.0	0.2	14.3	2
75	0.0	0.0	0.3	14.8	2
100	0.0	0.0	0.5	15.3	2
150	0.0	0.0	0.8	16.1	2
156	0.0	0.0	0.9	16.2	2
156	0.0	0.0	1.7	18.1	2
200	0.0	0.0	2.0	18.8	2
250	0.0	0.0	2.3	19.5	2
300	0.0	0.0	2.7	20.1	2

IN(OH)++ (AQ)

T(C)	DELTA H	DELTA F	ENTHALPY	ENTROPY	REF
25	-87.8	-74.8	-87.8	-31.0	1
50	-87.9	-73.7	-86.7	-27.2	3
75	-88.1	-72.6	-85.1	-22.4	3
100	-88.3	-71.5	-83.1	-16.9	3
150	-88.8	-69.1	-79.8	-8.4	3
156	-88.8	-68.9	-79.3	-7.2	2
156	-89.6	-68.9	-79.3	-7.2	2
200	-89.8	-66.8	-75.6	1.3	3
250	-90.6	-64.3	-71.1	10.4	3
300	-90.9	-61.7	-66.0	19.6	3

IN(OH)2+ (AQ)

T(C)	DELTA H	DELTA F	ENTHALPY	ENTROPY	REF
25	-147.1	-125.5	-147.1	1.0	1
50	-147.2	-123.7	-146.2	3.8	3
75	-147.3	-121.9	-145.1	7.2	3
100	-147.3	-120.0	-143.7	11.2	3
150	-147.6	-116.3	-141.4	17.0	3
156	-147.6	-115.9	-141.0	17.9	2
156	-148.4	-115.9	-141.0	17.9	2
200	-148.3	-112.6	-138.4	24.0	3
250	-148.4	-108.8	-135.1	30.5	3
300	-148.2	-105.0	-131.5	37.2	3

IN+++ (AQ)

T(C)	DELTA H	DELTA F	ENTHALPY	ENTROPY	REF
25	-24.3	-23.4	-24.3	-51.0	1
50	-24.5	-23.3	-22.9	-46.6	3
75	-24.9	-23.2	-21.0	-40.9	3
100	-25.5	-23.0	-18.7	-34.4	3
150	-26.6	-22.6	-14.7	-24.2	3
156	-26.7	-22.5	-14.2	-22.8	2
156	-27.5	-22.5	-14.2	-22.8	2
200	-28.3	-22.0	-9.9	-13.0	3
250	-30.3	-21.2	-4.5	-2.2	3
300	-31.7	-20.3	1.4	8.6	3

IR (C), IRIDIUM

T(C)	DELTA H	DELTA F	ENTHALPY	ENTROPY	REF
25	0.0	0.0	0.0	8.5	1
50	0.0	0.0	0.2	9.0	2
75	0.0	0.0	0.3	9.4	2
100	0.0	0.0	0.4	9.8	2
150	0.0	0.0	0.8	10.6	2
200	0.0	0.0	1.1	11.3	2
250	0.0	0.0	1.4	11.9	2
300	0.0	0.0	1.7	12.5	2

IRCL6-- (AQ)
  (EXTRAPOLATED AS CRISS & COBBLE TYPE 3)

T(C)	DELTA H	DELTA F	ENTHALPY	ENTROPY	REF
25	-154.9	-111.2	-154.9	63.0	5
50	-154.8	-107.5	-155.0	62.8	3
75	-154.3	-103.9	-155.1	62.4	3
100	-153.4	-100.4	-155.2	62.0	3
150	-152.5	-93.3	-156.0	59.9	3
200	-150.4	-86.7	-155.9	60.3	3
250	-147.6	-80.1	-156.2	59.7	3
300	-145.0	-73.7	-156.6	58.9	3

IRCL6--- (AQ)
  (EXTRAPOLATED AS CRISS & COBBLE TYPE 3)

T(C)	DELTA H	DELTA F	ENTHALPY	ENTROPY	REF
25	-186.0	-134.7	-186.0	58.0	5
50	-185.7	-130.4	-186.3	57.0	3
75	-184.7	-126.2	-186.7	55.8	3
100	-183.2	-122.1	-187.2	54.6	3
150	-181.2	-114.1	-188.4	51.4	3
200	-178.0	-106.6	-189.0	50.2	3
250	-173.6	-99.3	-190.0	48.1	3
300	-169.4	-92.3	-191.3	45.8	3

IRO2 (C), IRIDIUM OXIDE

T(C)	DELTA H	DELTA F	ENTHALPY	ENTROPY	REF
25	-57.4	-44.3	-57.4	13.7	2
50	-57.4	-43.3	-57.1	14.8	2
75	-57.3	-42.2	-56.7	15.9	2
100	-57.3	-41.1	-56.3	16.9	2
150	-57.2	-38.9	-55.6	18.8	2
200	-57.1	-36.7	-54.8	20.6	2
250	-57.0	-34.6	-53.9	22.3	2
300	-56.8	-32.5	-53.1	23.9	2

K (C,L), POTASSIUM
  MELTING PT = 63C

T(C)	DELTA H	DELTA F	ENTHALPY	ENTROPY	REF
25	0.0	0.0	0.0	15.5	2
50	0.0	0.0	0.2	16.0	2
63	0.0	0.0	0.3	16.3	2
63	0.0	0.0	0.8	18.0	2
75	0.0	0.0	0.9	18.3	2
100	0.0	0.0	1.1	18.8	2
150	0.0	0.0	1.5	19.8	2
200	0.0	0.0	1.9	20.6	2
250	0.0	0.0	2.2	21.3	2
300	0.0	0.0	2.6	22.0	2

K+ (AQ)

T(C)	DELTA H	DELTA F	ENTHALPY	ENTROPY	REF
25	-60.1	-67.5	-60.1	19.5	5
50	-60.1	-68.1	-59.4	21.7	3
63	-60.1	-68.4	-59.0	22.9	2
63	-60.7	-68.4	-59.0	22.9	2
75	-60.7	-68.7	-58.6	24.3	3
100	-60.7	-69.3	-57.5	27.4	3
150	-61.0	-70.4	-55.8	31.6	3
200	-60.8	-71.5	-53.4	37.2	3
250	-61.0	-72.7	-50.9	42.2	3
300	-60.9	-73.8	-48.1	47.3	3

K.33AL2.33SI3.67O10(OH)2 (C), POTASSIUM-
MONTMORILLONITE

T(C)	DELTA H	DELTA F	ENTHALPY	ENTROPY	REF
25	-1368.9	-1279.7	-1368.9	63.4	10
50	-1369.1	-1272.2	-1367.0	69.5	10
63	-1369.2	-1268.3	-1366.0	72.6	2
63	-1369.4	-1268.3	-1366.0	72.6	2
75	-1369.4	-1264.7	-1365.0	75.4	10
100	-1369.5	-1257.2	-1363.0	81.1	10
150	-1369.5	-1242.1	-1358.6	92.1	10
200	-1369.4	-1227.1	-1354.1	102.2	10
250	-1369.1	-1212.1	-1349.3	111.7	10
300	-1368.8	-1197.1	-1344.4	120.7	10

K.6MG.25AL2.3SI3.5O10(OH)2 (C), ILLITE

T(C)	DELTA H	DELTA F	ENTHALPY	ENTROPY	REF
25	-1390.8	-1277.1	-1390.8	66.4	10
50	-1391.3	-1267.5	-1388.9	72.7	10
63	-1391.6	-1262.4	-1387.8	76.0	2
63	-1391.9	-1262.4	-1387.8	76.0	2
75	-1392.1	-1257.9	-1386.8	78.9	10
100	-1392.6	-1248.2	-1384.7	84.8	10
150	-1393.3	-1228.8	-1380.2	96.0	10
200	-1393.8	-1209.4	-1375.5	106.5	10
250	-1394.3	-1189.9	-1370.6	116.3	10
300	-1394.7	-1170.3	-1365.6	125.5	10

K2CO3 (C), POTASSIUM CARBONATE

T(C)	DELTA H	DELTA F	ENTHALPY	ENTROPY	REF
25	-274.9	-254.4	-274.9	37.2	2
50	-274.9	-252.7	-274.2	39.4	2
63	-274.9	-251.8	-273.8	40.6	2
63	-276.0	-251.8	-273.8	40.6	2
75	-276.0	-251.0	-273.5	41.5	2
100	-276.0	-249.2	-272.8	43.6	2
150	-275.9	-245.6	-271.2	47.4	2
200	-275.7	-242.0	-269.6	51.0	2
250	-275.5	-238.5	-267.9	54.4	2
300	-275.3	-234.9	-266.2	57.6	2

K2O (C), POTASSIUM OXIDE

T(C)	DELTA H	DELTA F	ENTHALPY	ENTROPY	REF
25	-86.8	-77.0	-86.8	22.5	2
50	-86.7	-76.2	-86.3	24.1	2
63	-86.7	-75.7	-86.0	25.0	2
63	-87.8	-75.7	-86.0	25.0	2
75	-87.8	-75.3	-85.8	25.7	2
100	-87.8	-74.4	-85.3	27.1	2
150	-87.6	-72.6	-84.2	29.8	2
200	-87.5	-70.9	-83.1	32.2	2
250	-87.3	-69.1	-82.0	34.4	2
300	-87.1	-67.4	-80.9	36.5	2

K2O2 (C), POTASSIUM PEROXIDE

T(C)	DELTA H	DELTA F	ENTHALPY	ENTROPY	REF
25	-118.5	-102.7	-118.5	27.0	2
50	-118.4	-101.4	-117.9	29.0	2
63	-118.4	-100.7	-117.6	29.9	2
63	-119.5	-100.7	-117.6	29.9	2
75	-119.5	-100.1	-117.3	30.8	2
100	-119.4	-98.6	-116.6	32.5	2
150	-119.3	-95.9	-115.4	35.7	2
200	-119.1	-93.1	-114.1	38.7	2
250	-118.8	-90.4	-112.7	41.4	2
300	-118.5	-87.7	-111.3	44.0	2

K2SO4 (C), POTASSIUM SULFATE

T(C)	DELTA H	DELTA F	ENTHALPY	ENTROPY	REF
25	-358.0	-324.0	-358.0	42.0	2
50	-358.1	-321.1	-357.3	44.6	2
63	-358.1	-319.6	-356.8	45.9	2
63	-359.2	-319.6	-356.8	45.9	2
75	-359.2	-318.3	-356.4	47.0	2
100	-359.2	-315.3	-355.6	49.4	2
150	-359.1	-309.4	-353.8	53.8	2
200	-358.9	-303.6	-351.9	58.0	2
250	-358.7	-297.7	-350.0	61.9	2
300	-358.3	-291.9	-348.0	65.6	2

KAL3SI3O10(OH)2 (C), MUSCOVITE

T(C)	DELTA H	DELTA F	ENTHALPY	ENTROPY	REF
25	-1420.9	-1329.8	-1420.9	69.0	11
50	-1421.1	-1322.2	-1418.9	75.4	10
63	-1421.2	-1318.1	-1417.8	78.7	2
63	-1421.8	-1318.1	-1417.8	78.7	2
75	-1421.9	-1314.5	-1416.8	81.6	10
100	-1421.9	-1306.8	-1414.6	87.7	10
150	-1422.0	-1291.3	-1410.0	99.3	10
200	-1421.8	-1275.9	-1405.2	110.0	10
250	-1421.6	-1260.5	-1400.2	120.1	10
300	-1421.2	-1245.1	-1395.0	129.6	10

KALSI2O6 (C), LEUCITE

T(C)	DELTA H	DELTA F	ENTHALPY	ENTROPY	REF
25	-721.7	-681.6	-721.7	44.1	10
50	-721.8	-678.3	-720.7	47.2	10
63	-721.8	-676.5	-720.2	48.8	2
63	-722.4	-676.5	-720.2	48.8	2
75	-722.5	-674.9	-719.7	50.1	10
100	-722.6	-671.5	-718.7	52.8	10
150	-723.0	-664.6	-716.8	57.8	10
200	-723.3	-657.7	-714.8	62.2	10
250	-723.8	-650.7	-712.8	66.1	10
300	-724.2	-643.7	-710.9	69.7	10

KALSI3O8 (C), MICROCLINE

T(C)	DELTA H	DELTA F	ENTHALPY	ENTROPY	REF
25	-946.0	-892.6	-946.0	52.5	11
50	-946.1	-888.1	-944.8	56.5	12
63	-946.2	-885.7	-944.1	58.6	2
63	-946.8	-885.7	-944.1	58.6	2
75	-946.8	-883.5	-943.4	60.4	12
100	-946.9	-879.0	-942.0	64.3	12
150	-946.9	-869.9	-939.1	71.6	12
200	-946.8	-860.8	-936.1	78.4	12
250	-946.7	-851.7	-932.9	84.8	12
300	-946.5	-842.7	-929.7	90.7	12

KALSI3O8 (C), ADULARIA

T(C)	DELTA H	DELTA F	ENTHALPY	ENTROPY	REF
25	-945.0	-892.6	-945.0	56.0	11
50	-945.1	-888.2	-943.8	60.0	10
63	-945.2	-885.9	-943.1	62.1	2
63	-945.8	-885.9	-943.1	62.1	2
75	-945.8	-883.8	-942.4	64.0	10
100	-945.9	-879.3	-941.0	67.8	10
150	-945.9	-870.4	-938.1	75.1	10
200	-945.8	-861.5	-935.1	81.9	10
250	-945.7	-852.6	-931.9	88.3	10
300	-945.5	-843.7	-928.7	94.3	10

KALSI3O8 (C), HIGH SANIDINE

T(C)	DELTA H	DELTA F	ENTHALPY	ENTROPY	REF
25	-944.4	-892.3	-944.4	56.9	11
50	-944.5	-887.9	-943.1	61.0	10
63	-944.6	-885.6	-942.4	63.1	2
63	-945.1	-885.6	-942.4	63.1	2
75	-945.2	-883.5	-941.8	64.9	10
100	-945.3	-879.1	-940.4	68.8	10
150	-945.3	-870.2	-937.5	76.1	10
200	-945.2	-861.3	-934.4	82.9	10
250	-945.1	-852.4	-931.3	89.3	10
300	-944.9	-843.6	-928.0	95.2	10

KBR (C), POTASSIUM BROMIDE

T(C)	DELTA H	DELTA F	ENTHALPY	ENTROPY	REF
25	-94.1	-90.9	-94.1	22.9	2
50	-94.2	-90.6	-93.8	23.9	2
58	-94.2	-90.6	-93.7	24.3	2
58	-97.8	-90.6	-93.7	24.3	2
63	-97.8	-90.5	-93.6	24.5	2
63	-98.3	-90.5	-93.6	24.5	2
75	-98.3	-90.2	-93.5	24.9	2
100	-98.3	-89.6	-93.2	25.8	2
150	-98.3	-88.4	-92.5	27.4	2
200	-98.2	-87.3	-91.9	28.9	2
250	-98.1	-86.1	-91.2	30.2	2
300	-98.1	-85.0	-90.6	31.4	2

KBR (AQ), UNDISSOCIATED COMPLEX

T(C)	DELTA H	DELTA F	ENTHALPY	ENTROPY	REF
25	-89.0	-92.0	-89.0	43.8	5
50	-90.1	-92.3	-89.7	41.7	6
58	-90.4	-92.3	-89.9	41.0	2
58	-93.9	-92.3	-89.9	41.0	2
63	-94.1	-92.3	-90.0	40.7	2
63	-94.7	-92.3	-90.0	40.7	2
75	-95.1	-92.2	-90.2	40.0	6
100	-95.9	-91.9	-90.8	38.5	6
150	-98.2	-91.2	-92.5	34.0	6
200	-100.5	-90.3	-94.4	30.2	6

KCL (C), POTASSIUM CHLORIDE

T(C)	DELTA H	DELTA F	ENTHALPY	ENTROPY	REF
25	-104.4	-97.7	-104.4	19.7	2
50	-104.3	-97.1	-104.1	20.7	2
63	-104.3	-96.9	-103.9	21.2	2
63	-104.9	-96.9	-103.9	21.2	2
75	-104.9	-96.6	-103.8	21.7	2
100	-104.9	-96.0	-103.4	22.5	2
150	-104.8	-94.8	-102.8	24.1	2
200	-104.8	-93.6	-102.2	25.5	2
250	-104.7	-92.4	-101.5	26.8	2
300	-104.6	-91.2	-100.9	28.0	2

KCL (AQ), UNDISSOCIATED COMPLEX

T(C)	DELTA H	DELTA F	ENTHALPY	ENTROPY	REF
25	-92.2	-98.8	-92.2	37.7	5
50	-93.0	-99.3	-92.8	35.8	6
63	-93.4	-99.6	-93.1	34.8	2
63	-93.9	-99.6	-93.1	34.8	2
75	-94.3	-99.8	-93.4	34.0	6
100	-95.1	-100.1	-93.9	32.4	6
150	-97.1	-100.7	-95.6	28.1	6
200	-99.2	-101.0	-97.2	24.4	6

KF (C), POTASSIUM FLUORIDE

T(C)	DELTA H	DELTA F	ENTHALPY	ENTROPY	REF
25	-135.6	-128.5	-135.6	15.9	2
50	-135.6	-127.9	-135.3	16.9	2
63	-135.6	-127.6	-135.1	17.4	2
63	-136.1	-127.6	-135.1	17.4	2
75	-136.1	-127.3	-135.0	17.8	2
100	-136.1	-126.7	-134.7	18.6	2
150	-136.1	-125.4	-134.1	20.2	2
200	-136.0	-124.2	-133.4	21.6	2
250	-135.9	-122.9	-132.8	22.9	2
300	-135.9	-121.7	-132.2	24.0	2

KI (C), POTASSIUM IODIDE

T(C)	DELTA H	DELTA F	ENTHALPY	ENTROPY	REF
25	-78.4	-77.2	-78.4	25.4	2
50	-78.4	-77.1	-78.1	26.5	2
63	-78.4	-77.1	-77.9	27.0	2
63	-79.0	-77.1	-77.9	27.0	2
75	-79.0	-77.0	-77.7	27.4	2
100	-79.1	-76.8	-77.4	28.3	2
114	-79.1	-76.8	-77.2	28.7	2
114	-80.9	-76.8	-77.2	28.7	2
150	-81.1	-76.4	-76.8	29.9	2
185	-81.3	-76.0	-76.3	30.9	2
185	-86.3	-76.0	-76.3	30.9	2
200	-86.2	-75.6	-76.1	31.3	2
250	-86.2	-74.5	-75.5	32.7	2
300	-86.1	-73.4	-74.8	33.9	2

KI (AQ), UNDISSOCIATED COMPLEX

T(C)	DELTA H	DELTA F	ENTHALPY	ENTROPY	REF
25	-73.5	-79.8	-73.5	50.6	5
50	-74.5	-80.3	-74.2	48.3	6
63	-75.1	-80.5	-74.5	47.3	2
63	-75.6	-80.5	-74.5	47.3	2
75	-76.0	-80.7	-74.8	46.6	6
100	-76.9	-81.0	-75.3	45.3	6
114	-77.6	-81.1	-75.7	44.0	2
114	-79.4	-81.1	-75.7	44.0	2
150	-81.4	-81.2	-77.0	40.7	6
185	-83.2	-81.1	-78.2	37.9	2
185	-88.2	-81.1	-78.2	37.9	2
200	-88.9	-80.9	-78.7	36.7	6

KMG3ALSI3O10F2 (C), FLUORPHLOGOPITE

T(C)	DELTA H	DELTA F	ENTHALPY	ENTROPY	REF
25	-1498.1	-1415.6	-1498.1	75.9	10
50	-1498.2	-1408.7	-1496.0	82.6	10
63	-1498.3	-1405.0	-1494.9	86.1	2
63	-1498.8	-1405.0	-1494.9	86.1	2
75	-1498.8	-1401.7	-1493.8	89.2	10
100	-1498.9	-1394.8	-1491.6	95.4	10
150	-1498.8	-1380.8	-1486.2	107.3	10
200	-1498.6	-1366.9	-1481.9	118.2	10
250	-1498.4	-1353.0	-1476.9	128.3	10
300	-1498.1	-1339.1	-1471.8	137.7	10

KNO3 (C), POTASSIUM NITRATE
TRANS PT OF KNO3 AT 128C

T(C)	DELTA H	DELTA F	ENTHALPY	ENTROPY	REF
25	-117.7	-93.8	-117.7	31.8	2
50	-117.6	-91.8	-117.1	33.6	2
63	-117.6	-90.8	-116.8	34.6	2
63	-118.2	-90.8	-116.8	34.6	2
75	-118.1	-89.8	-116.5	35.4	2
100	-118.1	-87.8	-115.9	37.2	2
128	-118.0	-85.5	-115.2	39.0	2
128	-116.8	-85.5	-114.0	42.0	2
150	-116.6	-83.8	-113.3	43.6	2
200	-116.3	-79.9	-111.9	46.8	2
250	-115.9	-76.1	-110.4	49.7	2
300	-115.6	-72.3	-109.0	52.3	2

KNO3 (AQ), UNDISSOCIATED COMPLEX

T(C)	DELTA H	DELTA F	ENTHALPY	ENTROPY	REF
25	-106.5	-93.9	-106.5	69.5	5
50	-107.2	-92.8	-106.6	69.1	6
63	-107.5	-92.2	-106.7	69.0	2
63	-108.0	-92.2	-106.7	69.0	2
75	-108.2	-91.6	-106.6	69.2	6
100	-108.6	-90.4	-106.4	69.7	6
150	-109.1	-87.9	-105.9	71.1	6
200	-108.3	-85.5	-104.2	75.5	6

KO2 (C), POTASSIUM DIOXIDE

T(C)	DELTA H	DELTA F	ENTHALPY	ENTROPY	REF
25	-67.6	-56.7	-67.6	27.9	2
50	-67.5	-55.8	-67.1	29.4	2
63	-67.4	-55.3	-66.9	30.1	2
63	-68.0	-55.3	-66.9	30.1	2
75	-67.9	-54.9	-66.7	30.8	2
100	-67.8	-53.9	-66.2	32.1	2
150	-67.6	-52.1	-65.2	34.6	2
200	-67.4	-50.3	-64.2	36.8	2
250	-67.1	-48.5	-63.2	38.8	2
300	-66.8	-46.7	-62.2	40.8	2

KOH (C), POTASSIUM HYDROXIDE
TRANS PT OF KOH AT 249C

T(C)	DELTA H	DELTA F	ENTHALPY	ENTROPY	REF
25	-101.5	-90.6	-101.5	19.0	2
50	-101.5	-89.7	-101.1	20.2	2
63	-101.4	-89.2	-100.9	20.9	2
63	-102.0	-89.2	-100.9	20.9	2
75	-102.0	-88.7	-100.7	21.5	2
100	-101.9	-87.8	-100.3	22.6	2
150	-101.8	-85.9	-99.4	24.8	2
200	-101.6	-84.0	-98.5	26.9	2
249	-101.4	-82.2	-97.5	28.8	2
249	-99.8	-82.2	-96.0	31.7	2
250	-99.8	-82.2	-96.0	31.7	2
300	-99.6	-80.5	-95.1	33.4	2

KOH (AQ), UNDISSOCIATED COMPLEX

T(C)	DELTA H	DELTA F	ENTHALPY	ENTROPY	REF
25	-115.1	-105.1	-115.1	22.0	5
50	-115.9	-104.2	-115.5	20.5	6
63	-116.4	-103.7	-115.8	19.6	2
63	-116.9	-103.7	-115.8	19.6	2
75	-117.4	-103.2	-116.1	18.8	6
100	-118.4	-102.2	-116.8	16.9	6
150	-120.8	-99.8	-118.3	12.9	6
200	-123.0	-97.2	-119.8	9.5	6

LA (C), LANTHANUM

T(C)	DELTA H	DELTA F	ENTHALPY	ENTROPY	REF
25	0.0	0.0	0.0	13.6	1
50	0.0	0.0	0.2	14.1	2
75	0.0	0.0	0.3	14.6	2
100	0.0	0.0	0.5	15.1	2
150	0.0	0.0	0.8	16.0	2
200	0.0	0.0	1.2	16.7	2
250	0.0	0.0	1.5	17.4	2
300	0.0	0.0	1.9	18.1	2

LA+++ (AQ)

T(C)	DELTA H	DELTA F	ENTHALPY	ENTROPY	REF
25	-169.0	-163.4	-169.0	-67.0	1
50	-169.0	-162.9	-167.5	-62.1	3
75	-169.2	-162.4	-165.4	-55.7	3
100	-169.5	-161.9	-162.8	-48.4	3
150	-170.1	-160.9	-158.2	-36.9	3
200	-170.5	-159.7	-152.8	-24.3	3
250	-171.8	-158.5	-146.9	-12.3	3
300	-172.5	-157.2	-140.2	-0.1	3

LA2O3 (C), LANTHANUM OXIDE

T(C)	DELTA H	DELTA F	ENTHALPY	ENTROPY	REF
25	-428.7	-407.8	-428.7	30.4	1
50	-428.6	-406.0	-428.0	32.6	2
75	-428.6	-404.2	-427.4	34.6	2
100	-428.5	-402.5	-426.7	36.5	2
150	-428.3	-399.0	-425.3	40.0	2
200	-428.1	-395.6	-423.8	43.2	2
250	-427.9	-392.2	-422.4	46.1	2
300	-427.7	-388.7	-420.9	48.8	2

LACL3 (C), LANTHANUM CHLORIDE

T(C)	DELTA H	DELTA F	ENTHALPY	ENTROPY	REF
25	-255.9	-238.3	-255.9	34.5	2
50	-255.8	-236.8	-255.3	36.5	2
75	-255.6	-235.4	-254.7	38.4	2
100	-255.5	-233.9	-254.0	40.1	2
150	-255.2	-231.1	-252.8	43.3	2
200	-254.9	-228.2	-251.5	46.1	2
250	-254.6	-225.4	-250.2	48.7	2
300	-254.3	-222.6	-248.9	51.1	2

LI (C,L), LITHIUM
MELTING PT = 181C

T(C)	DELTA H	DELTA F	ENTHALPY	ENTROPY	REF
25	0.0	0.0	0.0	6.9	2
50	0.0	0.0	0.2	7.4	2
75	0.0	0.0	0.3	7.9	2
100	0.0	0.0	0.5	8.3	2
150	0.0	0.0	0.8	9.1	2
181	0.0	0.0	1.0	9.6	2
181	0.0	0.0	1.7	11.2	2
200	0.0	0.0	1.9	11.5	2
250	0.0	0.0	2.2	12.2	2
300	0.0	0.0	2.6	12.9	2

LI+ (AQ)

T(C)	DELTA H	DELTA F	ENTHALPY	ENTROPY	REF
25	-66.6	-70.2	-66.6	-1.6	5
50	-66.4	-70.5	-65.8	1.2	3
75	-66.0	-70.9	-64.6	4.8	3
100	-65.6	-71.2	-63.1	8.9	3
150	-65.2	-72.0	-60.7	14.9	3
181	-64.6	-72.5	-58.9	19.3	2
181	-65.3	-72.5	-58.9	19.3	2
200	-64.9	-72.9	-57.6	22.2	3
250	-64.3	-73.7	-54.3	28.9	3
300	-63.2	-74.7	-50.5	35.7	3

LI2CO3 (C), LITHIUM CARBONATE

T(C)	DELTA H	DELTA F	ENTHALPY	ENTROPY	REF
25	-290.6	-270.6	-290.6	21.5	2
50	-290.6	-268.9	-290.0	23.4	2
75	-290.7	-267.2	-289.4	25.3	2
100	-290.7	-265.5	-288.8	27.0	2
150	-290.7	-262.1	-287.4	30.4	2
181	-290.6	-260.1	-286.5	32.4	2
181	-292.1	-260.1	-286.5	32.4	2
200	-292.1	-258.7	-286.0	33.7	2
250	-292.0	-255.2	-284.4	36.8	2
300	-291.8	-251.7	-282.7	39.8	2

LI2O (C), DILITHIUM MONOXIDE

T(C)	DELTA H	DELTA F	ENTHALPY	ENTROPY	REF
25	-143.1	-134.4	-143.1	9.1	2
50	-143.1	-133.6	-142.8	10.1	2
75	-143.2	-132.9	-142.4	11.2	2
100	-143.2	-132.1	-142.0	12.2	2
150	-143.3	-130.6	-141.3	14.1	2
181	-143.4	-129.7	-140.8	15.3	2
181	-144.8	-129.7	-140.8	15.3	2
200	-144.8	-129.1	-140.5	15.9	2
250	-144.9	-127.4	-139.6	17.6	2
300	-144.9	-125.7	-138.8	19.2	2

LI2O2 (C), LITHIUM PEROXIDE
THERMAL DECOMP TEMP = 195C

T(C)	DELTA H	DELTA F	ENTHALPY	ENTROPY	REF
25	-151.2	-136.5	-151.2	13.5	2
50	-151.2	-135.2	-150.8	14.9	2
75	-151.3	-134.0	-150.3	16.2	2
100	-151.3	-132.8	-149.9	17.5	2
150	-151.3	-130.3	-148.9	20.0	2
181	-151.4	-128.7	-148.2	21.5	2
181	-152.8	-128.7	-148.2	21.5	2
195	-152.8	-128.0	-147.9	22.1	2

LIBR (C), LITHIUM BROMIDE

T(C)	DELTA H	DELTA F	ENTHALPY	ENTROPY	REF
25	-83.7	-81.5	-83.7	17.7	2
50	-83.8	-81.3	-83.4	18.7	2
58	-83.8	-81.2	-83.3	18.9	2
58	-87.3	-81.2	-83.3	18.9	2
75	-87.3	-80.9	-83.1	19.5	2
100	-87.3	-80.5	-82.8	20.3	2
150	-87.2	-79.6	-82.2	21.9	2
181	-87.2	-79.0	-81.8	22.7	2
181	-87.9	-79.0	-81.8	22.7	2
200	-87.9	-78.6	-81.6	23.3	2
250	-87.9	-77.6	-81.0	24.5	2
300	-87.8	-76.7	-80.3	25.7	2

LIBR (AQ), UNDISSOCIATED COMPLEX

T(C)	DELTA H	DELTA F	ENTHALPY	ENTROPY	REF
25	-95.5	-94.8	-95.5	22.7	5
50	-96.3	-94.7	-95.9	21.3	6
58	-96.6	-94.6	-96.1	20.8	2
58	-100.1	-94.6	-96.1	20.8	6
75	-100.4	-94.4	-96.2	20.5	6
100	-100.9	-93.9	-96.4	20.0	6
150	-102.4	-92.8	-97.4	17.3	6
181	-103.3	-92.1	-98.0	16.0	2
181	-104.0	-92.1	-98.0	16.0	2
200	-104.7	-91.6	-98.4	15.2	6

LICL (C), LITHIUM CHLORIDE

T(C)	DELTA H	DELTA F	ENTHALPY	ENTROPY	REF
25	-97.6	-91.8	-97.6	14.2	2
50	-97.5	-91.3	-97.3	15.1	2
75	-97.5	-90.8	-97.0	16.0	2
100	-97.5	-90.4	-96.7	16.8	2
150	-97.4	-89.4	-96.1	18.3	2
181	-97.4	-88.8	-95.7	19.2	2
181	-98.1	-88.8	-95.7	19.2	2
200	-98.1	-88.4	-95.5	19.7	2
250	-98.0	-87.4	-94.8	21.0	2
300	-97.9	-86.4	-94.2	22.2	2

LICL (AQ), UNDISSOCIATED COMPLEX

T(C)	DELTA H	DELTA F	ENTHALPY	ENTROPY	REF
25	-114.6	-101.6	-114.6	16.6	5
50	-115.3	-100.4	-115.0	15.4	6
75	-116.0	-99.3	-115.3	14.5	6
100	-116.5	-98.1	-115.5	14.0	6
150	-118.3	-95.4	-116.4	11.4	6
181	-119.3	-93.7	-117.0	10.2	2
181	-120.0	-93.7	-117.0	10.2	2
200	-120.6	-92.6	-117.3	9.4	6

LIF (C), LITHIUM FLUORIDE

T(C)	DELTA H	DELTA F	ENTHALPY	ENTROPY	REF
25	-146.5	-139.8	-146.5	8.5	2
50	-146.5	-139.2	-146.3	9.3	2
75	-146.5	-138.6	-146.0	10.1	2
100	-146.5	-138.0	-145.7	10.9	2
150	-146.4	-136.9	-145.2	12.3	2
181	-146.4	-136.2	-144.8	13.0	2
181	-147.1	-136.2	-144.8	13.0	2
200	-147.1	-135.8	-144.6	13.5	2
250	-147.1	-134.6	-144.0	14.7	2
300	-147.1	-133.4	-143.4	15.8	2

LIF (AQ), UNDISSOCIATED COMPLEX

T(C)	DELTA H	DELTA F	ENTHALPY	ENTROPY	REF
25	-145.3	-136.3	-145.3	1.1	5
50	-145.7	-135.5	-145.5	0.4	6
75	-146.2	-134.7	-145.7	-0.3	6
100	-146.7	-133.9	-146.0	-1.0	6
150	-147.9	-132.0	-146.7	-2.8	6
181	-148.6	-130.9	-147.0	-3.5	2
181	-149.3	-130.9	-147.0	-3.5	2
200	-149.7	-130.1	-147.2	-3.9	6

LII (C), LITHIUM IODIDE

T(C)	DELTA H	DELTA F	ENTHALPY	ENTROPY	REF
25	-64.6	-64.4	-64.6	20.5	2
50	-64.6	-64.4	-64.3	21.5	2
75	-64.6	-64.4	-63.9	22.4	2
100	-64.6	-64.4	-63.6	23.3	2
114	-64.6	-64.4	-63.5	23.7	2
114	-66.5	-64.4	-63.5	23.7	2
150	-66.6	-64.2	-63.0	24.8	2
181	-66.7	-64.0	-62.6	25.8	2
181	-67.4	-64.0	-62.6	25.8	2
185	-67.4	-64.0	-62.5	25.9	2
185	-72.5	-64.0	-62.5	25.9	2
200	-72.4	-63.7	-62.3	26.3	2
250	-72.4	-62.8	-61.7	27.6	2
300	-72.3	-61.9	-61.0	28.9	2

LII (AQ), UNDISSOCIATED COMPLEX

T(C)	DELTA H	DELTA F	ENTHALPY	ENTROPY	REF
25	-80.0	-82.6	-80.0	29.5	5
50	-80.8	-82.8	-80.5	27.9	6
75	-81.4	-82.9	-80.8	27.1	6
100	-81.9	-83.0	-80.9	26.8	6
114	-82.3	-83.0	-81.2	26.0	2
114	-84.2	-83.0	-81.2	26.0	2
150	-85.6	-82.9	-82.0	24.0	6
181	-86.7	-82.6	-82.5	22.6	2
181	-87.4	-82.6	-82.5	22.6	2
185	-87.6	-82.6	-82.6	22.4	2
185	-92.6	-82.6	-82.6	22.4	2
200	-93.1	-82.2	-82.9	21.7	6

LINO3 (AQ), UNDISSOCIATED COMPLEX

T(C)	DELTA H	DELTA F	ENTHALPY	ENTROPY	REF
25	-116.0	-96.6	-116.0	39.4	5
50	-116.8	-95.0	-116.3	37.4	6
75	-117.4	-93.3	-116.4	37.0	6
100	-118.0	-91.5	-116.5	36.9	6
150	-119.4	-87.9	-116.8	36.0	6
181	-120.1	-85.6	-116.9	36.0	2
181	-120.8	-85.6	-116.9	36.0	2
200	-121.3	-84.1	-116.9	36.0	6

LIOH (C), LITHIUM HYDROXIDE

T(C)	DELTA H	DELTA F	ENTHALPY	ENTROPY	REF
25	-115.8	-104.9	-115.8	10.2	2
50	-115.9	-103.9	-115.5	11.2	2
75	-115.9	-103.0	-115.2	12.1	2
100	-115.9	-102.1	-114.9	13.0	2
150	-115.9	-100.2	-114.2	14.8	2
181	-115.9	-99.1	-113.8	15.8	2
181	-116.6	-99.1	-113.8	15.8	2
200	-116.6	-98.4	-113.5	16.4	2
250	-116.5	-96.4	-112.7	17.9	2
300	-116.5	-94.5	-111.9	19.4	2

LIOH (AQ), UNDISSOCIATED COMPLEX

T(C)	DELTA H	DELTA F	ENTHALPY	ENTROPY	REF
25	-121.6	-107.8	-121.6	0.9	5
50	-122.2	-106.6	-121.9	0.0	6
75	-122.8	-105.4	-122.1	-0.8	6
100	-123.4	-104.1	-122.4	-1.6	6
150	-124.9	-101.4	-123.3	-3.8	6
181	-125.8	-99.7	-123.7	-4.8	2
181	-126.5	-99.7	-123.7	-4.8	2
200	-127.1	-98.5	-124.0	-5.5	6

MG (C), MAGNESIUM

T(C)	DELTA H	DELTA F	ENTHALPY	ENTROPY	REF
25	0.0	0.0	0.0	7.8	1
50	0.0	0.0	0.2	8.3	2
75	0.0	0.0	0.3	8.7	2
100	0.0	0.0	0.4	9.2	2
150	0.0	0.0	0.8	10.0	2
200	0.0	0.0	1.1	10.7	2
250	0.0	0.0	1.4	11.3	2
300	0.0	0.0	1.7	11.9	2

MG(NO3)2 (C), MAGNESIUM NITRATE

T(C)	DELTA H	DELTA F	ENTHALPY	ENTROPY	REF
25	-189.0	-140.9	-189.0	39.2	1
50	-189.0	-136.8	-188.1	42.0	2
75	-188.9	-132.8	-187.2	44.7	2
100	-188.8	-128.8	-186.3	47.3	2
150	-188.6	-120.7	-184.3	52.3	2
200	-188.2	-112.7	-182.1	57.2	2
250	-187.7	-104.8	-179.7	61.9	2
300	-187.0	-96.9	-177.2	66.5	2

MG(OH)2 (C), MAGNESIUM HYDROXIDE
THERMAL DECOMP TEMP = 268C

T(C)	DELTA H	DELTA F	ENTHALPY	ENTROPY	REF
25	-221.0	-199.2	-221.0	15.1	1
50	-221.0	-197.4	-220.5	16.6	2
75	-221.0	-195.6	-220.0	18.1	2
100	-221.0	-193.8	-219.5	19.5	2
150	-221.0	-190.1	-218.5	22.1	2
200	-220.9	-186.5	-217.4	24.6	2
250	-220.8	-182.8	-216.2	27.0	2
268	-220.7	-181.5	-215.7	27.8	2

MG++ (AQ)

T(C)	DELTA H	DELTA F	ENTHALPY	ENTROPY	REF
25	-111.6	-108.7	-111.6	-43.0	1
50	-111.4	-108.5	-110.3	-38.8	3
75	-111.1	-108.3	-108.5	-33.5	3
100	-110.9	-108.1	-106.3	-27.4	3
150	-110.7	-107.7	-102.6	-17.9	3
200	-110.1	-107.4	-98.0	-7.3	3
250	-110.0	-107.1	-93.0	2.8	3
300	-109.5	-106.8	-87.4	13.0	3

MG.167AL2.33SI3.67O10(OH)2 (C), MAGNESIUM-
MONTMORILLONITE

T(C)	DELTA H	DELTA F	ENTHALPY	ENTROPY	REF
25	-1364.1	-1275.4	-1364.1	61.2	10
50	-1364.3	-1268.0	-1362.3	67.1	10
75	-1364.5	-1260.5	-1360.4	72.9	10
100	-1364.6	-1253.0	-1358.3	78.5	10
150	-1364.6	-1238.1	-1354.1	89.2	10
200	-1364.5	-1223.1	-1349.6	99.1	10
250	-1364.3	-1208.2	-1345.0	108.5	10
300	-1363.9	-1193.3	-1340.2	117.2	10

MG2SIO4 (C), FORSTERITE

T(C)	DELTA H	DELTA F	ENTHALPY	ENTROPY	REF
25	-519.6	-491.2	-519.6	22.7	1
50	-519.7	-488.8	-518.9	25.1	10
75	-519.7	-486.4	-518.1	27.3	10
100	-519.7	-484.0	-517.3	29.5	10
150	-519.7	-479.2	-515.7	33.7	10
200	-519.6	-474.4	-513.9	37.5	10
250	-519.5	-469.7	-512.2	41.1	10
300	-519.3	-464.9	-510.3	44.4	10

MG3(PO4)2 (C), MAGNESIUM ORTHOPHOSPHATE

T(C)	DELTA H	DELTA F	ENTHALPY	ENTROPY	REF
25	-903.6	-845.8	-903.6	45.2	1
44	-903.7	-842.1	-902.6	48.4	2
44	-904.0	-842.1	-902.6	48.4	2
50	-904.1	-840.9	-902.3	49.4	2
75	-904.2	-836.1	-901.0	53.3	2
100	-904.3	-831.2	-899.6	57.2	2
150	-904.5	-821.3	-896.8	64.3	2
200	-904.6	-811.5	-893.8	71.0	2
250	-904.6	-801.7	-890.6	77.3	2
274	-904.5	-797.0	-889.1	80.2	2
274	-910.3	-797.0	-889.1	80.2	2
300	-910.2	-791.6	-887.4	83.3	2

MG3SI2O5(OH)4 (C), CHRYSOTILE

T(C)	DELTA H	DELTA F	ENTHALPY	ENTROPY	REF
25	-1043.4	-965.2	-1043.4	52.9	1
50	-1043.6	-958.6	-1041.7	58.3	10
75	-1043.6	-952.0	-1039.9	63.6	10
100	-1043.7	-945.4	-1038.1	68.7	10
150	-1043.6	-932.3	-1034.2	78.4	10
200	-1043.4	-919.1	-1030.2	87.5	10
250	-1043.1	-906.0	-1026.0	96.0	10
300	-1042.7	-892.9	-1021.6	103.9	10

MG3SI4O10(OH)2 (C), TALC

T(C)	DELTA H	DELTA F	ENTHALPY	ENTROPY	REF
25	-1415.5	-1324.8	-1415.5	62.3	1
50	-1415.7	-1317.1	-1413.6	68.5	10
75	-1416.0	-1309.5	-1411.7	74.2	10
100	-1416.3	-1301.8	-1409.7	79.6	10
150	-1417.0	-1286.5	-1405.9	89.2	10
200	-1417.8	-1271.0	-1402.0	97.8	10
250	-1418.6	-1255.4	-1398.2	105.5	10
300	-1419.5	-1239.8	-1394.4	112.6	10

MG5AL2SI3O10(OH)8 (C), MAGNESIUM-CHLORITE

T(C)	DELTA H	DELTA F	ENTHALPY	ENTROPY	REF
25	-2109.8	-1954.8	-2109.8	112.0	10
50	-2111.4	-1941.8	-2107.8	118.7	10
75	-2113.0	-1928.6	-2105.6	125.2	10
100	-2114.5	-1915.3	-2103.3	131.6	10
150	-2117.3	-1888.4	-2098.5	143.6	10
200	-2120.0	-1861.2	-2093.4	154.9	10
250	-2122.6	-1833.7	-2088.2	165.5	10
300	-2125.1	-1806.0	-2082.7	175.4	10

MGAL2O4 (C), SPINEL

T(C)	DELTA H	DELTA F	ENTHALPY	ENTROPY	REF
25	-552.8	-523.0	-552.8	19.3	11
50	-552.9	-520.4	-552.1	21.6	12
75	-552.9	-517.9	-551.3	23.8	12
100	-552.9	-515.4	-550.5	26.0	12
150	-552.9	-510.4	-548.9	30.2	12
200	-552.9	-505.4	-547.1	34.1	12
250	-552.8	-500.3	-545.3	37.7	12
300	-552.7	-495.3	-543.5	41.1	12

MGBR2 (C), MAGNESIUM BROMIDE

T(C)	DELTA H	DELTA F	ENTHALPY	ENTROPY	REF
25	-125.3	-120.5	-125.3	28.0	1
50	-125.4	-120.1	-124.9	29.4	2
58	-125.5	-119.9	-124.7	29.9	2
58	-132.6	-119.9	-124.7	29.9	2
75	-132.5	-119.3	-124.4	30.8	2
100	-132.5	-118.4	-124.0	32.0	2
150	-132.3	-116.5	-123.1	34.3	2
200	-132.1	-114.6	-122.1	36.3	2
250	-132.0	-112.8	-121.2	38.2	2
300	-131.8	-111.0	-120.3	39.9	2

MN2O3 (C), MANGANESE(3) OXIDE

T(C)	DELTA H	DELTA F	ENTHALPY	ENTROPY	REF
25	-229.2	-210.6	-229.2	26.4	1
50	-229.2	-209.0	-228.6	28.3	2
75	-229.2	-207.5	-228.0	30.2	2
100	-229.1	-205.9	-227.4	31.9	2
150	-229.1	-202.8	-226.0	35.2	2
200	-229.0	-199.7	-224.7	38.2	2
250	-228.9	-196.6	-223.3	41.0	2
300	-228.8	-193.6	-221.9	43.5	2

MN+++ (AQ)

T(C)	DELTA H	DELTA F	ENTHALPY	ENTROPY	REF
25	-26.9	-19.6	-27.0	-79.0	5
50	-26.9	-19.0	-25.3	-73.1	3
75	-26.9	-18.4	-23.1	-66.8	3
100	-27.0	-17.7	-20.3	-58.9	3
150	-27.2	-16.5	-15.4	-46.3	3
200	-27.2	-15.2	-9.5	-32.6	3
250	-28.0	-13.9	-3.1	-19.9	3
300	-28.2	-12.5	4.2	-6.7	3

MN++ (AQ)

T(C)	DELTA H	DELTA F	ENTHALPY	ENTROPY	REF
25	-52.7	-54.5	-52.7	-27.6	1
50	-52.7	-54.7	-51.6	-23.9	3
75	-52.7	-54.8	-50.0	-19.2	3
100	-52.8	-54.9	-48.1	-13.9	3
150	-53.1	-55.2	-44.9	-5.7	3
200	-53.0	-55.5	-40.9	3.7	3
250	-53.6	-55.7	-36.5	12.5	3
300	-53.8	-55.9	-31.6	21.5	3

MN (C), MANGANESE

T(C)	DELTA H	DELTA F	ENTHALPY	ENTROPY	REF
25	0.0	0.0	0.0	7.6	1
50	0.0	0.0	0.2	8.2	2
75	0.0	0.0	0.3	8.6	2
100	0.0	0.0	0.5	9.1	2
150	0.0	0.0	0.8	10.0	2
200	0.0	0.0	1.2	10.8	2
250	0.0	0.0	1.5	11.5	2
300	0.0	0.0	1.9	12.1	2

MGSO4 (AQ), UNDISSOCIATED COMPLEX

T(C)	DELTA H	DELTA F	ENTHALPY	ENTROPY	REF
25	-339.4	-299.2	-339.4	-1.7	1
50	-340.1	-295.8	-339.6	-2.0	6
75	-340.3	-292.4	-339.2	-0.9	6
100	-340.2	-289.0	-338.4	1.3	6
150	-338.3	-282.3	-335.5	9.4	6
200	-334.3	-275.7	-330.6	20.7	6

MGSO4 (C), MAGNESIUM SULFATE

T(C)	DELTA H	DELTA F	ENTHALPY	ENTROPY	REF
25	-322.4	-289.2	-322.4	21.9	1
50	-322.5	-286.5	-321.8	23.8	2
75	-322.5	-283.7	-321.2	25.6	2
100	-322.4	-281.0	-320.6	27.4	2
150	-322.3	-275.4	-319.3	30.7	2
200	-322.2	-269.7	-317.9	33.8	2
250	-322.1	-264.3	-316.5	36.7	2
300	-321.9	-258.8	-315.0	39.4	2

MGSIO3 (C), MAGNESIUM METASILICATE

T(C)	DELTA H	DELTA F	ENTHALPY	ENTROPY	REF
25	-370.2	-349.3	-370.2	16.2	1
50	-370.2	-347.6	-369.7	17.8	10
75	-370.3	-346.0	-369.2	19.3	10
100	-370.3	-344.2	-368.7	20.8	10
150	-370.3	-340.7	-367.6	23.6	10
200	-370.3	-337.2	-366.4	26.2	10
250	-370.2	-333.7	-365.2	28.7	10
300	-370.2	-330.3	-363.9	31.0	10

MGS (C), MAGNESIUM SULFIDE

T(C)	DELTA H	DELTA F	ENTHALPY	ENTROPY	REF
25	-98.0	-91.2	-98.0	12.0	1
50	-98.0	-90.6	-97.8	12.8	2
75	-98.0	-90.0	-97.6	13.6	2
100	-98.0	-89.4	-97.3	14.3	2
150	-98.0	-88.3	-96.8	15.6	2
200	-98.1	-87.1	-96.3	16.7	2
250	-98.1	-86.0	-95.7	17.8	2
300	-98.1	-84.8	-95.2	18.8	2

MGO (C), MAGNESIUM OXIDE

T(C)	DELTA H	DELTA F	ENTHALPY	ENTROPY	REF
25	-143.8	-136.1	-143.8	6.4	1
50	-143.8	-135.5	-143.6	7.2	2
75	-143.8	-134.7	-143.3	7.9	2
100	-143.8	-134.1	-143.1	8.6	2
150	-143.8	-132.9	-142.6	9.9	2
200	-143.8	-131.6	-142.0	11.0	2
250	-143.7	-130.3	-141.5	12.1	2
300	-143.7	-129.0	-140.9	13.2	2

MGFE2O4 (C), MAGNESIOFERRITE

T(C)	DELTA H	DELTA F	ENTHALPY	ENTROPY	REF
25	-341.4	-314.8	-341.4	29.6	1
50	-341.5	-312.5	-340.7	31.7	12
75	-341.6	-310.3	-340.0	33.9	12
100	-341.7	-308.0	-339.3	36.0	12
150	-341.7	-303.5	-337.6	40.2	12
200	-341.6	-299.0	-335.7	44.3	12
250	-341.4	-294.4	-333.1	48.3	12
300	-341.1	-290.1	-331.6	52.2	12

MGF2 (C), MAGNESIUM FLUORIDE

T(C)	DELTA H	DELTA F	ENTHALPY	ENTROPY	REF
25	-268.5	-255.8	-268.5	13.7	1
50	-268.5	-254.6	-268.1	14.9	2
75	-268.4	-253.7	-267.1	16.1	2
100	-268.3	-252.6	-266.3	17.2	2
150	-268.2	-250.2	-265.6	19.3	2
200	-268.1	-248.1	-265.0	21.2	2
250	-268.0	-246.2	-264.4	22.9	2
300	-267.8	-244.3	-263.9	24.5	2

MGCO3 (C), MAGNESIUM CARBONATE

T(C)	DELTA H	DELTA F	ENTHALPY	ENTROPY	REF
25	-261.9	-241.9	-261.9	15.7	1
50	-261.9	-240.3	-261.4	17.2	2
75	-261.9	-238.6	-261.0	18.7	2
100	-261.9	-236.9	-260.4	20.1	2
150	-261.8	-233.6	-259.4	22.9	2
200	-261.7	-230.2	-258.2	25.3	2
250	-261.6	-226.9	-257.0	27.7	2
300	-261.4	-223.6	-255.8	30.0	2

MGCL2 (C), MAGNESIUM CHLORIDE

T(C)	DELTA H	DELTA F	ENTHALPY	ENTROPY	REF
25	-153.3	-141.5	-153.3	21.4	1
50	-153.2	-140.5	-152.9	22.8	2
75	-153.1	-139.6	-152.6	24.1	2
100	-153.0	-138.5	-152.0	25.3	2
150	-152.9	-136.6	-151.1	27.6	2
200	-152.7	-134.7	-150.1	29.7	2
250	-152.5	-132.8	-149.2	31.6	2
300	-152.3	-130.9	-148.2	33.3	2

MN3O4 (C), MANGANESE(2,3) OXIDE

T(C)	DELTA H	DELTA F	ENTHALPY	ENTROPY	REF
25	-331.7	-306.7	-331.7	37.2	1
50	-331.6	-304.6	-330.8	40.1	2
75	-331.6	-302.5	-329.9	42.8	2
100	-331.5	-300.5	-329.0	45.3	2
150	-331.4	-296.3	-327.1	50.1	2
200	-331.3	-292.2	-325.2	54.3	2
250	-331.1	-288.1	-323.2	58.3	2
300	-331.0	-284.0	-321.2	61.9	2

MNBR2 (C), MANGANESE(2) BROMIDE

T(C)	DELTA H	DELTA F	ENTHALPY	ENTROPY	REF
25	-92.0	-88.7	-92.0	33.0	2
50	-92.1	-88.4	-91.6	34.5	2
58	-92.2	-88.3	-91.4	34.9	2
58	-99.3	-88.3	-91.4	34.9	2
75	-99.2	-87.8	-91.1	35.8	2
100	-99.2	-87.0	-90.6	37.1	2
150	-99.0	-85.4	-89.7	39.4	2
200	-98.9	-83.8	-88.8	41.5	2
250	-98.7	-82.2	-87.8	43.5	2
300	-98.5	-80.6	-86.8	45.2	2

MNCL2 (C), MANGANESE(2) CHLORIDE

T(C)	DELTA H	DELTA F	ENTHALPY	ENTROPY	REF
25	-115.0	-105.3	-115.0	28.3	1
50	-114.9	-104.5	-114.6	29.7	2
75	-114.9	-103.7	-114.1	31.0	2
100	-114.8	-102.9	-113.7	32.3	2
150	-114.6	-101.3	-112.8	34.6	2
200	-114.5	-99.7	-111.8	36.7	2
250	-114.3	-98.1	-110.9	38.6	2
300	-114.1	-96.6	-109.9	40.4	2

MNCO3 (C), MANGANESE CARBONATE

T(C)	DELTA H	DELTA F	ENTHALPY	ENTROPY	REF
25	-213.7	-195.2	-213.7	20.5	1
50	-213.7	-193.7	-213.2	22.1	2
75	-213.6	-192.1	-212.7	23.7	2
100	-213.6	-190.6	-212.1	25.2	2
150	-213.5	-187.5	-211.0	28.0	2
200	-213.4	-184.4	-209.8	30.7	2
250	-213.2	-181.4	-209.6	33.2	2
300	-213.1	-178.3	-207.3	35.5	2

MNF2 (C), MANGANESE(2) FLUORIDE

T(C)	DELTA H	DELTA F	ENTHALPY	ENTROPY	REF
25	-190.0	-179.9	-190.0	22.0	2
50	-189.9	-179.0	-189.6	23.4	2
75	-189.9	-178.2	-189.2	24.6	2
100	-189.8	-177.3	-188.8	25.7	2
150	-189.7	-175.6	-187.9	27.9	2
200	-189.6	-174.0	-187.1	29.8	2
250	-189.5	-172.3	-186.2	31.5	2
300	-189.4	-170.7	-185.3	33.2	2

MNI2 (C), MANGANESE(2) IODIDE

T(C)	DELTA H	DELTA F	ENTHALPY	ENTROPY	REF
25	-58.0	-58.2	-58.0	36.0	2
50	-58.0	-58.2	-57.6	37.5	2
75	-58.1	-58.2	-57.1	38.8	2
100	-58.2	-58.2	-56.6	40.1	2
114	-58.2	-58.2	-56.4	40.8	2
114	-61.9	-58.2	-56.4	40.8	2
150	-62.2	-57.8	-55.7	42.4	2
185	-62.4	-57.5	-55.0	43.9	2
185	-72.4	-57.5	-55.0	43.9	2
200	-72.4	-57.0	-54.8	44.6	2
250	-72.3	-55.4	-53.8	46.5	2
300	-72.1	-53.8	-52.8	48.3	2

MNO (C), MANGANESE(2) OXIDE

T(C)	DELTA H	DELTA F	ENTHALPY	ENTROPY	REF
25	-92.1	-86.7	-92.1	14.3	1
50	-92.1	-86.3	-91.8	15.1	2
75	-92.0	-85.9	-91.5	16.0	2
100	-92.0	-85.4	-91.3	16.7	2
150	-92.0	-84.5	-90.7	18.2	2
200	-91.9	-83.6	-90.1	19.4	2
250	-91.9	-82.8	-89.5	20.6	2
300	-91.9	-81.9	-88.9	21.7	2

MNO2 (C), MANGANESE(4) OXIDE

T(C)	DELTA H	DELTA F	ENTHALPY	ENTROPY	REF
25	-124.3	-111.2	-124.3	12.7	1
50	-124.3	-110.1	-124.0	13.8	2
75	-124.3	-109.0	-123.6	14.8	2
100	-124.3	-107.9	-123.3	15.8	2
150	-124.2	-105.7	-122.5	17.7	2
200	-124.1	-103.5	-121.7	19.5	2
250	-124.1	-101.3	-120.9	21.1	2
300	-124.0	-99.2	-120.1	22.6	2

MNO4- (AQ)

T(C)	DELTA H	DELTA F	ENTHALPY	ENTROPY	REF
25	-129.4	-106.9	-129.4	50.7	1
50	-130.1	-105.0	-130.1	48.6	3
75	-130.7	-103.0	-130.9	46.3	3
100	-131.2	-101.0	-131.7	43.8	3
150	-132.5	-96.9	-133.6	39.1	3
200	-133.4	-92.7	-135.2	35.4	3
250	-134.4	-89.3	-137.4	31.1	3
300	-135.5	-83.9	-139.8	26.7	3

MNO4-- (AQ)

T(C)	DELTA H	DELTA F	ENTHALPY	ENTROPY	REF
25	-156.3	-119.7	-156.3	24.0	1
50	-157.9	-116.6	-158.3	17.7	3
75	-159.1	-113.3	-160.5	11.2	3
100	-160.3	-110.0	-162.9	4.4	3
150	-162.3	-103.2	-167.0	-5.9	3
200	-165.2	-96.0	-172.4	-18.5	3
250	-167.8	-88.6	-178.6	-30.8	3
300	-171.0	-80.8	-185.3	-43.2	3

MNOH+ (AQ)

T(C)	DELTA H	DELTA F	ENTHALPY	ENTROPY	REF
25	-107.6	-96.8	-107.6	-9.0	1
50	-107.4	-95.9	-106.6	-5.9	3
75	-107.2	-95.0	-105.3	-2.0	3
100	-106.9	-94.2	-103.8	2.4	3
150	-106.5	-92.5	-101.1	9.1	3
200	-105.7	-90.9	-97.8	16.9	3
250	-105.1	-89.4	-94.1	24.2	3
300	-104.1	-88.0	-90.0	31.7	3

MNS (C), MANGANESE(2) SULFIDE

T(C)	DELTA H	DELTA F	ENTHALPY	ENTROPY	REF
25	-66.5	-61.7	-66.5	18.7	1
50	-66.5	-61.3	-66.2	19.7	2
75	-66.5	-60.9	-65.9	20.6	2
100	-66.4	-60.5	-65.6	21.4	2
150	-66.4	-59.7	-65.0	22.9	2
200	-66.3	-58.9	-64.4	24.3	2
250	-66.3	-58.2	-63.8	25.5	2
300	-66.2	-57.4	-63.2	26.7	2

MNSO4 (C), MANGANESE(2) SULFATE

T(C)	DELTA H	DELTA F	ENTHALPY	ENTROPY	REF
25	-269.9	-238.3	-269.9	26.8	1
50	-269.9	-235.6	-269.3	28.8	2
75	-269.9	-233.0	-268.7	30.7	2
100	-269.8	-230.4	-268.0	32.6	2
150	-269.7	-225.1	-266.6	36.2	2
200	-269.5	-219.8	-265.1	39.5	2
250	-269.3	-214.6	-263.5	42.6	2
300	-269.0	-209.3	-262.0	45.5	2

MNSO4 (AQ), UNDISSOCIATED COMPLEX

T(C)	DELTA H	DELTA F	ENTHALPY	ENTROPY	REF
25	-282.1	-245.1	-282.1	8.7	1
50	-283.2	-241.9	-282.5	7.3	6
75	-283.9	-238.7	-282.7	7.0	6
100	-284.3	-235.5	-282.5	7.6	6
150	-283.7	-229.0	-280.8	12.4	6
200	-281.7	-222.5	-278.1	19.4	6

MO (C), MOLYBDENUM

T(C)	DELTA H	DELTA F	ENTHALPY	ENTROPY	REF
25	0.0	0.0	0.0	6.9	1
50	0.0	0.0	0.1	7.3	2
75	0.0	0.0	0.3	7.7	2
100	0.0	0.0	0.4	8.1	2
150	0.0	0.0	0.7	8.9	2
200	0.0	0.0	1.0	9.5	2
250	0.0	0.0	1.3	10.1	2
300	0.0	0.0	1.6	10.7	2

MO2S3 (C), MOLYBDENUM SESQUISULFIDE

T(C)	DELTA H	DELTA F	ENTHALPY	ENTROPY	REF
25	-138.5	-118.6	-139.5	28.5	2
50	-138.4	-116.9	-137.8	30.8	2
75	-138.3	-115.2	-137.1	32.9	2
100	-138.1	-113.6	-136.4	35.0	2
150	-137.8	-110.3	-134.9	38.8	2
200	-137.5	-107.1	-133.3	42.3	2
250	-137.1	-103.9	-131.7	45.5	2
300	-136.6	-100.7	-130.0	48.6	2

MOO2 (C), MOLYBDENUM DIOXIDE

T(C)	DELTA H	DELTA F	ENTHALPY	ENTROPY	REF
25	-140.8	-127.4	-140.8	11.1	1
50	-140.7	-126.3	-140.4	12.2	2
75	-140.7	-125.2	-140.1	13.2	2
100	-140.7	-124.1	-139.7	14.2	2
150	-140.6	-121.8	-139.0	16.1	2
200	-140.5	-119.6	-138.2	17.8	2
250	-140.4	-117.4	-137.4	19.4	2
300	-140.2	-115.2	-136.6	20.9	2

MOO3 (C), MOLYBDENUM TRIOXIDE

T(C)	DELTA H	DELTA F	ENTHALPY	ENTROPY	REF
25	-178.1	-159.7	-178.1	18.6	1
50	-178.0	-158.1	-177.6	20.0	2
75	-178.0	-156.6	-177.2	21.4	2
100	-177.9	-155.0	-176.7	22.8	2
150	-177.8	-152.0	-175.7	25.3	2
200	-177.6	-149.0	-174.7	27.5	2
250	-177.4	-145.9	-173.6	29.6	2
300	-177.2	-142.9	-172.5	31.6	2

MOO4-- (AQ)

T(C)	DELTA H	DELTA F	ENTHALPY	ENTROPY	REF
25	-238.5	-199.9	-238.5	16.5	1
50	-240.4	-196.6	-240.8	9.1	3
75	-242.1	-193.1	-243.4	1.3	3
100	-243.6	-189.6	-246.3	-6.6	3
150	-246.2	-182.3	-251.0	-18.6	3
200	-250.1	-174.4	-257.5	-33.7	3
250	-253.8	-166.2	-264.8	-48.2	3
300	-258.1	-157.6	-272.8	-62.8	3

MOS2 (C), MOLYBDENUM DISULFIDE

T(C)	DELTA H	DELTA F	ENTHALPY	ENTROPY	REF
25	-86.9	-73.1	-86.9	15.0	1
50	-86.8	-71.9	-86.5	16.2	2
75	-86.8	-70.7	-86.1	17.4	2
100	-86.7	-69.6	-85.7	18.5	2
150	-86.6	-67.3	-84.9	20.6	2
200	-86.4	-65.0	-84.0	22.5	2
250	-86.3	-62.8	-83.1	24.3	2
300	-86.1	-60.5	-82.2	26.0	2

MOS3 (C), MOLYBDENUM TRISULFIDE

T(C)	DELTA H	DELTA F	ENTHALPY	ENTROPY	REF
25	-107.5	-85.8	-107.5	15.9	2
50	-107.5	-84.0	-107.1	17.2	2
75	-107.5	-82.2	-106.7	18.5	2
100	-107.6	-80.4	-106.2	19.7	2
150	-107.6	-76.7	-105.3	22.0	2
200	-107.5	-73.1	-104.4	24.1	2
250	-107.4	-69.4	-103.4	26.2	2
300	-107.3	-65.8	-102.3	28.1	2

N2 (G), NITROGEN (GAS)

T(C)	DELTA H	DELTA F	ENTHALPY	ENTROPY	REF
25	0.0	0.0	0.0	45.8	1
50	0.0	0.0	0.2	46.3	2
75	0.0	0.0	0.4	46.8	2
100	0.0	0.0	0.5	47.3	2
150	0.0	0.0	0.9	48.2	2
200	0.0	0.0	1.2	49.0	2
250	0.0	0.0	1.6	49.7	2
300	0.0	0.0	2.0	50.4	2

N2 (AQ)

T(C)	DELTA H	DELTA F	ENTHALPY	ENTROPY	REF
25	-2.5	4.4	-2.5	22.7	14
50	-1.4	4.8	-1.1	27.2	14
75	-0.4	5.2	0.1	30.7	14
100	0.4	5.4	1.2	33.8	14
150	2.1	6.2	3.2	38.6	14
200	3.7	6.2	5.4	43.6	14
250	5.9	6.4	7.9	48.8	14

N2H5+ (AQ)

T(C)	DELTA H	DELTA F	ENTHALPY	ENTROPY	REF
25	-1.8	19.7	-1.8	31.0	1
50	-2.3	21.5	-1.3	32.8	3
75	-2.9	23.4	-0.5	35.0	3
100	-3.6	25.3	0.3	37.5	3
150	-5.1	29.3	1.7	40.8	3
200	-6.1	33.4	3.6	45.3	3
250	-7.6	37.7	5.6	49.4	3
300	-9.0	42.1	7.9	53.6	3

N2H5OH (AQ), UNDISSOCIATED COMPLEX

T(C)	DELTA H	DELTA F	ENTHALPY	ENTROPY	REF
25	-60.2	-26.1	-60.2	49.7	1
50	-60.9	-23.2	-60.1	49.8	6
75	-61.6	-20.3	-60.1	49.9	6
100	-62.3	-17.3	-60.0	50.1	6
150	-63.2	-11.1	-59.4	51.7	6
200	-62.1	-4.9	-57.2	57.5	6

N2O (G), DINITROGEN MONOXIDE (GAS)

T(C)	DELTA H	DELTA F	ENTHALPY	ENTROPY	REF
25	19.6	24.9	19.6	52.5	1
50	19.6	25.3	19.8	53.3	2
75	19.5	25.8	20.1	54.0	2
100	19.5	26.2	20.3	54.6	2
150	19.5	27.2	20.8	55.9	2
200	19.5	28.0	21.3	57.1	2
250	19.5	29.0	21.9	58.2	2
300	19.5	29.9	22.5	59.2	2

N2O2-- (AQ)

T(C)	DELTA H	DELTA F	ENTHALPY	ENTROPY	REF
25	-2.4	33.2	-2.4	16.6	5
50	-4.1	36.3	-4.7	9.2	3
75	-5.6	39.4	-7.3	1.4	3
100	-7.0	42.7	-10.1	-6.5	3
150	-9.3	49.4	-14.9	-18.4	3
200	-12.9	56.7	-21.4	-33.5	3
250	-16.2	64.2	-28.6	-48.0	3
300	-20.2	72.1	-36.6	-62.5	3

N2O3 (G), DINITROGEN TRIOXIDE (GAS)

T(C)	DELTA H	DELTA F	ENTHALPY	ENTROPY	REF
25	20.0	33.3	20.0	74.6	1
50	20.0	34.4	20.4	75.9	2
75	19.9	35.6	20.8	77.1	2
100	19.9	36.7	21.2	78.2	2
150	19.8	38.9	22.1	80.4	2
200	19.8	41.2	23.0	82.4	2
250	19.8	43.5	23.9	84.3	2
300	19.9	45.7	24.9	86.0	2

N2O4 (G), DINITROGEN TETRAOXIDE (GAS)

T(C)	DELTA H	DELTA F	ENTHALPY	ENTROPY	REF
25	2.2	23.4	2.2	72.7	1
50	2.1	25.2	2.7	74.2	2
75	2.1	26.9	3.1	75.6	2
100	2.0	28.7	3.6	77.0	2
150	2.0	32.3	4.7	79.6	2
200	2.0	35.9	5.8	82.1	2
250	2.1	39.5	6.9	84.4	2
300	2.2	43.1	8.1	86.6	2

N2O5 (G), DINITROGEN PENTOXIDE (GAS)

T(C)	DELTA H	DELTA F	ENTHALPY	ENTROPY	REF
25	2.7	27.5	2.7	85.0	1
50	2.7	29.6	3.3	86.9	2
75	2.7	31.7	3.9	88.7	2
100	2.7	33.8	4.6	90.5	2
150	2.8	37.9	5.9	93.9	2
200	2.9	42.1	7.3	97.1	2
250	3.1	46.2	8.8	100.0	2
300	3.2	50.3	10.3	102.7	2

N3- (AQ)
(EXTRAPOLATED AS CRISS & COBBLE TYPE 2)

T(C)	DELTA H	DELTA F	ENTHALPY	ENTROPY	REF
25	65.8	83.2	65.8	30.8	1
50	64.6	84.7	64.4	26.5	3
75	63.6	86.3	63.0	22.1	3
100	62.7	87.9	61.4	17.8	3
150	60.4	91.5	58.1	9.2	3
200	57.8	95.4	54.1	0.0	3
250	55.3	99.5	49.8	-8.6	3
300	51.2	103.9	44.0	-19.3	3

NH3 (G), AMMONIA (GAS)

T(C)	DELTA H	DELTA F	ENTHALPY	ENTROPY	REF
25	-11.0	-3.9	-11.0	46.0	1
50	-11.1	-3.3	-10.8	46.7	2
75	-11.3	-2.7	-10.6	47.3	2
100	-11.4	-2.1	-10.4	47.9	2
150	-11.6	-0.9	-9.9	49.1	2
200	-11.9	0.4	-9.4	50.2	2
250	-12.0	1.7	-8.9	51.2	2
300	-12.2	3.1	-8.4	52.1	2

NH3 (AQ)

T(C)	DELTA H	DELTA F	ENTHALPY	ENTROPY	REF
25	-19.2	-6.4	-19.2	26.6	1
50	----	-5.2	----	----	15
75	----	-4.1	----	----	15
100	----	-3.1	----	----	15
150	----	-0.5	----	----	15
200	----	1.8	----	----	15
250	----	3.9	----	----	15
300	----	6.1	----	----	15

NH4+ (AQ)

T(C)	DELTA H	DELTA F	ENTHALPY	ENTROPY	REF
25	-31.7	-19.0	-31.7	22.1	1
50	-31.9	-17.9	-31.0	24.2	3
75	-32.2	-16.8	-30.2	26.8	3
100	-32.5	-15.7	-29.1	29.7	3
150	-33.4	-13.3	-27.5	33.7	3
200	-33.8	-11.0	-25.3	39.0	3
250	-34.6	-8.5	-22.9	43.8	3
300	-35.1	-6.0	-20.2	48.7	3

NH4CL (C), AMMONIUM CHLORIDE
PHASE TRANS FOR NH4CL AT 185C

T(C)	DELTA H	DELTA F	ENTHALPY	ENTROPY	REF
25	-75.1	-48.5	-75.1	22.6	1
50	-75.1	-46.3	-74.6	24.3	2
75	-75.1	-44.0	-74.1	26.0	2
100	-75.1	-41.8	-73.5	27.6	2
150	-74.9	-37.4	-72.3	30.6	2
185	-74.8	-34.3	-71.4	32.7	2
185	-73.8	-34.3	-70.4	34.8	2
200	-73.9	-33.0	-70.1	35.4	2
250	-73.9	-28.6	-69.0	37.6	2
300	-73.8	-24.3	-67.9	39.7	2

NH4OH (AQ), UNDISSOCIATED COMPLEX

T(C)	DELTA H	DELTA F	ENTHALPY	ENTROPY	REF
25	-87.5	-63.0	-87.5	43.3	1
50	-87.9	-61.0	-87.3	44.0	6
75	-88.2	-58.9	-87.0	44.8	6
100	-88.5	-56.7	-86.7	45.8	6
150	-88.3	-52.4	-85.4	49.3	6
200	-85.8	-48.3	-82.2	57.3	6

NO (G), NITRIC OXIDE (GAS)

T(C)	DELTA H	DELTA F	ENTHALPY	ENTROPY	REF
25	21.6	20.7	21.6	50.3	1
50	21.6	20.6	21.8	50.9	2
75	21.6	20.5	21.9	51.4	2
100	21.6	20.5	22.1	51.9	2
150	21.6	20.3	22.5	52.9	2
200	21.6	20.2	22.8	53.7	2
250	21.6	20.0	23.2	54.4	2
300	21.6	19.9	23.6	55.1	2

NO2 (G), NITROGEN DIOXIDE (GAS)

T(C)	DELTA H	DELTA F	ENTHALPY	ENTROPY	REF
25	7.9	12.3	7.9	57.3	1
50	7.9	12.6	8.1	58.1	2
75	7.9	13.0	8.4	58.8	2
100	7.8	13.4	8.6	59.4	2
150	7.8	14.1	9.1	60.6	2
200	7.7	14.9	9.6	61.8	2
250	7.7	15.6	10.1	62.8	2
300	7.7	16.4	10.7	63.8	2

NO2- (AQ)

T(C)	DELTA H	DELTA F	ENTHALPY	ENTROPY	REF
25	-25.0	-8.9	-25.0	38.5	1
50	-26.0	-7.5	-26.3	34.5	3
75	-27.0	-6.0	-27.7	30.2	3
100	-28.0	-4.5	-29.2	25.8	3
150	-29.8	-1.3	-32.1	18.5	3
200	-31.9	2.3	-35.5	10.8	3
250	-34.1	6.0	-39.5	2.8	3
300	-36.7	9.9	-43.9	-5.2	3

NO3 (G), NITROGEN TRIOXIDE (GAS)

T(C)	DELTA H	DELTA F	ENTHALPY	ENTROPY	REF
25	17.0	27.7	17.0	60.4	2
50	16.9	28.6	17.3	61.3	2
75	16.9	29.5	17.6	62.2	2
100	16.8	30.5	17.9	63.1	2
150	16.8	32.3	18.6	64.8	2
200	16.8	34.1	19.3	66.3	2
250	16.8	35.9	20.0	67.8	2
300	16.8	37.8	20.8	69.3	2

NO3- (AQ)

T(C)	DELTA H	DELTA F	ENTHALPY	ENTROPY	REF
25	-49.6	-26.6	-49.6	40.0	1
50	-50.6	-24.6	-50.8	36.2	3
75	-51.6	-22.6	-52.1	32.2	3
100	-52.6	-20.5	-53.6	28.0	3
150	-54.4	-16.1	-56.3	21.1	3
200	-56.5	-11.4	-59.5	13.8	3
250	-58.7	-6.5	-63.2	6.3	3
300	-61.2	-1.4	-67.4	-1.3	3

NA (C,L), SODIUM
MELTING PT = 97.8C

T(C)	DELTA H	DELTA F	ENTHALPY	ENTROPY	REF
25	0.0	0.0	0.0	12.2	2
50	0.0	0.0	0.2	12.8	2
75	0.0	0.0	0.4	13.3	2
98	0.0	0.0	0.5	13.8	2
98	0.0	0.0	1.1	15.4	2
100	0.0	0.0	1.2	15.5	2
150	0.0	0.0	1.5	16.4	2
200	0.0	0.0	1.9	17.3	2
250	0.0	0.0	2.3	18.0	2
300	0.0	0.0	2.6	18.7	2

NA+ (AQ)

T(C)	DELTA H	DELTA F	ENTHALPY	ENTROPY	REF
25	-57.3	-62.6	-57.3	9.4	5
50	-57.2	-63.0	-56.5	11.9	3
75	-57.0	-63.5	-55.5	15.0	3
98	-56.9	-63.9	-54.3	18.2	2
98	-57.5	-63.9	-54.3	18.2	2
100	-57.5	-64.0	-54.2	18.5	3
150	-57.4	-64.8	-52.2	23.6	3
200	-56.9	-65.8	-49.5	30.0	3
250	-56.7	-66.7	-46.6	35.8	3
300	-56.1	-67.7	-43.3	41.8	3

NA.33AL2.33SI3.67O10(OH)2 (C), SODIUM
MONTMORILLONITE

T(C)	DELTA H	DELTA F	ENTHALPY	ENTROPY	REF
25	-1366.8	-1277.8	-1366.8	62.8	10
50	-1367.0	-1270.3	-1365.0	68.9	10
75	-1367.2	-1262.8	-1363.0	74.7	10
98	-1367.3	-1255.9	-1361.1	79.9	2
98	-1367.4	-1255.9	-1361.1	79.9	2
100	-1367.5	-1255.3	-1360.9	80.4	10
150	-1367.5	-1240.3	-1356.6	91.2	10
200	-1367.4	-1225.2	-1352.1	101.3	10
250	-1367.2	-1210.2	-1347.4	110.8	10
300	-1366.9	-1195.2	-1342.6	119.6	10

NA2CO3 (C), SODIUM CARBONATE

T(C)	DELTA H	DELTA F	ENTHALPY	ENTROPY	REF
25	-270.3	-250.5	-270.3	33.2	2
50	-270.3	-248.9	-269.6	35.3	2
75	-270.2	-247.2	-268.9	37.4	2
98	-270.2	-245.7	-268.3	39.2	2
98	-271.5	-245.7	-268.3	39.2	2
100	-271.5	-245.6	-268.2	39.3	2
150	-271.5	-242.1	-266.7	43.0	2
200	-271.3	-238.6	-265.1	46.6	2
250	-271.1	-235.2	-263.4	50.0	2
300	-270.7	-231.8	-261.6	53.3	2

NA2O (C), SODIUM OXIDE

T(C)	DELTA H	DELTA F	ENTHALPY	ENTROPY	REF
25	-99.9	-90.6	-99.9	17.9	2
50	-99.9	-89.9	-99.5	19.3	2
75	-99.9	-89.1	-99.1	20.6	2
98	-99.9	-88.4	-98.6	21.7	2
98	-101.2	-88.4	-98.6	21.7	2
100	-101.2	-88.3	-98.6	21.8	2
150	-101.2	-86.6	-97.7	24.1	2
200	-101.2	-84.9	-96.8	26.2	2
250	-101.1	-83.1	-95.8	28.2	2
300	-101.1	-81.4	-94.8	30.0	2

NA2O2 (C), SODIUM PEROXIDE

T(C)	DELTA H	DELTA F	ENTHALPY	ENTROPY	REF
25	-122.7	-107.5	-122.7	22.7	2
50	-122.6	-106.2	-122.1	24.4	2
75	-122.6	-105.0	-121.6	26.1	2
98	-122.6	-103.8	-121.0	27.5	2
98	-123.8	-103.8	-121.0	27.5	2
100	-123.8	-103.7	-121.0	27.6	2
150	-123.8	-101.0	-119.8	30.5	2
200	-123.7	-98.3	-118.6	33.2	2
250	-123.6	-95.6	-117.4	35.7	2
300	-123.4	-93.0	-116.2	37.9	2

NA2S (C), DISODIUM MONOSULFIDE

T(C)	DELTA H	DELTA F	ENTHALPY	ENTROPY	REF
25	-104.3	-95.9	-104.3	23.4	2
50	-104.3	-95.2	-103.9	24.9	2
75	-104.3	-94.5	-103.4	26.4	2
98	-104.3	-93.9	-102.9	27.6	2
98	-105.5	-93.9	-102.9	27.6	2
100	-105.5	-93.8	-102.9	27.7	2
150	-105.5	-92.2	-101.9	30.1	2
200	-105.5	-90.6	-101.0	32.3	2
250	-105.4	-89.1	-100.0	34.3	2
300	-105.4	-87.5	-99.0	36.1	2

NA2SIO3 (C), SODIUM METASILICATE

T(C)	DELTA H	DELTA F	ENTHALPY	ENTROPY	REF
25	-373.2	-350.8	-373.2	27.2	2
50	-373.2	-348.9	-372.5	29.4	2
75	-373.3	-347.0	-371.8	31.5	2
98	-373.3	-345.3	-371.1	33.4	2
98	-374.5	-345.3	-371.1	33.4	2
100	-374.5	-345.1	-371.0	33.6	2
150	-374.5	-341.1	-369.5	37.5	2
200	-374.5	-337.2	-367.9	41.1	2
250	-374.4	-333.3	-366.2	44.4	2
300	-374.3	-329.3	-364.5	47.5	2

NA2SO4 (C), SODIUM SULFATE
PHASE TRANS FOR NA2SO4 AT 249C

T(C)	DELTA H	DELTA F	ENTHALPY	ENTROPY	REF
25	-346.9	-312.9	-346.9	35.8	2
50	-346.9	-310.1	-346.1	38.3	2
75	-346.9	-307.2	-345.3	40.7	2
98	-346.9	-304.6	-344.6	42.8	2
98	-348.2	-304.6	-344.6	42.8	2
100	-348.2	-304.3	-344.5	42.9	2
150	-348.1	-298.5	-342.8	47.3	2
200	-348.0	-292.6	-341.0	51.3	2
249	-347.8	-287.0	-339.1	55.0	2
249	-345.2	-287.0	-336.5	60.0	2
250	-345.2	-286.8	-336.5	60.1	2
300	-344.8	-281.2	-334.4	63.9	2

NAALSI206.H2O (C), ANALCIME

T(C)	DELTA H	DELTA F	ENTHALPY	ENTROPY	REF
25	-786.3	-734.2	-786.3	56.0	11
50	-786.4	-729.8	-785.1	60.1	10
75	-786.5	-725.5	-783.8	63.8	10
98	-786.6	-721.5	-782.7	67.0	2
98	-787.3	-721.5	-782.7	67.0	2
100	-787.3	-721.1	-782.5	67.3	10
150	-787.6	-712.2	-780.0	73.6	10
200	-788.0	-703.3	-777.5	79.2	10
250	-788.4	-694.3	-775.0	84.2	10
300	-788.8	-685.3	-772.5	88.8	10

NAALSI3O8 (C), LOW ALBITE

T(C)	DELTA H	DELTA F	ENTHALPY	ENTROPY	REF
25	-937.3	-884.1	-937.3	50.2	11
50	-937.4	-879.7	-936.0	54.3	10
75	-937.5	-875.2	-934.7	58.2	10
98	-937.6	-871.1	-933.5	61.7	2
98	-938.2	-871.1	-933.5	61.7	2
100	-938.2	-870.7	-933.3	62.0	10
150	-938.3	-861.7	-930.4	69.3	10
200	-938.2	-852.6	-927.4	76.0	10
250	-938.1	-843.6	-924.3	82.3	10
300	-937.9	-834.6	-921.1	88.2	10

NAALSIO4 (C), NEPHELINE

T(C)	DELTA H	DELTA F	ENTHALPY	ENTROPY	REF
25	-497.0	-469.7	-497.0	29.7	11
50	-497.1	-467.4	-496.3	32.0	10
75	-497.2	-465.1	-495.6	34.3	10
98	-497.2	-463.0	-494.8	36.3	2
98	-497.8	-463.0	-494.8	36.3	2
100	-497.8	-462.8	-494.8	36.5	10
150	-497.7	-458.1	-493.0	40.9	10
200	-497.5	-453.4	-491.1	45.1	10
250	-497.2	-448.7	-489.0	49.3	10
300	-496.6	-444.1	-486.8	53.5	10

NABR (C), SODIUM BROMIDE

T(C)	DELTA H	DELTA F	ENTHALPY	ENTROPY	REF
25	-86.4	-83.5	-86.4	20.8	2
50	-86.4	-83.3	-86.1	21.8	2
58	-86.5	-83.2	-86.0	22.1	2
58	-90.0	-83.2	-86.0	22.1	2
75	-90.0	-82.8	-85.8	22.7	2
98	-90.0	-82.4	-85.5	23.5	2
98	-90.6	-82.4	-85.5	23.5	2
100	-90.6	-82.3	-85.4	23.6	2
150	-90.6	-81.2	-84.8	25.2	2
200	-90.5	-80.1	-84.2	26.6	2
250	-90.4	-79.0	-83.5	27.9	2
300	-90.4	-77.9	-82.9	29.1	2

NABR (AQ), UNDISSOCIATED COMPLEX

T(C)	DELTA H	DELTA F	ENTHALPY	ENTROPY	REF
25	-86.2	-87.2	-86.2	33.7	5
50	-87.1	-87.2	-86.7	31.9	6
58	-87.4	-87.2	-86.9	31.4	2
58	-91.0	-87.2	-86.9	31.4	2
75	-91.4	-87.0	-87.2	30.7	6
98	-92.0	-86.7	-87.5	29.8	2
98	-92.6	-86.7	-87.5	29.8	2
100	-92.7	-86.7	-87.5	29.7	6
150	-94.7	-85.7	-88.9	26.0	6
200	-96.7	-84.6	-90.2	23.0	6

NACL (C), SODIUM CHLORIDE

T(C)	DELTA H	DELTA F	ENTHALPY	ENTROPY	REF
25	-98.3	-91.8	-98.3	17.2	2
50	-98.2	-91.3	-98.0	18.2	2
75	-98.2	-90.7	-97.6	19.1	2
98	-98.2	-90.2	-97.4	19.9	2
98	-98.8	-90.2	-97.4	19.9	2
100	-98.8	-90.2	-97.3	20.0	2
150	-98.8	-89.0	-96.7	21.6	2
200	-98.7	-87.9	-96.1	23.0	2
250	-98.6	-86.8	-95.4	24.3	2
300	-98.6	-85.6	-94.8	25.5	2

NACL (AQ), UNDISSOCIATED COMPLEX

T(C)	DELTA H	DELTA F	ENTHALPY	ENTROPY	REF
25	-97.3	-93.9	-97.3	27.6	5
50	-98.1	-93.6	-97.8	26.0	6
75	-98.8	-93.3	-98.2	24.7	6
98	-99.4	-92.9	-98.6	23.7	2
98	-100.0	-92.9	-98.6	23.7	2
100	-100.1	-92.8	-98.6	23.6	6
150	-102.1	-91.7	-100.0	20.1	6
200	-104.0	-90.4	-101.2	17.2	6

NACN (AQ), UNDISSOCIATED COMPLEX

T(C)	DELTA H	DELTA F	ENTHALPY	ENTROPY	REF
25	-19.1	-14.7	-19.1	21.8	5
50	-21.0	-14.3	-20.7	16.7	6
75	-23.0	-13.7	-22.3	11.7	6
98	-25.0	-13.0	-24.0	7.1	2
98	-25.6	-13.0	-24.0	7.1	2
100	-25.8	-12.9	-24.2	6.6	6
150	-30.7	-10.9	-28.3	-4.1	6
200	-37.9	-8.1	-34.4	-18.5	6

NAF (C), SODIUM FLUORIDE

T(C)	DELTA H	DELTA F	ENTHALPY	ENTROPY	REF
25	-137.1	-129.9	-137.1	12.3	2
50	-137.1	-129.3	-136.8	13.2	2
75	-137.1	-128.7	-136.5	14.0	2
98	-137.1	-128.1	-136.3	14.8	2
98	-137.7	-128.1	-136.3	14.8	2
100	-137.7	-128.1	-136.2	14.8	2
150	-137.7	-126.8	-135.6	16.3	2
200	-137.6	-125.5	-135.0	17.6	2
250	-137.6	-124.2	-134.5	18.9	2
300	-137.6	-122.9	-133.8	20.0	2

NAF (AQ), UNDISSOCIATED COMPLEX

T(C)	DELTA H	DELTA F	ENTHALPY	ENTROPY	REF
25	-135.9	-128.7	-135.9	12.1	5
50	-136.5	-128.0	-136.3	11.0	6
75	-137.2	-127.4	-136.6	9.9	6
98	-137.8	-126.7	-137.0	8.8	2
98	-138.5	-126.7	-137.0	8.8	2
100	-138.5	-126.6	-137.1	8.7	6
150	-140.2	-124.9	-138.2	5.9	6
200	-141.7	-123.0	-139.0	3.9	6

NAHCO3 (C), SODIUM BICARBONATE

T(C)	DELTA H	DELTA F	ENTHALPY	ENTROPY	REF
25	-226.4	-203.0	-226.4	24.4	2
50	-226.4	-201.1	-225.9	26.1	2
75	-226.5	-199.1	-225.3	27.8	2
98	-226.5	-197.3	-224.8	29.3	2
98	-227.1	-197.3	-224.8	29.3	2
100	-227.1	-197.2	-224.7	29.4	2
150	-227.1	-193.1	-223.5	32.5	2
200	-227.1	-189.1	-222.2	35.4	2

NAHCO3 (AQ), UNDISSOCIATED COMPLEX

T(C)	DELTA H	DELTA F	ENTHALPY	ENTROPY	REF
25	-220.0	-202.9	-220.0	37.1	5
50	-220.7	-201.4	-219.5	38.7	6
75	-221.2	-199.9	-218.5	41.6	6
98	-221.5	-198.6	-217.3	45.1	2
98	-222.8	-198.6	-217.3	45.1	2
100	-222.8	-198.4	-217.2	45.4	6
150	-226.0	-194.9	-217.0	45.5	6
200	-228.3	-191.3	-215.7	48.4	6

NAI (C), SODIUM IODIDE

T(C)	DELTA H	DELTA F	ENTHALPY	ENTROPY	REF
25	-68.8	-68.0	-68.8	23.5	2
50	-68.8	-68.0	-68.5	24.5	2
75	-68.9	-67.9	-68.2	25.5	2
98	-68.9	-67.8	-67.9	26.3	2
98	-69.5	-67.8	-67.9	26.3	2
100	-69.5	-67.8	-67.9	26.3	2
114	-69.6	-67.8	-67.7	26.8	2
114	-71.4	-67.8	-67.7	26.8	2
150	-71.6	-67.4	-67.2	28.0	2
185	-71.7	-67.1	-66.8	29.0	2
185	-76.7	-67.1	-66.8	29.0	2
200	-76.7	-66.7	-66.6	29.4	2
250	-76.6	-65.7	-65.9	30.7	2
300	-76.6	-64.6	-65.2	31.9	2

NAI (AQ), UNDISSOCIATED COMPLEX

T(C)	DELTA H	DELTA F	ENTHALPY	ENTROPY	REF
25	-66.6	-70.9	-66.6	40.5	5
50	-67.6	-71.3	-67.3	38.6	6
75	-68.4	-71.5	-67.7	37.3	6
98	-69.0	-71.7	-68.0	36.5	2
98	-69.6	-71.7	-68.0	36.5	2
100	-69.7	-71.7	-68.0	36.4	6
114	-70.3	-71.8	-68.4	35.4	2
114	-72.1	-71.8	-68.4	35.4	2
150	-73.9	-71.7	-69.4	32.7	6
185	-75.4	-71.5	-70.4	30.5	2
185	-80.4	-71.5	-70.4	30.5	2
200	-81.0	-71.1	-70.8	29.6	6

NANO3 (C), SODIUM NITRATE

T(C)	DELTA H	DELTA F	ENTHALPY	ENTROPY	REF
25	-111.5	-87.4	-111.5	27.8	2
50	-111.4	-85.4	-110.9	29.6	2
75	-111.4	-83.4	-110.3	31.5	2
98	-111.3	-91.5	-109.7	33.1	2
98	-111.9	-81.5	-109.7	33.1	2
100	-111.9	-81.4	-109.7	33.2	2
150	-111.6	-77.3	-108.3	36.7	2
200	-111.2	-73.3	-106.8	40.1	2
250	-110.7	-69.3	-105.1	43.4	2
300	-110.0	-65.3	-103.3	46.7	2

NANO3 (AQ), UNDISSOCIATED COMPLEX

T(C)	DELTA H	DELTA F	ENTHALPY	ENTROPY	REF
25	-106.7	-89.0	-106.7	49.4	5
50	-107.6	-87.5	-107.1	48.1	6
75	-108.4	-85.9	-107.4	47.2	6
98	-109.1	-84.4	-107.6	46.6	2
98	-109.8	-84.4	-107.6	46.6	2
100	-109.8	-84.3	-107.6	46.6	6
150	-111.6	-80.7	-108.3	44.7	6
200	-113.2	-77.0	-108.8	43.8	6

NAO2 (C), SODIUM DIOXIDE

T(C)	DELTA H	DELTA F	ENTHALPY	ENTROPY	REF
25	-62.3	-52.3	-62.3	27.7	2
50	-62.2	-51.5	-61.9	29.1	2
75	-62.1	-50.6	-61.4	30.4	2
98	-62.1	-49.9	-61.0	31.5	2
98	-62.7	-49.9	-61.0	31.5	2
100	-62.7	-49.8	-61.0	31.7	2
150	-62.5	-48.1	-60.1	33.9	2
200	-62.3	-46.4	-59.1	36.0	2
250	-62.1	-44.7	-58.2	37.9	2
300	-61.8	-43.1	-57.2	39.8	2

NAOH (C), SODIUM HYDROXIDE
PHASE TRANS FOR NAOH AT 293C

T(C)	DELTA H	DELTA F	ENTHALPY	ENTROPY	REF
25	-102.3	-91.3	-102.3	15.4	2
50	-102.3	-90.4	-101.9	16.6	2
75	-102.3	-89.4	-101.6	17.6	2
98	-102.3	-88.6	-101.2	18.6	2
98	-102.9	-88.6	-101.2	18.6	2
100	-102.9	-88.5	-101.2	18.7	2
150	-102.8	-86.6	-100.4	20.6	2
200	-102.7	-84.7	-99.6	22.5	2
250	-102.6	-82.8	-98.7	24.3	2
293	-102.4	-81.2	-97.9	25.8	2
293	-100.8	-81.2	-96.4	28.5	2
300	-100.8	-80.9	-96.2	28.8	2

NAOH (AQ), UNDISSOCIATED COMPLEX

T(C)	DELTA H	DELTA F	ENTHALPY	ENTROPY	REF
25	-112.2	-100.2	-112.2	11.9	5
50	-113.0	-99.1	-112.6	10.7	6
75	-113.7	-98.0	-113.0	9.4	6
98	-114.5	-97.0	-113.5	8.2	2
98	-115.1	-97.0	-113.5	8.2	2
100	-115.2	-96.9	-113.5	8.1	6
150	-117.2	-94.3	-114.8	4.9	6
200	-119.1	-91.5	-115.9	2.3	6

NB (C), NIOBIUM

T(C)	DELTA H	DELTA F	ENTHALPY	ENTROPY	REF
25	0.0	0.0	0.0	8.7	1
50	0.0	0.0	0.2	9.2	2
75	0.0	0.0	0.3	9.6	2
100	0.0	0.0	0.4	10.0	2
150	0.0	0.0	0.8	10.8	2
200	0.0	0.0	1.1	11.5	2
250	0.0	0.0	1.4	12.1	2
300	0.0	0.0	1.7	12.7	2

ND (C), NEODYMIUM

T(C)	DELTA H	DELTA F	ENTHALPY	ENTROPY	REF
25	0.0	0.0	0.0	17.1	1
50	0.0	0.0	0.2	17.6	2
75	0.0	0.0	0.3	18.1	2
100	0.0	0.0	0.5	18.6	2
150	0.0	0.0	0.8	19.4	2
200	0.0	0.0	1.2	20.2	2
250	0.0	0.0	1.5	20.9	2
300	0.0	0.0	1.9	21.6	2

ND(SO4)2- (AQ)
(EXTRAPOLATED AS CRISS & COBBLE TYPE 4)

T(C)	DELTA H	DELTA F	ENTHALPY	ENTROPY	REF
25	-626.6	-542.5	-626.6	6.0	1
50	-629.7	-535.3	-629.1	-2.0	3
75	-632.8	-527.8	-631.9	-10.4	3
100	-636.1	-520.2	-634.9	-18.9	3
150	-643.3	-504.1	-641.5	-35.7	3
200	-650.8	-487.0	-648.7	-52.2	3
250	-659.1	-469.3	-656.9	-68.8	3
300	-668.1	-450.8	-666.0	-85.4	3

ND+++ (AQ)

T(C)	DELTA H	DELTA F	ENTHALPY	ENTROPY	REF
25	-166.4	-160.5	-166.4	-64.4	1
50	-166.4	-160.0	-164.9	-59.5	3
75	-166.6	-159.5	-162.8	-53.3	3
100	-167.0	-159.0	-160.3	-46.1	3
150	-167.7	-157.8	-155.8	-34.8	3
200	-168.2	-156.6	-150.5	-22.5	3
250	-169.6	-155.3	-144.6	-10.7	3
300	-170.4	-153.9	-139.0	1.3	3

NDSO4+ (AQ)

T(C)	DELTA H	DELTA F	ENTHALPY	ENTROPY	REF
25	-395.5	-352.7	-395.5	-22.0	1
50	-395.5	-349.1	-394.5	-18.5	3
75	-395.4	-345.5	-393.0	-14.1	3
100	-395.1	-342.0	-391.2	-9.0	3
150	-394.9	-334.8	-388.1	-1.2	3
200	-394.2	-327.8	-384.3	7.7	3
250	-393.7	-320.9	-380.1	16.0	3
300	-392.7	-314.0	-375.5	24.5	3

NI (C), NICKEL

T(C)	DELTA H	DELTA F	ENTHALPY	ENTROPY	REF
25	0.0	0.0	0.0	7.1	1
50	0.0	0.0	0.2	7.6	2
75	0.0	0.0	0.3	8.1	2
100	0.0	0.0	0.5	8.6	2
150	0.0	0.0	0.8	9.4	2
200	0.0	0.0	1.2	10.2	2
250	0.0	0.0	1.5	10.9	2
300	0.0	0.0	1.9	11.6	2

NI(CN)4-- (AQ)
((EXTRAPOLATED AS CRISS & COBBLE TYPE 3)

T(C)	DELTA H	DELTA F	ENTHALPY	ENTROPY	REF
25	87.9	112.8	87.9	62.0	1
50	88.0	114.9	87.8	61.6	3
75	88.5	116.9	87.6	61.1	3
100	89.4	118.9	87.4	60.5	3
150	90.1	122.8	86.6	58.2	3
200	92.0	126.4	86.5	58.2	3
250	94.4	129.9	86.1	57.3	3
300	96.7	133.2	85.5	56.3	3

NI(NH3)+ (AQ)

T(C)	DELTA H	DELTA F	ENTHALPY	ENTROPY	REF
25	----	-21.1	----	----	15
50	----	-19.8	----	----	15
75	----	-19.6	----	----	15
100	----	-17.4	----	----	15

NI(NH3)2++ (AQ)

T(C)	DELTA H	DELTA F	ENTHALPY	ENTROPY	REF
25	-58.9	-30.6	-58.9	10.4	1
50	-59.9	-28.2	-58.1	12.9	3
75	-61.2	-25.7	-57.1	15.9	3
100	-62.6	-23.0	-55.9	19.4	3
150	-65.5	-17.5	-53.9	24.4	3
200	-68.2	-11.7	-51.2	30.7	3
250	-71.7	-5.5	-48.3	36.4	3
300	-75.0	1.0	-45.1	42.3	3

NI(NH3)3++ (AQ)

T(C)	DELTA H	DELTA F	ENTHALPY	ENTROPY	REF
25	----	-39.3	----	----	15
50	----	-35.7	----	----	15
75	----	-32.0	----	----	15
100	----	-28.7	----	----	15

NI(NH3)4++ (AQ)

T(C)	DELTA H	DELTA F	ENTHALPY	ENTROPY	REF
25	-104.9	-47.4	-104.9	51.8	1
50	-107.0	-42.5	-104.6	52.9	3
75	-109.6	-37.4	-104.1	54.2	3
100	-112.4	-32.0	-103.6	55.7	3
150	-118.1	-20.8	-103.0	57.2	3
200	-123.6	-9.0	-101.7	60.1	3
250	-130.2	3.5	-100.6	62.5	3
300	-136.7	16.6	-99.2	65.0	3

NI(NH3)5++ (AQ)

T(C)	DELTA H	DELTA F	ENTHALPY	ENTROPY	REF
25	----	-54.8	----	----	15
50	----	-48.6	----	----	15
75	----	-42.2	----	----	15
100	----	-36.4	----	----	15

NI(NH3)6++ (AQ)

T(C)	DELTA H	DELTA F	ENTHALPY	ENTROPY	REF
25	-150.6	-61.2	-150.6	84.3	1
50	-153.7	-53.6	-150.6	84.4	3
75	-157.4	-45.7	-150.6	84.3	3
100	-161.5	-37.4	-150.6	84.1	3
150	-169.6	-20.2	-151.1	83.0	3
200	-177.7	-2.0	-151.0	83.2	3
250	-187.0	17.1	-151.1	83.0	3
300	-196.4	37.0	-151.2	82.8	3

NI++ (AQ)

T(C)	DELTA H	DELTA F	ENTHALPY	ENTROPY	REF
25	-12.9	-10.9	-12.9	-40.8	1
50	-12.7	-10.7	-11.6	-36.7	3
75	-12.6	-10.6	-9.9	-31.5	3
100	-12.4	-10.5	-7.8	-25.4	3
150	-12.3	-10.2	-4.1	-16.1	3
200	-11.8	-10.0	0.4	-5.7	3
250	-11.8	-9.8	5.3	4.2	3
300	-11.4	-9.6	10.8	14.2	3

NI3S2 (C), TRINICKEL DISULFIDE

T(C)	DELTA H	DELTA F	ENTHALPY	ENTROPY	REF
25	-79.2	-66.1	-79.2	32.0	1
50	-79.0	-65.0	-78.4	34.6	2
75	-78.9	-63.9	-77.6	37.1	2
100	-78.8	-62.8	-76.7	39.3	2
150	-78.6	-60.7	-75.1	43.5	2
200	-78.3	-58.6	-73.4	47.2	2
250	-78.1	-56.6	-71.7	50.6	2
300	-77.9	-54.5	-70.1	53.6	2

NICL2 (C), NICKEL CHLORIDE

T(C)	DELTA H	DELTA F	ENTHALPY	ENTROPY	REF
25	-73.0	-61.9	-73.0	23.3	1
50	-72.9	-61.0	-72.5	24.7	2
75	-72.8	-60.1	-72.1	26.0	2
100	-72.8	-59.2	-71.7	27.3	2
150	-72.6	-57.3	-70.8	29.5	2
200	-72.5	-55.6	-69.9	31.6	2
250	-72.3	-53.8	-68.9	33.4	2
300	-72.2	-52.0	-68.0	35.2	2

NIF2 (C), NICKEL FLUORIDE

T(C)	DELTA H	DELTA F	ENTHALPY	ENTROPY	REF
25	-155.7	-144.4	-155.7	17.6	1
50	-155.6	-143.4	-155.3	19.0	2
75	-155.5	-142.5	-154.8	20.4	2
100	-155.4	-141.6	-154.3	21.7	2
150	-155.2	-139.7	-153.4	23.9	2
200	-155.1	-137.9	-152.5	26.0	2
250	-154.9	-136.1	-151.6	27.8	2
300	-154.7	-134.3	-150.7	29.5	2

NIO (C), NICKEL MONOXIDE
PHASE TRANS FOR NIO AT 252C AND 292C

T(C)	DELTA H	DELTA F	ENTHALPY	ENTROPY	REF
25	-57.3	-50.6	-57.3	9.1	1
50	-57.3	-50.0	-57.0	9.9	2
75	-57.3	-49.4	-56.8	10.8	2
100	-57.2	-48.9	-56.5	11.6	2
150	-57.1	-47.8	-55.8	13.1	2
200	-57.0	-46.7	-55.2	14.7	2
250	-56.7	-45.6	-54.4	16.2	2
252	-56.7	-45.6	-54.4	16.3	2
252	-56.7	-45.6	-54.4	16.3	2
292	-56.6	-44.7	-53.8	17.3	2
292	-56.6	-44.7	-53.8	17.3	2
300	-56.6	-44.6	-53.7	17.5	2

NIOH+ (AQ)

T(C)	DELTA H	DELTA F	ENTHALPY	ENTROPY	REF
25	-68.9	-54.4	-68.9	-22.0	1
50	-68.6	-53.2	-67.8	-19.5	3
75	-68.2	-52.0	-66.3	-14.1	3
100	-67.6	-50.9	-64.6	-9.0	3
150	-66.9	-48.7	-61.5	-1.2	3
200	-65.6	-46.7	-57.7	7.7	3
250	-64.4	-44.8	-53.5	16.0	3
300	-62.8	-43.0	-48.8	24.5	3

NIP207-- (AQ)
(EXTRAPOLATED AS CRISS & COBBLE TYPE 3)

T(C)	DELTA H	DELTA F	ENTHALPY	ENTROPY	REF
25	-559.9	-479.1	-559.9	-31.5	1
44	-563.5	-473.8	-563.4	-42.9	2
44	-563.8	-473.8	-563.4	-42.9	2
50	-565.0	-472.1	-564.5	-46.4	3
75	-569.8	-464.8	-569.7	-61.8	3
100	-574.7	-457.0	-575.3	-77.5	3
150	-582.6	-440.9	-584.2	-99.5	3
200	-594.6	-423.0	-597.5	-130.6	3
250	-606.7	-404.3	-611.9	-159.6	3
274	-613.0	-394.9	-619.2	-173.4	2
274	-618.7	-394.9	-619.2	-173.4	2
300	-626.1	-384.3	-627.7	-188.5	3

NIS (C), NICKEL MONOSULFIDE

T(C)	DELTA H	DELTA F	ENTHALPY	ENTROPY	REF
25	-34.9	-28.5	-34.9	12.7	1
50	-34.9	-27.9	-34.6	13.7	2
75	-34.8	-27.4	-34.3	14.7	2
100	-34.7	-26.9	-33.9	15.7	2
150	-34.5	-25.8	-33.2	17.5	2
200	-34.3	-24.8	-32.5	19.2	2
250	-34.1	-23.8	-31.7	20.7	2
300	-33.9	-22.8	-30.9	22.2	2

NISO4 (C), NICKEL SULFATE

T(C)	DELTA H	DELTA F	ENTHALPY	ENTROPY	REF
25	-224.0	-191.0	-224.0	22.0	1
50	-223.8	-188.3	-223.1	24.7	2
75	-223.5	-185.6	-222.3	27.2	2
100	-223.3	-182.9	-221.5	29.5	2
150	-222.9	-177.5	-219.8	33.8	2
200	-222.5	-172.1	-218.0	37.6	2
250	-222.0	-166.8	-216.3	41.2	2
300	-221.6	-161.5	-214.5	44.4	2

NISO4 (AQ), UNDISSOCIATED COMPLEX

T(C)	DELTA H	DELTA F	ENTHALPY	ENTROPY	REF
25	-242.2	-201.5	-242.2	-4.3	1
50	-243.1	-198.0	-242.5	-5.2	6
75	-243.7	-194.5	-242.5	-4.9	6
100	-243.8	-191.0	-242.0	-3.7	6
150	-242.8	-184.0	-239.9	2.3	6
200	-240.3	-177.1	-236.6	10.5	6

O2 (G), OXYGEN (GAS)

T(C)	DELTA H	DELTA F	ENTHALPY	ENTROPY	REF
25	0.0	0.0	0.0	49.0	1
50	0.0	0.0	0.2	49.6	2
75	0.0	0.0	0.4	50.1	2
100	0.0	0.0	0.5	50.6	2
150	0.0	0.0	0.9	51.5	2
200	0.0	0.0	1.3	52.3	2
250	0.0	0.0	1.7	53.1	2
300	0.0	0.0	2.0	53.8	2

O2 (AQ)

T(C)	DELTA H	DELTA F	ENTHALPY	ENTROPY	REF
25	-2.8	3.9	-2.8	26.5	1
50	-1.6	4.5	-1.5	30.8	14
75	-0.7	4.9	-0.4	34.0	14
100	0.5	5.5	0.8	37.3	14
150	2.4	6.0	3.1	43.1	14
200	4.1	6.0	5.5	48.3	14
250	6.0	6.2	7.7	52.8	14
300	7.6	5.8	10.0	57.0	14

OH- (AQ)

T(C)	DELTA H	DELTA F	ENTHALPY	ENTROPY	REF
25	-55.0	-37.6	-55.0	2.4	1
50	-55.8	-36.1	-56.1	-1.3	3
75	-56.8	-34.5	-57.6	-5.7	3
100	-57.8	-32.9	-59.3	-10.6	3
150	-59.8	-29.4	-62.6	-18.9	3
200	-62.2	-25.6	-66.4	-27.8	3
250	-64.4	-21.6	-70.7	-36.3	3
300	-68.2	-17.3	-76.4	-46.8	3

OS (C), OSMIUM

T(C)	DELTA H	DELTA F	ENTHALPY	ENTROPY	REF
25	0.0	0.0	0.0	7.8	1
50	0.0	0.0	0.2	8.3	2
75	0.0	0.0	0.3	8.7	2
100	0.0	0.0	0.4	9.1	2
150	0.0	0.0	0.8	9.9	2
200	0.0	0.0	1.0	10.6	2
250	0.0	0.0	1.3	11.2	2
300	0.0	0.0	1.7	11.7	2

P (C,L,P4 GAS), PHOSPHORUS

	T(C)	DELTA H	DELTA F	ENTHALPY	ENTROPY	REF
P(C)	25	0.0	0.0	0.0	9.8	1
	44	0.0	0.0	0.1	10.2	2
P(L)	44	0.0	0.0	0.3	10.7	2
	50	0.0	0.0	0.3	10.8	2
	75	0.0	0.0	0.5	11.3	2
	100	0.0	0.0	0.6	11.7	2
	150	0.0	0.0	0.9	12.5	2
	200	0.0	0.0	1.3	13.2	2
	250	0.0	0.0	1.6	13.8	2
	274	0.0	0.0	1.7	14.1	2
P4(G)	274	0.0	0.0	18.5	77.5	2
	300	0.0	0.0	19.0	78.4	2

P207---- (AQ)

T(C)	DELTA H	DELTA F	ENTHALPY	ENTROPY	REF
25	-542.7	-458.7	-542.7	-8.0	1
44	-544.6	-453.3	-545.3	-16.6	2
44	-544.9	-453.3	-545.3	-16.6	2
50	-545.5	-451.5	-546.1	-19.2	2
75	-547.5	-444.1	-550.0	-30.9	3
100	-549.0	-436.8	-554.3	-42.8	3
150	-551.4	-421.7	-561.2	-59.9	3
200	-555.9	-405.9	-571.1	-83.2	3
250	-559.5	-389.8	-582.0	-105.1	3
274	-561.7	-382.1	-587.6	-115.5	2
274	-567.4	-382.1	-587.6	-115.5	2
300	-570.1	-373.4	-594.0	-126.9	3

PO4--- (AQ)

T(C)	DELTA H	DELTA F	ENTHALPY	ENTROPY	REF
25	-305.4	-243.5	-305.4	-38.0	1
44	-308.5	-239.4	-309.2	-50.2	2
44	-308.6	-239.4	-309.2	-50.2	2
50	-309.6	-238.1	-310.3	-53.9	3
75	-313.5	-232.5	-315.8	-70.3	3
100	-317.3	-226.5	-321.8	-87.1	3
150	-322.9	-214.2	-331.3	-110.5	3
200	-332.8	-200.3	-345.5	-143.8	3
250	-342.3	-185.8	-360.9	-174.7	3
274	-347.4	-179.6	-368.7	-189.4	2
274	-350.3	-179.6	-368.7	-189.4	2
300	-356.2	-170.4	-377.8	-205.5	3

PB (C), LEAD

T(C)	DELTA H	DELTA F	ENTHALPY	ENTROPY	REF
25	0.0	0.0	0.0	15.5	1
50	0.0	0.0	0.2	16.0	2
75	0.0	0.0	0.3	16.5	2
100	0.0	0.0	0.5	16.9	2
150	0.0	0.0	0.8	17.8	2
200	0.0	0.0	1.1	18.5	2
250	0.0	0.0	1.5	19.2	2
300	0.0	0.0	1.8	19.8	2

PB++ (AQ)

T(C)	DELTA H	DELTA F	ENTHALPY	ENTROPY	REF
25	-0.4	-5.8	-0.4	-7.5	1
50	-0.5	-6.3	0.5	-4.5	3
75	-0.9	-6.7	1.8	-0.6	3
100	-1.3	-7.1	3.4	3.7	3
150	-2.2	-7.8	5.9	10.3	3
200	-2.9	-8.4	9.3	18.0	3
250	-4.3	-8.9	12.8	25.2	3
300	-5.3	-9.3	16.8	32.5	3

PB3(OH)4++ (AQ)

T(C)	DELTA H	DELTA F	ENTHALPY	ENTROPY	REF
25	-248.1	-212.4	-248.1	46.0	1
50	-249.8	-209.3	-247.7	47.3	3
75	-251.9	-206.1	-247.2	48.9	3
100	-254.3	-202.7	-246.5	50.6	3
150	-259.0	-195.4	-245.7	52.6	3
200	-263.7	-187.6	-244.3	56.0	3
250	-269.3	-179.3	-242.9	58.9	3
300	-274.9	-170.4	-241.3	61.8	3

PB3O4 (C), TRILEAD TETRAOXIDE

T(C)	DELTA H	DELTA F	ENTHALPY	ENTROPY	REF
25	-171.7	-143.7	-171.7	50.5	1
50	-171.6	-141.3	-170.8	53.3	2
75	-171.6	-139.0	-169.9	56.0	2
100	-171.6	-136.6	-169.0	58.4	2
150	-171.5	-132.0	-167.3	63.0	2
200	-171.4	-127.3	-165.4	67.0	2
250	-171.3	-122.7	-163.6	70.7	2
300	-171.3	-118.0	-161.7	74.1	2

PB4(OH)4++++ (AQ)

T(C)	DELTA H	DELTA F	ENTHALPY	ENTROPY	REF
25	-254.8	-223.8	-254.8	36.0	1
50	-257.5	-221.1	-254.3	37.6	3
75	-261.0	-218.1	-253.6	39.6	3
100	-265.2	-214.8	-252.9	41.8	3
150	-273.1	-207.4	-251.7	44.7	3
200	-281.4	-199.0	-249.9	48.9	3
250	-291.6	-189.8	-248.1	52.6	3
300	-301.8	-179.5	-246.0	56.3	3

PB6(OH)8++++ (AQ)

T(C)	DELTA H	DELTA F	ENTHALPY	ENTROPY	REF
25	-499.6	-430.3	-499.6	99.0	1
50	-503.9	-424.3	-499.7	98.6	3
75	-509.4	-417.9	-499.9	97.9	3
100	-515.7	-411.0	-500.2	97.0	3
150	-527.7	-396.0	-501.0	94.6	3
200	-540.3	-379.5	-501.6	93.7	3
250	-555.2	-361.7	-502.3	92.3	3
300	-570.3	-342.5	-503.1	90.9	3

PBBR2 (C), LEAD BROMIDE

T(C)	DELTA H	DELTA F	ENTHALPY	ENTROPY	REF
25	-66.6	-62.6	-66.6	38.6	1
50	-66.7	-62.3	-66.1	40.2	2
58	-66.7	-62.2	-66.0	40.6	2
58	-73.8	-62.2	-66.0	40.6	2
75	-73.8	-61.5	-65.6	41.6	2
100	-73.7	-60.8	-65.1	42.9	2
150	-73.5	-59.1	-64.2	45.4	2
200	-73.3	-57.4	-63.2	47.6	2
250	-73.1	-55.7	-62.2	49.6	2
300	-72.9	-54.0	-61.2	51.3	2

PBCL2 (C), LEAD CHLORIDE

T(C)	DELTA H	DELTA F	ENTHALPY	ENTROPY	REF
25	-85.9	-75.1	-85.9	32.5	1
50	-85.8	-74.2	-85.3	33.9	2
75	-85.7	-73.3	-85.0	35.3	2
100	-85.7	-72.4	-84.6	36.5	2
150	-85.5	-70.6	-83.6	38.8	2
200	-85.3	-68.9	-82.7	40.9	2
250	-85.1	-67.1	-81.7	42.9	2
300	-84.9	-65.4	-80.7	44.7	2

PBCO3 (C), LEAD CARBONATE

T(C)	DELTA H	DELTA F	ENTHALPY	ENTROPY	REF
25	-167.1	-149.5	-167.1	31.3	1
50	-167.0	-148.0	-166.6	33.0	2
75	-167.0	-146.5	-166.0	34.7	2
100	-166.9	-145.1	-165.5	36.2	2
150	-166.7	-142.2	-164.3	39.2	2
200	-166.5	-139.3	-163.0	42.0	2
250	-166.3	-136.4	-161.7	44.7	2
300	-166.0	-133.6	-160.3	47.3	2

PBF2 (C), LEAD FLUORIDE

T(C)	DELTA H	DELTA F	ENTHALPY	ENTROPY	REF
25	-158.7	-147.5	-158.7	26.4	1
50	-158.6	-146.5	-158.3	27.8	2
75	-158.5	-145.6	-157.8	29.2	2
100	-158.4	-144.7	-157.3	30.5	2
150	-158.1	-142.9	-156.4	32.9	2
200	-157.9	-141.1	-155.4	35.2	2
250	-157.6	-139.4	-154.3	37.3	2
300	-157.3	-137.6	-153.2	39.2	2

PBI2 (C), LEAD IODIDE

T(C)	DELTA H	DELTA F	ENTHALPY	ENTROPY	REF
25	-41.9	-41.5	-41.9	41.8	1
50	-41.9	-41.5	-41.4	43.4	2
75	-41.9	-41.4	-41.0	44.8	2
100	-42.0	-41.4	-40.5	46.2	2
114	-42.0	-41.4	-40.2	46.9	2
114	-45.7	-41.4	-40.2	46.9	2
150	-45.9	-41.0	-39.5	48.7	2
185	-46.1	-40.5	-38.8	50.3	2
185	-56.2	-40.5	-38.8	50.3	2
200	-56.1	-40.0	-38.5	50.9	2
250	-55.9	-38.3	-37.5	53.0	2
300	-55.6	-36.7	-36.4	54.8	2

PBO (C), LEAD MONOXIDE (RED)

T(C)	DELTA H	DELTA F	ENTHALPY	ENTROPY	REF
25	-52.3	-45.2	-52.3	15.9	1
50	-52.3	-44.6	-52.1	16.8	2
75	-52.3	-44.0	-51.8	17.6	2
100	-52.3	-43.4	-51.5	18.4	2
150	-52.2	-42.2	-50.9	19.8	2
200	-52.1	-41.0	-50.4	21.1	2
250	-52.1	-39.8	-49.8	22.3	2
300	-52.0	-38.7	-49.2	23.4	2

PBO2 (C), LEAD DIOXIDE

T(C)	DELTA H	DELTA F	ENTHALPY	ENTROPY	REF
25	-66.3	-52.0	-66.3	16.4	1
50	-66.3	-50.9	-65.9	17.6	2
75	-66.2	-49.6	-65.5	18.9	2
100	-66.2	-48.4	-65.1	19.8	2
150	-66.1	-46.0	-64.4	21.8	2
200	-66.0	-43.6	-63.6	23.6	2
250	-65.9	-41.3	-62.7	25.3	2
300	-65.7	-38.9	-61.9	26.8	2

PBS (C), LEAD MONOSULFIDE

T(C)	DELTA H	DELTA F	ENTHALPY	ENTROPY	REF
25	-39.3	-33.1	-39.3	21.8	1
50	-39.3	-32.6	-39.0	22.8	2
75	-39.3	-32.1	-38.8	23.6	2
100	-39.2	-31.5	-38.4	24.5	2
150	-39.2	-30.5	-37.8	26.0	2
200	-39.1	-29.5	-37.2	27.3	2
250	-39.0	-28.5	-36.6	28.6	2
300	-38.9	-27.5	-36.0	29.7	2

PBSO4 (C), LEAD SULFATE

T(C)	DELTA H	DELTA F	ENTHALPY	ENTROPY	REF
25	-235.2	-203.8	-235.2	35.5	1
50	-235.2	-201.2	-234.6	37.5	2
75	-235.2	-198.6	-234.0	39.4	2
100	-235.2	-195.9	-233.3	41.2	2
150	-235.1	-190.7	-232.0	44.4	2
200	-235.1	-185.4	-230.7	47.4	2
250	-235.0	-180.2	-229.3	50.2	2
300	-234.8	-175.0	-227.8	52.9	2

PBSO4·PBO (C), DILEAD OXIDE SULFATE

T(C)	DELTA H	DELTA F	ENTHALPY	ENTROPY	REF
25	-295.3	-256.2	-295.3	49.4	1
50	-295.3	-252.9	-294.4	52.3	2
75	-295.3	-249.6	-293.5	55.0	2
100	-295.2	-246.3	-292.6	57.6	2
150	-295.1	-239.8	-290.7	62.3	2
200	-295.0	-233.3	-288.8	66.6	2
250	-294.8	-226.8	-286.8	70.7	2
300	-294.5	-220.3	-284.7	74.5	2

PBSO4·2PBO (C), TRILEAD OXIDE SULFATE

T(C)	DELTA H	DELTA F	ENTHALPY	ENTROPY	REF
25	-349.7	-303.5	-349.7	65.6	1
50	-349.7	-299.6	-348.6	69.4	2
75	-349.6	-295.7	-347.4	72.9	2
100	-349.5	-291.9	-346.2	76.3	2
150	-349.3	-284.2	-343.7	82.4	2
200	-349.1	-276.5	-341.2	89.1	2
250	-348.8	-268.8	-338.5	93.4	2
300	-348.5	-261.2	-335.8	98.4	2

PBSO4·4PBO (C), PENTALEAD OXIDE SULFATE

T(C)	DELTA H	DELTA F	ENTHALPY	ENTROPY	REF
25	-449.5	-390.9	-449.5	104.1	2
50	-449.3	-386.0	-447.8	109.7	2
75	-449.2	-381.0	-446.0	114.9	2
100	-449.1	-376.2	-444.2	119.8	2
150	-448.8	-366.4	-440.6	128.9	2
200	-449.4	-356.7	-436.9	137.2	2
250	-447.9	-347.0	-433.0	145.0	2
300	-447.4	-337.4	-429.0	152.2	2

PD (C), PALLADIUM

T(C)	DELTA H	DELTA F	ENTHALPY	ENTROPY	REF
25	0.0	0.0	0.0	9.0	1
50	0.0	0.0	0.2	9.5	2
75	0.0	0.0	0.3	10.0	2
100	0.0	0.0	0.5	10.4	2
150	0.0	0.0	0.8	11.2	2
200	0.0	0.0	1.1	11.9	2
250	0.0	0.0	1.4	12.5	2
300	0.0	0.0	1.8	13.1	2

PD++ (AQ)

T(C)	DELTA H	DELTA F	ENTHALPY	ENTROPY	REF
25	40.5	42.2	40.5	-38.0	1
50	40.6	42.3	41.7	-34.0	3
75	40.8	42.5	43.4	-28.9	3
100	40.9	42.6	45.5	-23.0	3
150	40.9	42.8	49.1	-13.9	3
200	41.4	43.0	53.5	-3.7	3
250	41.2	43.2	58.3	6.0	3
300	41.6	43.3	63.7	15.8	3

PDBR4-- (AQ)
(EXTRAPOLATED AS CRISS & COBBLE TYPE 3)

T(C)	DELTA H	DELTA F	ENTHALPY	ENTROPY	REF
25	-88.8	-76.0	-88.8	80.0	1
50	-88.1	-74.9	-88.1	82.4	3
58	-87.9	-74.6	-87.8	83.2	2
58	-102.0	-74.6	-87.8	83.2	2
75	-100.8	-73.3	-87.3	84.8	3
100	-98.7	-71.4	-86.4	87.1	3
150	-96.0	-67.8	-85.8	88.6	3
200	-90.9	-65.0	-83.3	94.6	3
250	-85.3	-62.5	-81.0	99.1	3
300	-79.4	-60.6	-78.7	103.4	3

PDCL4-- (AQ)
(EXTRAPOLATED AS CRISS & COBBLE TYPE 3)

T(C)	DELTA H	DELTA F	ENTHALPY	ENTROPY	REF
25	-129.7	-96.7	-129.7	46.0	5
50	-130.2	-93.9	-130.6	43.1	3
75	-130.4	-91.1	-131.6	40.1	3
100	-130.3	-88.3	-132.8	36.9	3
150	-130.5	-82.7	-135.0	31.2	3
200	-130.4	-77.0	-137.3	25.9	3
250	-129.8	-71.6	-140.2	20.2	3
300	-129.4	-66.0	-143.3	14.4	3

PDCL6-- (AQ)
(EXTRAPOLATED AS CRISS & COBBLE TYPE 3)

T(C)	DELTA H	DELTA F	ENTHALPY	ENTROPY	REF
25	-143.7	-99.6	-143.7	62.0	5
50	-143.7	-95.9	-143.9	61.6	3
75	-143.2	-92.2	-144.0	61.1	3
100	-142.4	-88.6	-144.2	60.5	3
150	-141.6	-81.5	-145.1	58.2	3
200	-139.6	-74.7	-145.2	58.2	3
250	-137.1	-68.0	-145.6	57.3	3
300	-134.6	-61.5	-146.2	56.3	3

PDO (C), PALLADIUM MONOXIDE

T(C)	DELTA H	DELTA F	ENTHALPY	ENTROPY	REF
25	-21.7	-15.7	-21.7	13.4	2
50	-21.8	-15.2	-21.5	14.0	2
75	-21.8	-14.7	-21.3	14.6	2
100	-21.8	-14.2	-21.1	15.2	2
150	-21.9	-13.2	-20.7	16.3	2
200	-21.9	-12.1	-20.2	17.4	2
250	-21.9	-11.1	-19.7	18.5	2
300	-21.9	-10.1	-19.1	19.5	2

PT (C), PLATINUM

T(C)	DELTA H	DELTA F	ENTHALPY	ENTROPY	REF
25	0.0	0.0	0.0	10.0	1
50	0.0	0.0	0.2	10.5	2
75	0.0	0.0	0.3	10.9	2
100	0.0	0.0	0.5	11.4	2
150	0.0	0.0	0.8	12.1	2
200	0.0	0.0	1.1	12.9	2
250	0.0	0.0	1.4	13.5	2
300	0.0	0.0	1.8	14.1	2

PTCL4-- (AQ)
(EXTRAPOLATED AS CRISS & COBBLE TYPE 3)

T(C)	DELTA H	DELTA F	ENTHALPY	ENTROPY	REF
25	-120.2	-88.1	-120.2	50.0	1
50	-120.5	-85.4	-120.9	47.8	3
75	-120.5	-82.7	-121.7	45.3	3
100	-120.2	-80.0	-122.6	42.8	3
150	-120.0	-74.7	-124.5	37.9	3
200	-119.3	-69.5	-126.3	34.0	3
250	-118.1	-64.3	-128.5	29.5	3
300	-117.1	-59.2	-131.0	24.9	3

PTCL6-- (AQ)
(EXTRAPOLATED AS CRISS & COBBLE TYPE 3)

T(C)	DELTA H	DELTA F	ENTHALPY	ENTROPY	REF
25	-161.3	-117.0	-161.3	62.6	1
50	-161.2	-113.3	-161.3	62.3	3
75	-160.7	-109.6	-161.5	61.9	3
100	-159.8	-106.0	-161.6	61.4	3
150	-159.0	-98.9	-162.5	59.2	3
200	-156.9	-92.1	-162.4	59.4	3
250	-154.3	-85.4	-162.8	58.7	3
300	-151.7	-78.9	-163.3	57.9	3

PTS (C), PLATINUM (2) SULFIDE

T(C)	DELTA H	DELTA F	ENTHALPY	ENTROPY	REF
25	-34.8	-27.7	-34.8	13.2	1
50	-34.8	-27.1	-34.5	14.1	2
75	-34.7	-26.5	-34.2	15.0	2
100	-34.7	-25.9	-33.9	15.9	2
150	-34.6	-24.7	-33.3	17.4	2
200	-34.5	-23.5	-32.7	18.8	2
250	-34.4	-22.4	-32.1	20.1	2
300	-34.3	-21.3	-31.4	21.2	2

PTS2 (C), PLATINUM (4) SULFIDE

T(C)	DELTA H	DELTA F	ENTHALPY	ENTROPY	REF
25	-56.7	-42.8	-56.7	17.8	1
50	-56.6	-41.6	-56.3	19.1	2
75	-56.6	-40.5	-55.9	20.4	2
100	-56.5	-39.3	-55.5	21.5	2
150	-56.4	-37.0	-54.6	23.6	2
200	-56.3	-34.7	-53.8	25.5	2
250	-56.2	-32.4	-52.9	27.3	2
300	-56.0	-30.2	-52.0	28.9	2

PU (C), PLUTONIUM
PHASE TRANSITIONS AT 122C AND 206C

T(C)	DELTA H	DELTA F	ENTHALPY	ENTROPY	REF
25	0.0	0.0	0.0	12.3	2
50	0.0	0.0	0.2	12.9	2
75	0.0	0.0	0.4	13.5	2
100	0.0	0.0	0.6	14.1	2
122	0.0	0.0	0.8	14.6	2
122	0.0	0.0	1.8	17.0	2
150	0.0	0.0	2.0	17.7	2
200	0.0	0.0	2.5	18.8	2
206	0.0	0.0	2.6	18.9	2
206	0.0	0.0	2.7	19.2	2
250	0.0	0.0	3.2	20.2	2
300	0.0	0.0	3.8	21.2	2

PU+++ (AQ)

T(C)	DELTA H	DELTA F	ENTHALPY	ENTROPY	REF
25	-141.8	-140.5	-141.8	-54.0	5
50	-142.0	-140.4	-140.4	-49.5	3
75	-142.4	-140.2	-138.5	-43.7	3
100	-143.0	-140.0	-136.2	-37.0	3
122	-143.5	-139.8	-134.4	-32.4	2
122	-144.4	-139.8	-134.4	-32.4	2
150	-145.1	-139.5	-132.0	-26.6	3
200	-146.1	-138.8	-127.1	-15.1	3
206	-146.4	-138.7	-126.5	-13.8	2
206	-146.5	-138.7	-126.5	-13.8	2
250	-148.3	-137.9	-121.6	-4.1	3
300	-149.8	-136.8	-115.5	7.0	3

RB (C,L), RUBIDIUM
MELTING POINT = 39C

T(C)	DELTA H	DELTA F	ENTHALPY	ENTROPY	REF
25	0.0	0.0	0.0	18.1	2
39	0.0	0.0	0.1	18.4	2
39	0.0	0.0	0.6	20.2	2
50	0.0	0.0	0.7	20.4	2
75	0.0	0.0	0.9	21.0	2
100	0.0	0.0	1.1	21.5	2
150	0.0	0.0	1.5	22.5	2
200	0.0	0.0	1.8	23.3	2
250	0.0	0.0	2.2	24.0	2
300	0.0	0.0	2.6	24.7	2

RB+ (AQ)

T(C)	DELTA H	DELTA F	ENTHALPY	ENTROPY	REF
25	-59.3	-67.4	-59.3	24.7	5
39	-59.4	-67.8	-59.0	25.8	2
39	-59.4	-67.8	-59.0	25.8	2
50	-59.9	-68.1	-58.7	26.7	3
75	-60.0	-68.7	-57.9	29.2	3
100	-60.1	-69.4	-56.9	31.9	3
150	-60.6	-70.6	-55.4	35.8	3
200	-60.6	-71.8	-53.2	40.9	3
250	-61.0	-72.9	-50.9	45.5	3
300	-61.1	-74.1	-48.4	50.1	3

RE (C), RHENIUM

T(C)	DELTA H	DELTA F	ENTHALPY	ENTROPY	REF
25	0.0	0.0	0.0	8.8	1
50	0.0	0.0	0.2	9.3	2
75	0.0	0.0	0.3	9.8	2
100	0.0	0.0	0.5	10.2	2
150	0.0	0.0	0.8	11.0	2
200	0.0	0.0	1.1	11.7	2
250	0.0	0.0	1.4	12.3	2
300	0.0	0.0	1.7	12.9	2

RH (C), RHODIUM

T(C)	DELTA H	DELTA F	ENTHALPY	ENTROPY	REF
25	0.0	0.0	0.0	7.5	1
50	0.0	0.0	0.2	8.0	2
75	0.0	0.0	0.3	8.5	2
100	0.0	0.0	0.4	8.9	2
150	0.0	0.0	0.8	9.7	2
200	0.0	0.0	1.1	10.4	2
250	0.0	0.0	1.4	11.0	2
300	0.0	0.0	1.7	11.6	2

RH2O (C), RHODIUM (1) OXIDE

T(C)	DELTA H	DELTA F	ENTHALPY	ENTROPY	REF
25	-22.7	-19.1	-22.7	27.4	2
50	-22.7	-18.8	-22.3	28.8	2
75	-22.6	-18.5	-21.8	30.1	2
100	-22.5	-18.2	-21.4	31.4	2
150	-22.4	-17.6	-20.5	33.7	2
200	-22.3	-17.0	-19.5	35.7	2
250	-22.2	-16.5	-18.6	37.6	2
300	-22.1	-15.9	-17.6	39.4	2

RH2O3 (C), RHODIUM (3) OXIDE

T(C)	DELTA H	DELTA F	ENTHALPY	ENTROPY	REF
25	-68.3	-49.8	-68.3	26.5	2
50	-68.2	-48.3	-67.7	28.5	2
75	-68.2	-46.7	-67.0	30.4	2
100	-68.1	-45.2	-66.4	32.2	2
150	-68.0	-42.1	-65.1	35.5	2
200	-67.8	-39.1	-63.7	38.5	2
250	-67.6	-36.0	-62.4	41.3	2
300	-67.4	-33.0	-60.9	43.8	2

RHCL6-- (AQ)
(EXTRAPOLATED AS CRISS & COBBLE TYPE 3)

T(C)	DELTA H	DELTA F	ENTHALPY	ENTROPY	REF
25	-202.6	-158.3	-202.6	60.0	5
50	-202.6	-154.6	-202.8	59.3	3
75	-202.3	-150.9	-203.1	58.5	3
100	-201.6	-147.3	-203.4	57.6	3
150	-201.0	-140.0	-204.5	54.8	3
200	-199.3	-133.1	-204.8	54.2	3
250	-197.0	-126.3	-205.5	52.7	3
300	-194.8	-119.6	-206.4	51.1	3

RHO (C), RHODIUM (2) OXIDE

T(C)	DELTA H	DELTA F	ENTHALPY	ENTROPY	REF
25	-21.7	-16.0	-21.7	12.9	2
50	-21.7	-15.5	-21.4	13.8	2
75	-21.6	-15.0	-21.1	14.7	2
100	-21.5	-14.6	-20.8	15.5	2
150	-21.4	-13.7	-20.2	17.0	2
200	-21.3	-12.7	-19.6	18.4	2
250	-21.2	-11.8	-19.0	19.6	2
300	-21.1	-11.0	-18.4	20.8	2

RU (C), RUTHENIUM

T(C)	DELTA H	DELTA F	ENTHALPY	ENTROPY	REF
25	0.0	0.0	0.0	6.8	1
50	0.0	0.0	0.1	7.3	2
75	0.0	0.0	0.3	7.7	2
100	0.0	0.0	0.4	8.1	2
150	0.0	0.0	0.7	8.9	2
200	0.0	0.0	1.0	9.5	2
250	0.0	0.0	1.3	10.1	2
300	0.0	0.0	1.6	10.7	2

S (RHOMBIC,MONOCLINIC,L), SULFUR SOL PHASE TRANSI-
TION AT 95C, SOL-LIQ TRANS AT 115C

T(C)	DELTA H	DELTA F	ENTHALPY	ENTROPY	REF
25	-15.3	-9.5	-15.3	7.6	1
50	-15.3	-9.0	-15.2	8.0	2
75	-15.3	-8.5	-15.1	8.5	2
95	-15.2	-8.1	-14.9	8.8	2
95	-15.1	-8.1	-14.9	9.1	2
100	-15.1	-8.0	-14.8	9.1	2
115	-15.1	-7.7	-14.7	9.4	2
115	-14.7	-7.7	-14.3	10.4	2
150	-14.5	-7.1	-14.0	11.1	2
200	-14.3	-6.3	-13.6	12.1	2
250	-14.0	-5.4	-13.1	13.0	2
300	-13.8	-4.6	-12.7	13.8	2

S-- (AQ)

T(C)	DELTA H	DELTA F	ENTHALPY	ENTROPY	REF
25	-7.4	11.0	-7.4	6.5	1
50	-7.8	12.6	-8.6	2.7	3
75	-7.9	14.2	-10.1	-1.7	3
100	-7.9	15.7	-11.8	-6.5	3
150	-8.2	18.9	-15.1	-14.9	3
200	-8.7	22.2	-19.0	-23.8	3
250	-8.5	25.4	-23.2	-32.3	3
300	-9.8	28.7	-29.0	-42.9	3

S2 (G), SULFUR (DIATOMIC GAS)

T(C)	DELTA H	DELTA F	ENTHALPY	ENTROPY	REF
25	0.0	0.0	0.0	54.5	1
50	0.0	0.0	0.2	55.1	2
75	0.0	0.0	0.4	55.7	2
100	0.0	0.0	0.6	56.3	2
150	0.0	0.0	1.0	57.3	2
200	0.0	0.0	1.4	58.2	2
250	0.0	0.0	1.8	59.1	2
300	0.0	0.0	2.3	59.8	2

S2-- (AQ)
(EXTRAPOLATED AS CRISS & COBBLE TYPE 2)

T(C)	DELTA H	DELTA F	ENTHALPY	ENTROPY	REF
25	-23.5	0.0	-23.5	16.8	1
50	-24.0	2.0	-24.7	12.8	3
75	-24.2	4.0	-26.2	8.4	3
100	-24.3	6.1	-27.8	3.8	3
150	-24.8	10.1	-31.2	-4.7	3
200	-25.5	14.3	-35.1	-13.7	3
250	-25.5	18.5	-39.3	-22.3	3
300	-27.0	22.8	-45.1	-32.9	3

S2O3-- (AQ)

T(C)	DELTA H	DELTA F	ENTHALPY	ENTROPY	REF
25	-188.0	-143.0	-188.0	18.0	5
50	-189.8	-139.1	-190.3	10.8	3
75	-191.4	-135.1	-192.8	3.3	3
100	-192.8	-131.1	-195.5	-4.4	3
150	-195.2	-122.7	-200.2	-16.0	3
200	-198.8	-113.8	-206.5	-30.6	3
250	-202.2	-104.7	-213.5	-44.7	3
300	-206.2	-95.2	-221.2	-58.9	3

S2O4-- (AQ)

T(C)	DELTA H	DELTA F	ENTHALPY	ENTROPY	REF
25	-210.7	-162.5	-210.7	32.0	1
50	-211.9	-158.4	-212.2	27.0	3
75	-212.8	-154.2	-214.0	21.7	3
100	-213.5	-150.0	-216.0	16.2	3
150	-214.8	-141.5	-219.4	7.6	3
200	-216.6	-132.7	-223.7	-2.4	3
250	-218.1	-123.7	-228.6	-12.3	3
300	-220.1	-114.6	-234.1	-22.2	3

S2O5-- (AQ)

T(C)	DELTA H	DELTA F	ENTHALPY	ENTROPY	REF
25	-262.6	-208.0	-262.6	35.0	5
50	-263.7	-203.3	-264.0	30.4	3
75	-264.5	-198.6	-265.6	25.6	3
100	-265.2	-193.9	-267.4	20.7	3
150	-266.5	-184.3	-270.6	12.6	3
200	-268.0	-174.5	-274.5	3.7	3
250	-269.2	-164.6	-278.9	-5.3	3
300	-270.9	-154.5	-283.9	-14.4	3

S206-- (AQ)

T(C)	DELTA H	DELTA F	ENTHALPY	ENTROPY	REF
25	-311.1	-250.0	-311.1	37.7	5
50	-312.2	-244.8	-312.4	33.6	3
75	-312.9	-239.6	-313.8	29.2	3
100	-313.5	-234.3	-315.4	24.7	3
150	-314.7	-223.7	-318.4	17.3	3
200	-316.1	-212.8	-321.9	9.2	3
250	-317.1	-201.9	-326.0	1.0	3
300	-318.5	-190.8	-330.5	-7.2	3

S208-- (AQ)

T(C)	DELTA H	DELTA F	ENTHALPY	ENTROPY	REF
25	-350.7	-284.4	-350.7	69.3	1
50	-350.4	-278.8	-350.5	70.0	3
75	-349.7	-273.3	-350.2	70.7	3
100	-348.6	-267.9	-350.0	71.3	3
150	-347.5	-257.1	-350.3	70.5	3
200	-344.8	-246.9	-349.3	73.0	3
250	-341.4	-236.7	-348.6	74.3	3
300	-338.0	-226.9	-348.0	75.4	3

S3-- (AQ)
(EXTRAPOLATED AS CRISS & COBBLE TYPE 2)

T(C)	DELTA H	DELTA F	ENTHALPY	ENTROPY	REF
25	-39.8	-10.8	-39.8	25.8	1
50	-40.5	-8.4	-41.1	21.6	3
75	-40.8	-5.9	-42.6	17.2	3
100	-40.9	-3.4	-44.2	12.8	3
150	-41.7	1.7	-47.5	4.2	3
200	-42.6	6.9	-51.4	-4.9	3
250	-42.8	12.1	-55.7	-13.5	3
300	-44.6	17.4	-61.6	-24.1	3

S306-- (AQ)

T(C)	DELTA H	DELTA F	ENTHALPY	ENTROPY	REF
25	-325.0	-257.4	-325.0	43.3	5
50	-325.9	-251.7	-326.0	40.0	3
75	-326.5	-246.0	-327.2	36.6	3
100	-326.8	-240.2	-328.5	32.9	3
150	-327.8	-228.6	-331.0	26.7	3
200	-328.6	-216.9	-333.7	20.5	3
250	-328.9	-205.1	-336.9	14.0	3
300	-329.6	-193.2	-340.5	7.4	3

S4-- (AQ)
(EXTRAPOLATED AS CRISS & COBBLE TYPE 2)

T(C)	DELTA H	DELTA F	ENTHALPY	ENTROPY	REF
25	-55.9	-21.4	-55.9	34.7	1
50	-56.7	-18.5	-57.2	30.3	3
75	-57.1	-15.5	-58.7	26.0	3
100	-57.2	-12.6	-60.2	21.7	3
150	-58.3	-6.5	-63.6	13.0	3
200	-59.4	-0.3	-67.6	3.8	3
250	-59.9	5.9	-71.8	-4.8	3
300	-61.9	12.3	-77.7	-15.5	3

S406-- (AQ)

T(C)	DELTA H	DELTA F	ENTHALPY	ENTROPY	REF
25	-357.1	-282.2	-357.1	46.0	5
50	-358.0	-275.9	-358.0	43.1	3
75	-358.5	-269.5	-359.0	40.1	3
100	-358.8	-263.2	-360.2	36.9	3
150	-359.8	-250.3	-362.4	31.2	3
200	-360.4	-237.4	-364.7	25.9	3
250	-360.5	-224.4	-367.6	20.2	3
300	-361.0	-211.4	-370.8	14.4	3

S5-- (AQ)
(EXTRAPOLATED AS CRISS & COBBLE TYPE 2)

T(C)	DELTA H	DELTA F	ENTHALPY	ENTROPY	REF
25	-71.6	-31.7	-71.6	43.6	1
50	-72.6	-28.3	-73.1	39.0	3
75	-73.1	-24.9	-74.5	34.7	3
100	-73.3	-21.5	-76.0	30.6	3
150	-74.5	-14.4	-79.4	21.8	3
200	-75.9	-7.2	-83.4	12.6	3
250	-76.6	0.1	-87.7	3.9	3
300	-78.9	7.5	-93.6	-6.8	3

S506-- (AQ)

T(C)	DELTA H	DELTA F	ENTHALPY	ENTROPY	REF
25	-357.7	-275.9	-357.7	50.1	5
50	-358.5	-269.0	-358.4	47.9	3
75	-358.9	-262.1	-359.2	45.5	3
100	-359.0	-255.2	-360.1	43.0	3
150	-359.8	-241.2	-362.0	38.2	3
200	-360.0	-227.3	-363.7	34.3	3
250	-359.8	-213.3	-365.9	29.8	3
300	-359.8	-199.3	-368.4	25.3	3

SO2 (G), SULFUR DIOXIDE (GAS)

T(C)	DELTA H	DELTA F	ENTHALPY	ENTROPY	REF
25	-86.3	-81.2	-86.3	59.3	1
50	-86.3	-80.8	-86.0	60.1	2
75	-86.3	-80.4	-85.8	60.8	2
100	-86.4	-79.9	-85.5	61.5	2
150	-86.4	-79.1	-85.0	62.8	2
200	-86.4	-78.2	-84.5	64.1	2
250	-86.5	-77.3	-83.9	65.2	2
300	-86.5	-76.5	-83.4	66.2	2

SO3 (G), SULFUR TRIOXIDE (GAS)

T(C)	DELTA H	DELTA F	ENTHALPY	ENTROPY	REF
25	-109.9	-98.2	-109.9	61.3	1
50	-110.0	-97.2	-109.6	62.3	2
75	-110.0	-96.2	-109.3	63.3	2
100	-110.1	-95.2	-108.9	64.2	2
150	-110.1	-93.2	-108.3	66.0	2
200	-110.1	-91.2	-107.5	67.6	2
250	-110.1	-89.2	-106.8	69.1	2
300	-110.1	-87.2	-106.0	70.6	2

SO3-- (AQ)

T(C)	DELTA H	DELTA F	ENTHALPY	ENTROPY	REF
25	-167.2	-125.8	-167.2	3.0	1
50	-169.6	-122.2	-170.2	-6.5	3
75	-171.9	-118.5	-173.5	-16.4	3
100	-174.0	-114.6	-177.1	-26.6	3
150	-177.7	-106.5	-183.0	-41.3	3
200	-183.0	-97.5	-191.4	-60.9	3
250	-188.5	-88.2	-200.7	-79.5	3
300	-194.7	-78.3	-210.9	-98.1	3

SO4-- (AQ)

T(C)	DELTA H	DELTA F	ENTHALPY	ENTROPY	REF
25	-232.7	-187.5	-232.7	14.8	1
50	-234.6	-183.6	-235.1	7.1	3
75	-236.3	-179.6	-237.7	-0.9	3
100	-237.9	-175.5	-240.7	-9.2	3
150	-240.5	-167.0	-245.6	-21.4	3
200	-244.6	-157.9	-252.3	-37.1	3
250	-248.4	-148.6	-259.8	-52.2	3
300	-253.0	-138.8	-268.1	-67.3	3

SB (C), ANTIMONY

T(C)	DELTA H	DELTA F	ENTHALPY	ENTROPY	REF
25	0.0	0.0	0.0	10.9	1
50	0.0	0.0	0.2	11.4	2
75	0.0	0.0	0.3	11.9	2
100	0.0	0.0	0.4	12.3	2
150	0.0	0.0	0.8	13.1	2
200	0.0	0.0	1.1	13.8	2
250	0.0	0.0	1.4	14.4	2
300	0.0	0.0	1.7	15.0	2

SB2O3 (C), ANTIMONY TRIOXIDE

T(C)	DELTA H	DELTA F	ENTHALPY	ENTROPY	REF
25	-169.4	-149.7	-169.4	29.4	2
50	-169.4	-148.1	-168.8	31.4	2
75	-169.3	-146.5	-168.2	33.2	2
100	-169.3	-144.8	-167.5	35.0	2
150	-169.1	-141.5	-166.2	38.2	2
200	-169.0	-138.3	-164.9	41.2	2
250	-168.8	-135.1	-163.5	44.0	2
300	-168.6	-131.9	-162.1	46.6	2

SB2O4 (C), ANTIMONY TETRAOXIDE

T(C)	DELTA H	DELTA F	ENTHALPY	ENTROPY	REF
25	-216.9	-190.2	-216.9	30.4	1
50	-216.9	-188.0	-216.2	32.6	12
75	-216.8	-185.8	-215.5	34.7	12
100	-216.8	-183.5	-214.8	36.7	12
150	-216.7	-179.1	-213.3	40.3	12
200	-216.5	-174.7	-211.9	43.7	12
250	-216.4	-170.2	-210.3	46.8	12
300	-216.2	-165.8	-208.7	49.6	12

SB2O5 (C), ANTIMONY PENTOXIDE

T(C)	DELTA H	DELTA F	ENTHALPY	ENTROPY	REF
25	-232.3	-198.2	-232.3	29.9	1
50	-232.3	-195.3	-231.6	32.2	2
75	-232.3	-192.5	-230.8	34.5	2
100	-232.3	-189.6	-230.0	36.7	2
150	-232.1	-183.9	-228.3	40.9	2
200	-231.8	-178.2	-226.5	45.0	2

SB2S3 (C), ANTIMONY TRISULFIDE

T(C)	DELTA H	DELTA F	ENTHALPY	ENTROPY	REF
25	-87.8	-69.9	-87.8	43.5	1
50	-87.7	-68.4	-87.1	45.8	2
75	-87.6	-66.9	-86.4	47.9	2
100	-87.5	-65.4	-85.7	49.9	2
150	-87.2	-62.5	-84.2	53.6	2
200	-87.0	-59.6	-82.7	57.0	2
250	-86.7	-56.7	-81.2	60.1	2
300	-86.4	-53.8	-79.6	62.9	2

SB2S4-- (AQ)
(EXTRAPOLATED AS CRISS & COBBLE TYPE 3)

T(C)	DELTA H	DELTA F	ENTHALPY	ENTROPY	REF
25	-113.8	-61.7	-113.8	-2.5	1
50	-116.8	-57.2	-117.0	-12.9	3
75	-119.6	-52.5	-120.6	-23.7	3
100	-122.5	-47.6	-124.5	-34.7	3
150	-127.1	-37.4	-131.0	-50.6	3
200	-134.1	-26.1	-140.1	-72.1	3
250	-141.0	-14.4	-150.2	-92.3	3
300	-148.9	-1.9	-161.3	-112.6	3

SBBR3 (C,L,G), ANTIMONY TRIBROMIDE
MELT PT = 95C; BOIL PT = 289C

T(C)	DELTA H	DELTA F	ENTHALPY	ENTROPY	REF
25	-62.0	-57.2	-62.0	49.5	1
50	-62.1	-56.8	-61.3	51.6	2
58	-62.2	-56.7	-61.1	52.3	2
58	-72.8	-56.7	-61.1	52.3	2
75	-72.7	-55.9	-60.7	53.6	2
95	-72.5	-55.0	-60.1	55.2	2
95	-69.0	-55.0	-56.6	64.7	2
100	-69.0	-54.8	-56.5	65.1	2
150	-68.4	-52.9	-55.0	68.9	2
200	-67.9	-51.1	-53.5	72.2	2
250	-67.4	-49.3	-52.0	75.2	2
289	-67.0	-48.0	-50.8	77.4	2
289	-50.3	-48.0	-34.1	107.1	2
300	-50.3	-48.0	-33.9	107.5	2

SBCL3 (C,L,G), ANTIMONY TRICHLORIDE
MELT PT = 73C; BOIL PT = 220C

T(C)	DELTA H	DELTA F	ENTHALPY	ENTROPY	REF
25	-91.3	-77.4	-91.3	44.0	1
50	-91.1	-76.2	-90.7	46.1	2
73	-90.9	-75.2	-90.1	48.0	2
73	-87.8	-75.2	-87.0	56.9	2
75	-87.8	-75.1	-86.9	57.1	2
100	-87.5	-74.2	-86.2	59.2	2
150	-87.0	-72.4	-84.7	62.9	2
200	-86.5	-70.7	-83.2	66.2	2
220	-86.3	-70.1	-82.6	67.4	2
220	-75.9	-70.1	-72.2	88.5	2
250	-75.9	-69.7	-71.6	89.6	2
300	-75.9	-69.1	-70.7	91.4	2

SBI3 (C,L), ANTIMONY TRIIODIDE
MELTING PT = 171C

T(C)	DELTA H	DELTA F	ENTHALPY	ENTROPY	REF
25	-23.0	-22.7	-23.0	51.5	2
50	-23.0	-22.7	-22.4	53.4	2
75	-23.1	-22.6	-21.8	55.2	2
100	-23.2	-22.6	-21.2	56.9	2
114	-23.2	-22.6	-20.8	57.8	2
114	-28.8	-22.6	-20.8	57.8	2
150	-29.2	-22.0	-19.9	60.1	2
171	-29.3	-21.6	-19.4	61.4	2
171	-23.9	-21.6	-13.9	73.6	2
185	-23.9	-21.5	-13.4	74.7	2
185	-38.9	-21.5	-13.4	74.7	2
200	-38.7	-21.0	-12.9	75.8	2
250	-38.0	-19.1	-11.2	79.3	2
300	-37.3	-17.3	-9.5	82.4	2

SBO2 (C), ANTIMONY DIOXIDE

T(C)	DELTA H	DELTA F	ENTHALPY	ENTROPY	REF
25	-113.5	-100.2	-113.5	15.2	2
50	-113.5	-99.1	-113.1	16.3	2
75	-113.5	-97.9	-112.8	17.4	2
100	-113.4	-96.8	-112.4	18.3	2
150	-113.4	-94.6	-111.7	20.2	2
200	-113.3	-92.4	-111.0	21.8	2
250	-113.3	-90.2	-110.2	23.4	2
300	-113.2	-88.0	-109.4	24.8	2

SBS3-- (AQ)
(EXTRAPOLATED AS CRISS & COBBLE TYPE 3)

T(C)	DELTA H	DELTA F	ENTHALPY	ENTROPY	REF
25	-50.0	-60.4	-50.0	168.8	5
50	-44.5	-61.5	-45.0	185.0	3
75	-38.0	-63.1	-39.5	201.5	3
100	-30.7	-65.3	-33.5	218.2	3
150	-20.1	-70.4	-25.2	238.4	3
200	-2.3	-78.4	-10.1	274.1	3
250	16.8	-87.4	5.4	305.2	3
300	37.4	-98.3	22.2	336.0	3

SE (BLACK,L), SELENIUM
SOL-LIQ TRANS AT 221C

T(C)	DELTA H	DELTA F	ENTHALPY	ENTROPY	REF
25	0.0	0.0	0.0	10.1	1
50	0.0	0.0	0.2	10.6	2
75	0.0	0.0	0.3	11.1	2
100	0.0	0.0	0.5	11.5	2
150	0.0	0.0	0.8	12.4	2
200	0.0	0.0	1.2	13.2	2
221	0.0	0.0	1.3	13.5	2
221	0.0	0.0	2.6	16.1	2
250	0.0	0.0	2.8	16.6	2
300	0.0	0.0	3.3	17.4	2

SE-- (AQ)

T(C)	DELTA H	DELTA F	ENTHALPY	ENTROPY	REF
25	30.3	42.6	30.3	10.1	5
50	29.9	43.7	29.1	6.2	3
75	29.7	44.7	27.6	1.8	3
100	29.7	45.8	26.0	-2.9	3
150	29.2	48.0	22.7	-11.3	3
200	28.6	50.2	18.8	-20.3	3
221	28.6	51.2	17.1	-23.8	2
221	27.3	51.2	17.1	-23.8	2
250	27.2	52.5	14.5	-28.8	3
300	25.7	55.0	8.8	-39.4	3

SEO3-- (AQ)

T(C)	DELTA H	DELTA F	ENTHALPY	ENTROPY	REF
25	-121.8	-88.4	-121.8	13.0	5
50	-123.7	-85.5	-124.2	5.0	3
75	-125.5	-82.5	-127.0	-3.3	3
100	-127.1	-79.4	-130.0	-11.8	3
150	-129.9	-72.9	-135.1	-24.5	3
200	-134.2	-65.7	-142.1	-40.7	3
221	-135.8	-62.7	-145.3	-47.3	2
221	-137.1	-62.7	-145.3	-47.3	2
250	-139.6	-58.3	-149.9	-56.3	3
300	-144.5	-50.3	-158.4	-72.0	3

SEO4-- (AQ)

T(C)	DELTA H	DELTA F	ENTHALPY	ENTROPY	REF
25	-143.2	-105.5	-143.2	22.9	1
50	-144.8	-102.3	-145.2	16.5	3
75	-146.1	-98.9	-147.5	9.7	3
100	-147.3	-95.5	-149.9	2.8	3
150	-149.4	-88.5	-154.1	-7.8	3
200	-152.4	-81.1	-159.7	-20.7	3
221	-153.6	-77.9	-162.3	-26.0	2
221	-154.9	-77.9	-162.3	-26.0	2
250	-156.6	-73.4	-166.0	-33.4	3
300	-160.0	-65.3	-173.0	-46.1	3

SI (C), SILICON

T(C)	DELTA H	DELTA F	ENTHALPY	ENTROPY	REF
25	0.0	0.0	0.0	4.5	1
50	0.0	0.0	0.1	4.9	2
75	0.0	0.0	0.3	5.3	2
100	0.0	0.0	0.4	5.6	2
150	0.0	0.0	0.6	6.3	2
200	0.0	0.0	0.9	6.9	2
250	0.0	0.0	1.2	7.4	2
300	0.0	0.0	1.5	8.0	2

SICL4 (L,G), SILICON TETRACHLORIDE
BOILING PT = 58C

T(C)	DELTA H	DELTA F	ENTHALPY	ENTROPY	REF
25	-164.2	-148.2	-164.2	57.3	1
50	-163.9	-146.8	-163.4	60.0	2
58	-163.8	-146.4	-163.1	60.8	2
58	-157.0	-146.4	-156.3	81.4	2
75	-157.0	-145.9	-155.9	82.5	2
100	-157.0	-145.1	-155.4	84.1	2
150	-156.9	-143.5	-154.2	87.0	2
200	-156.8	-141.9	-153.0	89.6	2
250	-156.8	-140.3	-151.8	92.1	2
300	-156.7	-138.8	-150.6	94.3	2

SIF6-- (AQ)
(EXTRAPOLATED AS CRISS & COBBLE TYPE 2)

T(C)	DELTA H	DELTA F	ENTHALPY	ENTROPY	REF
25	-571.0	-525.7	-571.0	39.2	1
50	-572.1	-521.9	-572.4	34.7	3
75	-572.8	-518.0	-573.8	30.4	3
100	-573.3	-514.1	-575.3	26.2	3
150	-574.9	-506.0	-578.7	17.5	3
200	-576.7	-497.7	-582.7	8.3	3
250	-577.9	-489.3	-587.0	-0.4	3
300	-580.6	-480.7	-592.9	-11.1	3

SIO2 (C), SILICONDIOXIDE (CRISTOBALITE)
TRANSITION PT FOR SIO2 AT 270C

T(C)	DELTA H	DELTA F	ENTHALPY	ENTROPY	REF
25	-217.4	-204.5	-217.4	10.2	1
50	-217.4	-203.4	-217.1	11.1	2
75	-217.4	-202.3	-216.8	12.0	2
100	-217.4	-201.2	-216.5	12.8	2
150	-217.4	-199.0	-215.9	14.4	2
200	-217.4	-196.9	-215.2	15.9	2
250	-217.3	-194.7	-214.5	17.3	2
270	-217.3	-193.8	-214.2	17.8	2
270	-217.0	-193.8	-213.9	18.4	2
300	-217.0	-192.5	-213.5	19.2	2

SIO2 (C), SILICON DIOXIDE (QUARTZ)

T(C)	DELTA H	DELTA F	ENTHALPY	ENTROPY	REF
25	-217.7	-204.8	-217.7	10.0	1
50	-217.7	-203.7	-217.5	10.9	2
75	-217.8	-202.6	-217.2	11.7	2
100	-217.8	-201.5	-216.9	12.6	2
150	-217.8	-199.3	-216.2	14.2	2
200	-217.7	-197.1	-215.5	15.7	2
250	-217.7	-194.9	-214.8	17.1	2
300	-217.6	-192.8	-214.1	18.5	2

SIS2 (C), SILICON DISULFIDE

T(C)	DELTA H	DELTA F	ENTHALPY	ENTROPY	REF
25	-81.4	-69.6	-81.4	16.0	2
50	-81.3	-67.5	-81.0	17.2	2
75	-81.3	-66.4	-80.6	18.3	2
100	-81.2	-65.4	-80.2	19.4	2
150	-81.1	-63.2	-79.4	21.4	2
200	-81.0	-61.1	-78.6	23.2	2
250	-80.9	-59.0	-77.8	24.8	2
300	-80.7	-57.0	-77.0	26.4	2

SM (C), SAMARIUM

T(C)	DELTA H	DELTA F	ENTHALPY	ENTROPY	REF
25	0.0	0.0	0.0	16.6	1
50	0.0	0.0	0.2	17.2	2
75	0.0	0.0	0.4	17.8	2
100	0.0	0.0	0.6	18.3	2
150	0.0	0.0	1.0	19.3	2
200	0.0	0.0	1.4	20.2	2
250	0.0	0.0	1.8	21.1	2
300	0.0	0.0	2.3	21.9	2

SM+++ (AQ)

T(C)	DELTA H	DELTA F	ENTHALPY	ENTROPY	REF
25	-165.4	-159.3	-165.4	-65.6	1
50	-165.5	-158.8	-163.9	-60.7	3
75	-165.7	-158.3	-161.8	-54.4	3
100	-166.0	-157.7	-159.2	-47.2	3
150	-166.7	-156.5	-154.7	-35.8	3
200	-167.2	-155.3	-149.4	-23.3	3
250	-168.7	-153.9	-143.4	-11.4	3
300	-169.5	-152.5	-136.8	0.6	3

SN (C,L) TIN
MELTING PT = 232C

T(C)	DELTA H	DELTA F	ENTHALPY	ENTROPY	REF
25	0.0	0.0	0.0	12.3	1
50	0.0	0.0	0.2	12.8	2
75	0.0	0.0	0.3	13.3	2
100	0.0	0.0	0.5	13.8	2
150	0.0	0.0	0.8	14.7	2
200	0.0	0.0	1.2	15.5	2
232	0.0	0.0	1.4	15.9	2
232	0.0	0.0	3.1	19.2	2
250	0.0	0.0	3.2	19.5	2
300	0.0	0.0	3.6	20.1	2

SN++ (AQ)

T(C)	DELTA H	DELTA F	ENTHALPY	ENTROPY	REF
25	-2.4	-6.3	-2.4	-15.9	5
50	-2.5	-6.6	-1.4	-12.6	3
75	-2.7	-6.9	0.0	-8.4	3
100	-3.0	-7.2	1.7	-3.6	3
150	-3.6	-7.7	4.6	3.6	3
200	-4.0	-8.1	8.2	12.0	3
232	-4.7	-8.4	10.6	17.0	2
232	-6.4	-8.4	10.6	17.0	2
250	-6.8	-8.5	12.1	19.9	3
300	-7.5	-8.7	16.5	27.9	3

SNBR4 (C,L,G), TIN(4) BROMIDE
MELT PT = 30C; BOIL PT = 207C

T(C)	DELTA H	DELTA F	ENTHALPY	ENTROPY	REF
25	-90.2	-83.7	-90.2	63.2	1
30	-90.2	-83.6	-90.0	63.8	2
30	-87.4	-83.6	-87.2	73.2	2
50	-87.4	-83.3	-86.4	75.6	2
58	-87.5	-83.2	-86.1	76.5	2
58	-101.7	-83.2	-86.1	76.5	2
75	-101.4	-82.3	-85.5	78.4	2
100	-101.1	-81.0	-84.5	81.0	2
150	-100.4	-78.3	-82.6	85.8	2
200	-99.8	-75.7	-80.8	90.0	2
207	-99.7	-75.4	-80.5	90.5	2
207	-89.8	-75.4	-70.6	110.4	2
232	-89.8	-74.3	-70.0	111.7	2
232	-91.5	-74.3	-70.0	111.7	2
250	-91.5	-73.7	-69.5	112.6	2
300	-91.4	-72.0	-68.3	114.9	2

SNCL+ (AQ)

T(C)	DELTA H	DELTA F	ENTHALPY	ENTROPY	REF
25	-39.5	-39.4	-39.5	18.0	1
50	-39.6	-39.4	-38.8	20.2	3
75	-39.6	-39.4	-37.9	23.0	3
100	-39.7	-39.3	-36.8	26.1	3
150	-40.1	-39.3	-35.1	30.5	3
200	-40.1	-39.2	-32.7	36.1	3
232	-40.3	-39.1	-31.0	39.4	2
232	-41.9	-39.1	-31.0	39.4	2
250	-42.1	-39.1	-30.1	41.2	3
300	-42.2	-38.8	-27.2	46.5	3

SNCL2 (C,L), TIN(2) CHLORIDE
MELTING PT = 247C

T(C)	DELTA H	DELTA F	ENTHALPY	ENTROPY	REF
25	-77.7	-67.4	-77.7	31.0	2
50	-77.6	-66.5	-77.2	32.5	2
75	-77.5	-65.7	-76.7	34.0	2
100	-77.4	-64.8	-76.3	35.3	2
150	-77.1	-63.2	-75.3	37.8	2
200	-76.9	-61.5	-74.2	40.1	2
232	-76.7	-60.5	-73.6	41.4	2
232	-78.4	-60.5	-73.6	41.4	2
247	-78.3	-60.0	-73.3	42.1	2
247	-75.3	-60.0	-70.2	47.9	2
250	-75.3	-59.9	-70.1	48.1	2
300	-74.9	-58.4	-69.1	50.1	2

SNCL4 (L,G), TIN(4) CHLORIDE
BOILING PT = 115C

T(C)	DELTA H	DELTA F	ENTHALPY	ENTROPY	REF
25	-122.2	-105.2	-122.2	61.8	1
50	-121.9	-103.8	-121.3	64.7	2
75	-121.6	-102.4	-120.4	67.4	2
100	-121.2	-101.0	-119.5	69.9	2
115	-121.1	-100.2	-119.0	71.3	2
115	-112.9	-100.2	-110.9	92.2	2
150	-112.9	-99.1	-110.0	94.3	2
200	-112.9	-97.4	-108.8	97.1	2
232	-112.9	-96.4	-108.0	98.7	2
232	-114.5	-96.4	-108.0	98.7	2
250	-114.5	-95.7	-107.5	99.6	2
300	-114.5	-93.9	-106.3	101.8	2

SNO (C), TIN(2) OXIDE

T(C)	DELTA H	DELTA F	ENTHALPY	ENTROPY	REF
25	-68.3	-61.3	-69.3	13.5	1
50	-69.3	-60.8	-68.0	14.4	2
75	-68.3	-60.2	-67.8	15.2	2
100	-68.3	-59.6	-67.5	15.9	2
150	-68.2	-58.4	-66.9	17.3	2
200	-68.2	-57.3	-66.4	18.5	2
232	-68.2	-56.6	-66.0	19.3	2
232	-69.9	-56.6	-66.0	19.3	2
250	-69.9	-56.1	-65.8	19.7	2
300	-69.8	-54.8	-65.3	20.7	2

SNO2 (C), TIN(4) OXIDE

T(C)	DELTA H	DELTA F	ENTHALPY	ENTROPY	REF
25	-138.8	-124.2	-138.8	12.5	1
50	-138.8	-123.0	-138.5	13.5	2
75	-138.8	-121.8	-138.1	14.6	2
100	-138.8	-120.6	-137.8	15.6	2
150	-138.7	-118.1	-137.0	17.5	2
200	-138.6	-115.7	-136.2	19.3	2
232	-138.6	-114.2	-135.7	20.4	2
232	-140.3	-114.2	-135.7	20.4	2
250	-140.2	-113.2	-135.4	21.0	2
300	-140.1	-110.7	-134.5	22.6	2

SNOH+ (AQ)

T(C)	DELTA H	DELTA F	ENTHALPY	ENTROPY	REF
25	-68.3	-60.9	-68.3	7.0	1
50	-68.3	-60.3	-67.5	9.6	3
75	-68.3	-59.7	-66.4	12.8	3
100	-68.3	-59.0	-65.1	16.4	3
150	-68.4	-57.8	-63.0	21.7	3
200	-68.2	-56.6	-60.2	28.3	3
232	-68.2	-55.8	-58.4	32.1	2
232	-69.9	-55.8	-58.4	32.1	2
250	-69.9	-55.4	-57.2	34.3	3
300	-69.6	-54.0	-53.9	40.4	3

SNS (C), TIN(2) SULFIDE

T(C)	DELTA H	DELTA F	ENTHALPY	ENTROPY	REF
25	-39.3	-33.0	-39.3	18.4	1
50	-39.3	-32.5	-39.1	19.3	2
75	-39.3	-32.0	-38.8	20.2	2
100	-39.2	-31.5	-38.4	21.1	2
150	-39.2	-30.4	-37.8	22.6	2
200	-39.1	-29.4	-37.2	24.0	2
232	-39.1	-28.7	-36.8	24.8	2
232	-40.8	-28.7	-36.8	24.8	2
250	-40.7	-28.3	-36.6	25.2	2
300	-40.7	-27.1	-35.9	26.4	2

SNS2 (C), TIN(4) SULFIDE

T(C)	DELTA H	DELTA F	ENTHALPY	ENTROPY	REF
25	-70.7	-57.0	-70.7	20.9	2
50	-70.6	-55.8	-70.3	22.3	2
75	-70.6	-54.7	-69.8	23.5	2
100	-70.5	-53.6	-69.4	24.7	2
150	-70.4	-51.3	-68.6	26.9	2
200	-70.3	-49.1	-67.7	28.8	2
232	-70.2	-47.6	-67.1	29.9	2
232	-71.9	-47.6	-67.1	29.9	2
250	-71.9	-46.8	-66.8	30.6	2
300	-71.7	-44.3	-65.9	32.2	2

SR (C), STRONTIUM

T(C)	DELTA H	DELTA F	ENTHALPY	ENTROPY	REF
25	0.0	0.0	0.0	12.5	1
50	0.0	0.0	0.2	13.0	2
75	0.0	0.0	0.3	13.5	2
100	0.0	0.0	0.5	13.9	2
150	0.0	0.0	0.8	14.8	2
200	0.0	0.0	1.2	15.5	2
250	0.0	0.0	1.5	16.2	2
300	0.0	0.0	1.9	16.9	2

SR++ (AQ)

T(C)	DELTA H	DELTA F	ENTHALPY	ENTROPY	REF
25	-130.5	-133.7	-130.5	-17.8	1
50	-130.5	-134.0	-129.4	-14.4	3
75	-130.7	-134.2	-128.0	-10.2	3
100	-130.9	-134.5	-126.3	-5.3	3
150	-131.5	-134.9	-123.4	2.1	3
200	-131.8	-135.3	-119.7	10.6	3
250	-132.8	-135.6	-115.7	18.7	3
300	-133.4	-135.8	-111.2	26.8	3

SRBR2 (C), STRONTIUM BROMIDE

T(C)	DELTA H	DELTA F	ENTHALPY	ENTROPY	REF
25	-171.5	-166.5	-171.5	32.3	1
50	-171.6	-166.1	-171.0	33.8	2
58	-171.6	-166.0	-170.9	34.3	2
58	-178.8	-165.0	-170.9	34.3	2
75	-178.7	-165.4	-170.5	35.3	2
100	-179.6	-164.4	-170.1	36.6	2
150	-178.4	-162.5	-169.1	39.0	2
200	-179.2	-160.7	-168.1	41.2	2
250	-179.0	-158.8	-167.1	43.2	2
300	-177.8	-157.0	-166.1	45.0	2

SRCL2 (C), STRONTIUM CHLORIDE

T(C)	DELTA H	DELTA F	ENTHALPY	ENTROPY	REF
25	-199.1	-186.7	-198.1	27.5	1
50	-198.0	-185.7	-197.6	29.0	2
75	-197.9	-184.8	-197.1	30.4	2
100	-197.8	-183.8	-196.7	31.7	2
150	-197.6	-182.0	-195.7	34.1	2
200	-197.4	-180.1	-194.8	36.3	2
250	-197.2	-178.3	-193.8	38.2	2
300	-197.0	-176.5	-192.8	40.0	2

SRCO3 (C), STRONTIUM CARBONATE

T(C)	DELTA H	DELTA F	ENTHALPY	ENTROPY	REF
25	-291.6	-272.5	-291.6	23.2	1
50	-291.6	-270.9	-291.1	24.8	2
75	-291.5	-269.3	-290.5	26.4	2
100	-291.5	-267.7	-290.0	28.0	2
150	-291.4	-264.5	-288.9	30.8	2
200	-291.2	-261.3	-287.7	33.4	2
250	-291.1	-258.2	-286.5	35.9	2
300	-291.0	-255.0	-285.2	38.2	2

SRF2 (C), STRONTIUM FLUORIDE

T(C)	DELTA H	DELTA F	ENTHALPY	ENTROPY	REF
25	-290.7	-278.4	-290.7	19.6	1
50	-290.6	-277.4	-290.2	21.1	2
75	-290.5	-276.3	-289.8	22.5	2
100	-290.4	-275.3	-289.3	23.8	2
150	-290.2	-273.3	-288.4	26.2	2
200	-289.9	-271.4	-287.4	28.3	2
250	-289.8	-269.4	-286.5	30.2	2
300	-289.6	-267.5	-285.5	32.0	2

SRI2 (C), STRONTIUM IODIDE

T(C)	DELTA H	DELTA F	ENTHALPY	ENTROPY	REF
25	-134.0	-133.7	-134.0	39.2	2
50	-134.0	-133.7	-133.5	40.8	2
75	-134.0	-133.6	-133.0	42.2	2
100	-134.0	-133.6	-132.5	43.6	2
114	-134.1	-133.6	-132.3	44.3	2
114	-137.8	-133.6	-132.3	44.3	2
150	-138.0	-133.2	-131.5	46.1	2
185	-138.2	-132.8	-130.8	47.7	2
185	-148.2	-132.8	-130.8	47.7	2
200	-148.2	-132.3	-130.5	48.3	2
250	-148.0	-130.6	-129.5	50.3	2
300	-147.8	-129.0	-128.5	52.2	2

SRO (C), STRONTIUM OXIDE

T(C)	DELTA H	DELTA F	ENTHALPY	ENTROPY	REF
25	-141.5	-134.3	-141.5	13.0	1
50	-141.5	-133.7	-141.2	13.9	2
75	-141.5	-133.1	-141.0	14.7	2
100	-141.4	-132.5	-140.7	15.5	2
150	-141.4	-131.4	-140.1	17.0	2
200	-141.3	-130.2	-139.5	18.3	2
250	-141.2	-129.0	-138.9	19.5	2
300	-141.1	-127.9	-138.3	20.6	2

SRO2 (C), STRONTIUM PEROXIDE

T(C)	DELTA H	DELTA F	ENTHALPY	ENTROPY	REF
25	-156.4	-141.9	-156.4	13.0	2
50	-156.3	-140.7	-155.9	14.5	2
75	-156.1	-139.5	-155.4	16.0	2
100	-156.0	-138.4	-155.0	17.3	2
150	-155.7	-136.0	-154.0	19.7	2
200	-155.4	-133.7	-153.0	21.9	2
250	-155.2	-131.4	-152.0	23.9	2
300	-154.9	-129.1	-151.0	25.8	2

SRSO4 (C), STRONTIUM SULFATE

T(C)	DELTA H	DELTA F	ENTHALPY	ENTROPY	REF
25	-362.6	-329.9	-362.6	28.0	1
50	-362.6	-327.2	-362.0	30.1	2
75	-362.6	-324.4	-361.3	32.0	2
100	-362.5	-321.7	-360.7	33.9	2
150	-362.4	-316.2	-359.3	37.3	2
200	-362.3	-310.8	-357.9	40.4	2
250	-362.2	-305.3	-356.5	43.3	2
300	-362.1	-299.9	-355.0	45.9	2

TA (C), TANTALUM

T(C)	DELTA H	DELTA F	ENTHALPY	ENTROPY	REF
25	0.0	0.0	0.0	9.9	1
50	0.0	0.0	0.2	10.4	2
75	0.0	0.0	0.3	10.9	2
100	0.0	0.0	0.5	11.3	2
150	0.0	0.0	0.8	12.1	2
200	0.0	0.0	1.1	12.8	2
250	0.0	0.0	1.4	13.4	2
300	0.0	0.0	1.7	14.0	2

TE (C), TELLURIUM

T(C)	DELTA H	DELTA F	ENTHALPY	ENTROPY	REF
25	0.0	0.0	0.0	11.9	1
50	0.0	0.0	0.2	12.4	2
75	0.0	0.0	0.3	12.9	2
100	0.0	0.0	0.5	13.3	2
150	0.0	0.0	0.8	14.1	2
200	0.0	0.0	1.2	14.9	2
250	0.0	0.0	1.5	15.6	2
300	0.0	0.0	1.9	16.3	2

TE(OH)3+ (AQ)

T(C)	DELTA H	DELTA F	ENTHALPY	ENTROPY	REF
25	-145.4	-118.6	-145.4	21.7	1
50	-145.9	-116.3	-144.8	23.8	3
75	-146.4	-114.0	-143.9	26.4	3
100	-147.0	-111.7	-142.9	29.3	3
150	-148.4	-106.8	-141.2	33.4	3
200	-149.3	-101.9	-139.0	38.7	3
250	-150.6	-96.8	-136.5	43.6	3
300	-151.8	-91.6	-133.8	48.5	3

TECL4 (C,L), TELLURIUM TETRACHLORIDE
MELTING PT = 224C

T(C)	DELTA H	DELTA F	ENTHALPY	ENTROPY	REF
25	-78.0	-57.6	-78.0	50.0	2
50	-77.7	-55.9	-77.2	52.7	2
75	-77.5	-54.2	-76.3	55.2	2
100	-77.2	-52.6	-75.5	57.4	2
150	-76.7	-49.3	-73.9	61.6	2
200	-76.3	-46.1	-72.2	65.3	2
224	-76.1	-44.5	-71.4	67.0	2
224	-71.6	-44.5	-66.9	76.0	2
250	-70.8	-43.1	-65.5	78.8	2
300	-69.4	-40.6	-62.8	83.6	2

TEO2 (C), TELLURIUM DIOXIDE

T(C)	DELTA H	DELTA F	ENTHALPY	ENTROPY	REF
25	-77.1	-64.6	-77.1	19.0	1
50	-77.1	-63.6	-76.7	20.2	2
75	-77.0	-62.5	-76.3	21.4	2
100	-76.9	-61.5	-75.9	22.5	2
150	-76.8	-59.4	-75.1	24.5	2
200	-76.7	-57.4	-74.3	26.4	2
250	-76.6	-55.3	-73.4	28.1	2
300	-76.5	-53.3	-72.6	29.7	2

TEO3-- (AQ)

T(C)	DELTA H	DELTA F	ENTHALPY	ENTROPY	REF
25	-143.4	-108.0	-143.4	8.0	5
50	-145.6	-104.9	-146.1	-0.8	3
75	-147.6	-101.7	-149.1	-9.9	3
100	-149.6	-98.4	-152.5	-19.2	3
150	-152.8	-91.4	-158.0	-32.9	3
200	-157.7	-83.7	-165.7	-50.8	3
250	-162.5	-75.6	-174.2	-67.9	3
300	-168.1	-67.0	-183.5	-85.1	3

TH (C), THORIUM

T(C)	DELTA H	DELTA F	ENTHALPY	ENTROPY	REF
25	0.0	0.0	0.0	12.8	2
50	0.0	0.0	0.2	13.3	2
75	0.0	0.0	0.3	13.8	2
100	0.0	0.0	0.5	14.3	2
150	0.0	0.0	0.8	15.1	2
200	0.0	0.0	1.2	15.9	2
250	0.0	0.0	1.5	16.6	2
300	0.0	0.0	1.9	17.2	2

TH(SO4)2 (C), THORIUM SULFATE

T(C)	DELTA H	DELTA F	ENTHALPY	ENTROPY	REF
25	-638.1	-570.1	-638.1	35.4	2
50	-638.1	-564.4	-637.0	38.8	2
75	-638.1	-558.7	-635.9	42.0	2
100	-638.1	-553.1	-634.8	45.2	2
150	-637.9	-541.7	-632.5	51.1	2
200	-637.7	-530.3	-630.0	56.6	2
250	-637.3	-519.0	-627.3	61.9	2
300	-636.8	-507.7	-624.6	66.9	2

TH++++ (AQ)

T(C)	DELTA H	DELTA F	ENTHALPY	ENTROPY	REF
25	-182.8	-175.2	-182.8	-95.0	5
50	-183.0	-174.6	-180.9	-89.2	3
75	-183.5	-173.9	-179.4	-81.6	3
100	-184.2	-173.1	-175.4	-72.9	3
150	-185.5	-171.5	-169.9	-59.0	3
200	-186.7	-169.8	-163.5	-44.2	3
250	-189.2	-167.9	-156.4	-30.0	3
300	-191.0	-165.8	-148.5	-15.5	3

THO2 (C), THORIUM DIOXIDE

T(C)	DELTA H	DELTA F	ENTHALPY	ENTROPY	REF
25	-293.2	-279.4	-293.2	15.6	2
50	-293.2	-278.3	-292.8	16.8	2
75	-293.1	-277.1	-292.4	18.0	2
100	-293.1	-276.0	-292.0	19.0	2
150	-293.0	-273.7	-291.2	21.1	2
200	-292.9	-271.4	-290.4	22.9	2
250	-292.8	-269.2	-289.6	24.6	2
300	-292.6	-266.9	-288.7	26.2	2

TI (C), TITANIUM

T(C)	DELTA H	DELTA F	ENTHALPY	ENTROPY	REF
25	0.0	0.0	0.0	7.3	1
50	0.0	0.0	0.2	7.8	2
75	0.0	0.0	0.3	8.3	2
100	0.0	0.0	0.5	8.7	2
150	0.0	0.0	0.8	9.5	2
200	0.0	0.0	1.1	10.2	2
250	0.0	0.0	1.4	10.9	2
300	0.0	0.0	1.8	11.5	2

TI2O3 (C), DITITANIUM TRIOXIDE
TRANSITION PT FOR TI2O3 AT 200C

T(C)	DELTA H	DELTA F	ENTHALPY	ENTROPY	REF
25	-363.5	-342.8	-363.5	18.8	1
50	-363.5	-341.1	-362.9	20.8	2
75	-363.4	-339.4	-362.3	22.7	2
100	-363.3	-337.7	-361.6	24.6	2
150	-363.0	-334.2	-360.1	28.2	2
200	-362.7	-330.9	-358.6	31.6	2
200	-362.5	-330.9	-358.4	32.0	2
250	-362.1	-327.5	-356.8	35.2	2
300	-361.8	-324.2	-355.2	38.1	2

TI3O5 (C), TRITITANIUM PENTOXIDE
TRANSITION PT FOR TI3O5 AT 177C

T(C)	DELTA H	DELTA F	ENTHALPY	ENTROPY	REF
25	-587.8	-553.9	-587.8	30.9	1
50	-587.7	-551.1	-586.8	34.0	2
75	-587.6	-548.3	-585.8	37.0	2
100	-587.5	-545.4	-584.8	39.9	2
150	-587.2	-539.8	-582.6	45.3	2
177	-587.0	-536.8	-581.4	48.1	2
177	-584.2	-536.8	-578.6	54.4	2
200	-584.0	-534.4	-577.6	56.6	2
250	-583.6	-529.2	-575.3	61.2	2
300	-583.3	-524.0	-572.9	65.4	2

TIBR2 (C), TITANIUM DIBROMIDE (C)

T(C)	DELTA H	DELTA F	ENTHALPY	ENTROPY	REF
25	-96.9	-91.6	-96.9	25.9	2
50	-97.0	-91.1	-96.4	27.4	2
58	-97.0	-91.0	-96.3	27.9	2
58	-104.1	-91.0	-96.3	27.9	2
75	-104.1	-90.4	-96.0	28.8	2
100	-104.0	-89.4	-95.5	30.2	2
150	-103.8	-87.4	-94.5	32.6	2
200	-103.6	-85.5	-93.6	34.7	2
250	-103.4	-83.6	-92.6	36.7	2
300	-103.2	-81.7	-91.6	38.4	2

TIBR3 (C), TITANIUM TRIBROMIDE

T(C)	DELTA H	DELTA F	ENTHALPY	ENTROPY	REF
25	-131.1	-125.2	-131.1	42.2	1
50	-131.3	-124.7	-130.5	44.2	2
58	-131.4	-124.6	-130.3	44.8	2
58	-142.0	-124.6	-130.3	44.8	2
75	-141.9	-123.7	-129.9	46.0	2
100	-141.8	-122.4	-129.3	47.7	2
150	-141.5	-119.8	-128.0	50.8	2
200	-141.2	-117.3	-126.7	53.8	2
250	-140.8	-114.8	-125.3	56.5	2
300	-140.4	-112.3	-123.9	59.1	2

TIBR4 (C,L,G), TITANIUM TETRABROMIDE
MELT PT = 38C;   BOIL PT = 231C

T(C)	DELTA H	DELTA F	ENTHALPY	ENTROPY	REF
25	-147.4	-140.9	-147.4	58.2	1
38	-147.5	-140.6	-147.0	59.6	2
38	-144.4	-140.6	-143.9	69.5	2
50	-144.5	-140.4	-143.5	70.8	2
58	-144.5	-140.3	-143.2	71.7	2
58	-158.7	-140.3	-143.2	71.7	2
75	-158.5	-139.5	-142.6	73.5	2
100	-158.2	-139.1	-141.7	76.0	2
150	-157.6	-135.5	-139.8	80.6	2
200	-157.0	-132.9	-138.0	84.6	2
231	-156.6	-131.3	-136.9	86.9	2
231	-145.8	-131.3	-126.1	108.4	2
250	-145.8	-130.8	-125.6	109.3	2
300	-145.7	-129.3	-124.4	111.6	2

TICL2 (C), TITANIUM DICHLORIDE

T(C)	DELTA H	DELTA F	ENTHALPY	ENTROPY	REF
25	-122.8	-111.0	-122.8	20.9	1
50	-122.7	-110.0	-122.4	22.3	2
75	-122.7	-109.0	-121.9	23.5	2
100	-122.6	-108.0	-121.5	24.7	2
150	-122.5	-106.1	-120.6	26.9	2
200	-122.3	-104.1	-119.8	28.9	2
250	-122.2	-102.2	-118.8	30.7	2
300	-122.0	-100.3	-117.9	32.4	2

TICL3 (C), TITANIUM TRICHLORIDE

T(C)	DELTA H	DELTA F	ENTHALPY	ENTROPY	REF
25	-172.3	-156.2	-172.3	33.4	1
50	-172.2	-154.9	-171.7	35.3	2
75	-172.0	-153.6	-171.1	37.0	2
100	-171.9	-152.3	-170.5	38.7	2
150	-171.7	-149.6	-169.4	41.6	2
200	-171.5	-147.0	-168.2	44.3	2
250	-171.2	-144.5	-167.0	46.7	2
300	-171.0	-141.9	-165.8	48.9	2

TICL4 (L,G), TITANIUM TETRACHLORIDE
BOILING PT = 136C

T(C)	DELTA H	DELTA F	ENTHALPY	ENTROPY	REF
25	-192.2	-176.2	-192.2	60.3	1
50	-191.9	-174.9	-191.3	63.1	2
75	-191.6	-173.6	-190.5	65.7	2
100	-191.3	-172.3	-189.6	68.1	2
136	-190.9	-170.5	-188.3	71.3	2
136	-182.3	-170.5	-179.8	92.2	2
150	-182.3	-170.1	-179.4	93.1	2
200	-182.2	-168.7	-179.2	95.8	2
250	-182.2	-167.2	-177.0	98.3	2
300	-182.1	-165.8	-175.8	100.5	2

TIF3 (C), TITANIUM TRIFLUORIDE

T(C)	DELTA H	DELTA F	ENTHALPY	ENTROPY	REF
25	-343.1	-325.5	-343.1	21.0	2
50	-343.0	-324.0	-342.5	22.8	2
75	-342.9	-322.6	-342.0	24.4	2
100	-342.8	-321.1	-341.4	26.0	2
150	-342.6	-318.2	-340.3	28.8	2
200	-342.4	-315.4	-339.2	31.3	2
250	-342.2	-312.5	-338.1	33.6	2
300	-342.0	-309.7	-336.9	35.7	2

TIF4 (C,G), TITANIUM TETRAFLUORIDE
SUBLIMATION PT = 286C

T(C)	DELTA H	DELTA F	ENTHALPY	ENTROPY	REF
25	-394.2	-372.7	-394.2	32.0	1
50	-394.0	-370.9	-393.5	34.3	2
75	-393.9	-369.1	-392.8	36.4	2
100	-393.7	-367.3	-392.1	38.4	2
150	-393.3	-363.8	-390.5	42.2	2
200	-392.8	-360.4	-389.0	45.7	2
250	-392.4	-357.0	-387.4	48.9	2
286	-392.0	-354.6	-386.2	51.1	2
286	-368.7	-354.6	-362.8	92.9	2
300	-368.7	-354.2	-362.5	93.5	2

TII2 (C), TITANIUM DIIODIDE

T(C)	DELTA H	DELTA F	ENTHALPY	ENTROPY	REF
25	-64.5	-64.6	-64.5	35.3	2
50	-64.5	-64.6	-64.0	37.0	2
75	-64.4	-64.6	-63.5	38.5	2
100	-64.4	-64.6	-62.9	39.9	2
114	-64.4	-64.6	-62.7	40.7	2
114	-68.1	-64.6	-62.7	40.7	2
150	-68.3	-64.3	-61.9	42.6	2
185	-68.5	-63.9	-61.2	44.2	2
185	-78.5	-63.9	-61.2	44.2	2
200	-78.4	-63.4	-60.9	44.9	2
250	-78.2	-61.9	-59.8	47.0	2
300	-77.9	-60.3	-58.8	48.9	2

TII3 (C), TITANIUM TRIIODIDE

T(C)	DELTA H	DELTA F	ENTHALPY	ENTROPY	REF
25	-81.4	-80.5	-81.4	46.0	2
50	-81.4	-80.4	-80.7	48.3	2
75	-81.3	-80.4	-80.0	50.3	2
100	-81.3	-80.3	-79.3	52.3	2
114	-81.3	-80.3	-78.9	53.3	2
114	-86.9	-80.3	-78.9	53.3	2
150	-87.1	-79.6	-77.9	55.8	2
185	-87.4	-79.0	-76.9	58.0	2
185	-102.4	-79.0	-76.9	58.0	2
200	-102.3	-78.3	-76.5	58.9	2
250	-101.9	-75.7	-75.1	61.8	2
300	-101.5	-73.3	-73.7	64.4	2

TII4 (C,L), TITANIUM TETRAIODIDE
TRANSITION PT = 106C; MELT PT = 155C

T(C)	DELTA H	DELTA F	ENTHALPY	ENTROPY	REF
25	-89.8	-88.8	-89.8	59.6	1
50	-89.9	-88.8	-89.0	62.1	2
75	-89.9	-88.7	-88.3	64.4	2
100	-90.0	-88.6	-87.4	66.6	2
106	-90.0	-88.6	-87.3	67.2	2
106	-87.6	-88.6	-84.9	73.4	2
114	-87.6	-88.6	-84.6	74.1	2
114	-95.0	-88.6	-84.6	74.1	2
150	-95.4	-88.0	-83.3	77.3	2
155	-95.4	-87.9	-83.1	77.7	2
155	-90.7	-87.9	-78.4	88.8	2
185	-90.9	-87.7	-77.3	91.3	2
185	-111.0	-87.7	-77.3	91.3	2
200	-110.8	-86.9	-76.7	92.5	2
250	-110.1	-84.4	-74.8	96.3	2
300	-109.5	-82.0	-73.0	99.7	2

TIO (C), TITANIUM MONOXIDE

T(C)	DELTA H	DELTA F	ENTHALPY	ENTROPY	REF
25	-124.2	-117.2	-124.2	8.3	1
50	-124.2	-116.6	-124.0	9.1	2
75	-124.2	-116.0	-123.7	9.9	2
100	-124.2	-115.4	-123.4	10.6	2
150	-124.1	-114.3	-122.9	11.9	2
200	-124.1	-113.1	-122.3	13.2	2
250	-124.0	-111.9	-121.8	14.4	2
300	-123.9	-110.8	-121.2	15.4	2

TIO2 (C), TITANIUM DIOXIDE (ANATASE)

T(C)	DELTA H	DELTA F	ENTHALPY	ENTROPY	REF
25	-224.6	-211.4	-224.6	11.9	1
50	-224.6	-210.3	-224.3	13.0	2
75	-224.6	-209.1	-223.9	14.1	2
100	-224.5	-208.0	-223.5	15.1	2
150	-224.5	-205.8	-222.8	17.0	2
200	-224.4	-203.6	-222.0	18.8	2
250	-224.2	-201.5	-221.2	20.4	2
300	-224.1	-199.3	-220.4	21.9	2

TIO2 (C), TITANIUM DIOXIDE (RUTILE)

T(C)	DELTA H	DELTA F	ENTHALPY	ENTROPY	REF
25	-225.8	-212.6	-225.8	12.0	1
50	-225.8	-211.5	-225.5	13.1	2
75	-225.8	-210.4	-225.1	14.1	2
100	-225.8	-209.3	-224.8	15.1	2
150	-225.7	-207.1	-224.0	17.0	2
200	-225.6	-204.9	-223.3	18.6	2
250	-225.6	-202.7	-222.5	20.2	2
300	-225.5	-200.5	-221.7	21.6	2

TIS2 (C), TITANIUM DISULFIDE
TRANSITION PT FOR TIS2 AT 147C

T(C)	DELTA H	DELTA F	ENTHALPY	ENTROPY	REF
25	-110.7	-97.8	-110.7	18.7	2
50	-110.6	-95.8	-110.3	20.1	2
75	-110.5	-95.7	-109.8	21.3	2
100	-110.4	-94.6	-109.4	22.6	2
147	-110.2	-92.6	-108.5	24.8	2
147	-110.2	-92.6	-108.5	24.8	2
150	-110.2	-92.5	-108.4	25.0	2
200	-110.1	-90.4	-107.6	26.9	2
250	-110.0	-88.4	-106.7	28.7	2
300	-109.8	-86.3	-105.8	30.3	2

TL (C), THALLIUM
TRANSITION PT AT 234C

T(C)	DELTA H	DELTA F	ENTHALPY	ENTROPY	REF
25	0.0	0.0	0.0	15.3	1
50	0.0	0.0	0.2	15.9	2
75	0.0	0.0	0.3	16.3	2
100	0.0	0.0	0.5	16.8	2
150	0.0	0.0	0.8	17.6	2
200	0.0	0.0	1.1	18.3	2
234	0.0	0.0	1.4	18.9	2
234	0.0	0.0	1.5	19.0	2
250	0.0	0.0	1.6	19.3	2
300	0.0	0.0	2.0	20.0	2

TL+ (AQ)

T(C)	DELTA H	DELTA F	ENTHALPY	ENTROPY	REF
25	1.3	-7.7	1.3	25.0	1
50	1.3	-8.5	1.9	27.0	3
75	1.2	-9.3	2.7	29.4	3
100	1.1	-10.0	3.7	32.2	3
150	0.7	-11.4	5.2	36.0	3
200	0.7	-12.9	7.4	41.1	3
234	0.5	-13.9	8.9	44.2	2
234	0.4	-13.9	8.9	44.2	2
250	0.3	-14.3	9.6	45.7	3
300	0.1	-15.7	12.2	50.3	3

TL+++ (AQ)

T(C)	DELTA H	DELTA F	ENTHALPY	ENTROPY	REF
25	47.0	51.3	47.0	-61.0	1
50	46.9	51.7	48.4	-56.3	3
75	46.6	52.1	50.5	-50.2	3
100	46.2	52.5	52.9	-43.1	3
150	45.4	53.4	57.3	-32.1	3
200	44.8	54.4	62.5	-20.1	3
234	43.8	55.1	66.3	-12.3	2
234	43.7	55.1	66.3	-12.3	2
250	43.2	55.5	68.2	-8.5	3
300	42.2	56.7	74.6	3.2	3

U (C), URANIUM

T(C)	DELTA H	DELTA F	ENTHALPY	ENTROPY	REF
25	0.0	0.0	0.0	12.0	2
50	0.0	0.0	0.2	12.6	2
75	0.0	0.0	0.3	13.1	2
100	0.0	0.0	0.5	13.5	2
150	0.0	0.0	0.9	14.4	2
200	0.0	0.0	1.2	15.2	2
250	0.0	0.0	1.6	16.0	2
300	0.0	0.0	2.0	16.7	2

U+++ (AQ)

T(C)	DELTA H	DELTA F	ENTHALPY	ENTROPY	REF
25	-123.0	-124.4	-123.0	-45.0	5
50	-123.2	-124.5	-121.7	-40.8	3
75	-123.7	-124.6	-119.9	-35.3	3
100	-124.4	-124.6	-117.7	-29.1	3
150	-125.7	-124.5	-113.9	-19.4	3
200	-126.9	-124.3	-109.2	-8.7	3
250	-129.1	-123.9	-104.1	1.5	3
300	-130.9	-123.3	-98.4	11.9	3

U++++ (AQ)

T(C)	DELTA H	DELTA F	ENTHALPY	ENTROPY	REF
25	-146.6	-138.4	-146.6	-98.0	5
50	-146.8	-137.7	-144.8	-92.1	3
75	-147.3	-137.0	-142.2	-84.4	3
100	-148.0	-136.2	-139.1	-75.6	3
150	-149.1	-134.5	-133.6	-61.4	3
200	-150.3	-132.7	-127.1	-46.4	3
250	-152.7	-130.7	-119.8	-31.8	3
300	-154.4	-128.5	-111.8	-17.1	3

U3O8 (C), 3-URANIUM 9-OXIDE

T(C)	DELTA H	DELTA F	ENTHALPY	ENTROPY	REF
25	-854.4	-805.3	-854.4	67.5	2
50	-854.2	-801.2	-852.9	72.2	2
75	-853.9	-797.1	-851.5	76.6	2
100	-853.6	-793.1	-849.9	80.9	2
150	-852.9	-785.0	-846.8	88.9	2
200	-852.2	-777.0	-843.5	96.2	2
250	-851.4	-769.1	-840.1	102.9	2
300	-850.7	-761.3	-836.7	109.1	2

U4O9 (C), 4-URANIUM 9-OXIDE

T(C)	DELTA H	DELTA F	ENTHALPY	ENTROPY	REF
25	-1078.1	-1021.9	-1078.1	80.3	2
50	-1077.8	-1017.3	-1076.3	86.0	2
75	-1077.4	-1012.6	-1074.5	91.6	2
100	-1077.0	-1007.9	-1072.6	96.8	2
150	-1076.1	-998.8	-1068.6	106.7	2
200	-1075.1	-989.7	-1064.6	115.7	2
250	-1074.2	-980.7	-1060.4	124.1	2
300	-1073.3	-971.8	-1056.2	131.8	2

UBR4 (C), URANIUM TETRABROMIDE

T(C)	DELTA H	DELTA F	ENTHALPY	ENTROPY	REF
25	-197.5	-188.9	-197.5	56.0	2
50	-197.8	-188.2	-196.8	58.4	2
58	-197.9	-187.9	-196.5	59.2	2
58	-212.1	-187.9	-196.5	59.2	2
75	-212.0	-186.8	-196.0	60.6	2
100	-211.8	-185.0	-195.2	62.8	2
150	-211.5	-191.4	-193.7	66.7	2
200	-211.1	-177.9	-192.1	70.3	2
250	-210.7	-174.4	-190.4	73.6	2
300	-210.4	-170.9	-188.8	76.6	2

UF3 (C), URANIUM TRIFLUORIDE

T(C)	DELTA H	DELTA F	ENTHALPY	ENTROPY	REF
25	-357.1	-340.2	-357.1	28.0	2
50	-357.0	-338.8	-356.5	29.9	2
75	-356.8	-337.4	-355.9	31.6	2
100	-356.7	-336.0	-355.4	33.2	2
150	-356.5	-333.2	-354.2	36.2	2
200	-356.3	-330.5	-353.0	38.9	2
250	-356.0	-327.8	-351.8	41.3	2
300	-355.8	-325.1	-350.5	43.6	2

UF4 (C), URANIUM TETRAFLUORIDE

T(C)	DELTA H	DELTA F	ENTHALPY	ENTROPY	REF
25	-453.7	-432.0	-453.7	36.3	2
50	-453.5	-430.2	-453.0	38.5	2
75	-453.4	-428.4	-452.3	40.6	2
100	-453.3	-426.7	-451.6	42.6	2
150	-453.0	-423.1	-450.2	46.2	2
200	-452.7	-419.6	-448.7	49.4	2
250	-452.4	-416.1	-447.3	52.3	2
300	-452.2	-412.7	-445.8	55.0	2

UF6 (C,G), URANIUM HEXAFLUORIDE
SUBLIMATION PT FOR UF6 AT 57C

T(C)	DELTA H	DELTA F	ENTHALPY	ENTROPY	REF
25	-523.0	-492.3	-523.0	54.4	2
50	-522.7	-489.7	-522.0	57.7	2
57	-522.6	-489.0	-521.7	58.6	2
57	-511.1	-489.0	-510.1	93.5	2
75	-511.0	-487.8	-509.5	95.3	2
100	-511.0	-486.2	-508.7	97.6	2
150	-510.8	-482.8	-507.0	101.8	2
200	-510.7	-479.5	-505.3	105.6	2
250	-510.5	-476.3	-503.6	109.1	2
300	-510.4	-473.0	-501.8	112.3	2

UI4 (C), URANIUM TETRAIODIDE

T(C)	DELTA H	DELTA F	ENTHALPY	ENTROPY	REF
25	-126.5	-126.3	-126.5	67.0	2
50	-126.5	-126.3	-125.7	69.5	2
75	-126.6	-126.3	-124.9	71.9	2
100	-126.6	-126.3	-124.1	74.2	2
114	-126.7	-126.3	-123.6	75.4	2
114	-134.1	-126.3	-123.6	75.4	2
150	-134.5	-125.5	-122.4	78.4	2
185	-134.9	-124.8	-121.2	81.1	2
185	-155.0	-124.8	-121.2	81.1	2
200	-154.9	-123.8	-120.7	82.3	2
250	-154.4	-120.5	-118.9	85.8	2
300	-153.9	-117.3	-117.2	89.0	2

UO2 (C), URANIUM DIOXIDE

T(C)	DELTA H	DELTA F	ENTHALPY	ENTROPY	REF
25	-259.3	-246.6	-259.3	18.6	2
50	-259.3	-245.6	-258.9	19.9	2
75	-259.2	-244.5	-258.5	21.1	2
100	-259.1	-243.5	-258.1	22.2	2
150	-259.0	-241.4	-257.2	24.4	2
200	-258.8	-239.3	-256.3	26.4	2
250	-258.6	-237.3	-255.4	28.3	2
300	-258.5	-235.3	-254.5	30.0	2

UO2+ (AQ)

T(C)	DELTA H	DELTA F	ENTHALPY	ENTROPY	REF
25	-247.6	-237.6	-247.6	7.0	5
50	-247.6	-236.8	-246.8	9.6	3
75	-247.6	-235.9	-245.7	12.8	3
100	-247.5	-235.1	-244.4	16.4	3
150	-247.7	-233.4	-242.3	21.7	3
200	-247.5	-231.8	-239.5	28.3	3
250	-247.5	-230.1	-236.5	34.3	3
300	-247.3	-228.5	-233.1	40.4	3

UO2++ (AQ)

T(C)	DELTA H	DELTA F	ENTHALPY	ENTROPY	REF
25	-250.4	-236.4	-250.4	-27.0	5
50	-250.5	-235.2	-249.2	-23.3	3
75	-250.7	-234.0	-247.7	-18.7	3
100	-251.0	-232.8	-245.8	-13.4	3
150	-251.7	-230.3	-242.6	-5.2	3
200	-252.0	-227.8	-238.6	4.1	3
250	-253.0	-225.2	-234.2	12.9	3
300	-253.6	-222.5	-229.3	21.8	3

UO3 (C), URANIUM TRIOXIDE

T(C)	DELTA H	DELTA F	ENTHALPY	ENTROPY	REF
25	-294.0	-275.5	-294.0	23.6	2
50	-293.9	-274.0	-293.5	25.2	2
75	-293.9	-272.5	-293.0	26.7	2
100	-293.8	-270.9	-292.5	28.1	2
150	-293.6	-267.9	-291.4	30.8	2
200	-293.4	-264.8	-290.3	33.2	2
250	-293.3	-261.8	-289.2	35.5	2
300	-293.1	-258.8	-288.1	37.5	2

UOH+++ (AQ)

T(C)	DELTA H	DELTA F	ENTHALPY	ENTROPY	REF
25	-204.0	-193.5	-204.0	-45.0	5
50	-204.5	-192.6	-202.7	-40.8	3
75	-205.1	-191.6	-200.9	-35.3	3
100	-206.0	-190.6	-198.7	-29.1	3
150	-207.6	-188.4	-194.9	-19.4	3
200	-209.2	-186.0	-190.3	-8.7	3
250	-211.8	-183.5	-185.2	1.5	3
300	-213.9	-180.6	-179.5	11.9	3

V (C), VANADIUM

T(C)	DELTA H	DELTA F	ENTHALPY	ENTROPY	REF
25	0.0	0.0	0.0	6.9	1
50	0.0	0.0	0.2	7.4	2
75	0.0	0.0	0.3	7.8	2
100	0.0	0.0	0.4	8.2	2
150	0.0	0.0	0.8	9.0	2
200	0.0	0.0	1.1	9.7	2
250	0.0	0.0	1.4	10.3	2
300	0.0	0.0	1.7	10.9	2

V2O3 (C), VANADIUM TRIOXIDE

T(C)	DELTA H	DELTA F	ENTHALPY	ENTROPY	REF
25	-293.5	-274.5	-293.5	23.5	1
50	-293.4	-272.9	-292.9	25.5	2
75	-293.3	-271.3	-292.2	27.5	2
100	-293.2	-269.7	-291.5	29.3	2
150	-293.0	-266.6	-290.2	32.8	2
200	-292.7	-263.5	-288.7	36.1	2
250	-292.5	-260.4	-287.2	39.0	2
300	-292.1	-257.3	-285.7	41.8	2

V2O5 (C), VANADIUM PENTOXIDE

T(C)	DELTA H	DELTA F	ENTHALPY	ENTROPY	REF
25	-370.6	-339.3	-370.6	31.3	1
50	-370.5	-336.7	-369.8	33.8	2
75	-370.5	-334.0	-369.0	36.3	2
100	-370.3	-331.4	-368.1	38.8	2
150	-370.0	-326.2	-366.3	43.4	2
200	-369.7	-321.1	-364.4	47.6	2
250	-369.3	-316.0	-362.4	51.6	2
300	-368.8	-310.9	-360.4	55.2	2

VCL2 (C), VANADIUM DICHLORIDE

T(C)	DELTA H	DELTA F	ENTHALPY	ENTROPY	REF
25	-108.0	-97.0	-108.0	23.2	1
50	-107.9	-96.1	-107.6	24.6	2
75	-107.8	-95.1	-107.1	25.9	2
100	-107.8	-94.2	-106.7	27.1	2
150	-107.6	-92.4	-105.8	29.4	2
200	-107.4	-90.6	-104.9	31.4	2
250	-107.2	-88.9	-104.0	33.2	2
300	-107.1	-87.1	-103.0	34.9	2

VCL3 (C), VANADIUM TRICHLORIDE

T(C)	DELTA H	DELTA F	ENTHALPY	ENTROPY	REF
25	-138.8	-122.2	-138.8	31.3	1
50	-138.7	-120.9	-138.2	33.1	2
75	-138.6	-119.5	-137.7	34.8	2
100	-138.5	-118.1	-137.1	36.4	2
150	-138.2	-115.4	-135.9	39.4	2
200	-138.0	-112.7	-134.7	42.0	2
250	-137.7	-110.1	-133.5	44.5	2
300	-137.5	-107.4	-132.3	46.7	2

VCL4 (L,G), VANADIUM TETRACHLORIDE
BOILING PT = 152C

T(C)	DELTA H	DELTA F	ENTHALPY	ENTROPY	REF
25	-136.1	-120.4	-136.1	61.0	1
50	-135.7	-119.2	-135.1	64.1	2
75	-135.3	-117.9	-134.2	67.0	2
100	-134.9	-116.7	-133.2	69.7	2
150	-134.1	-114.3	-131.3	74.5	2
152	-134.1	-114.2	-131.2	74.7	2
152	-125.0	-114.2	-122.1	96.1	2
200	-125.0	-113.0	-121.0	98.6	2
250	-124.9	-111.7	-119.8	101.0	2
300	-124.9	-110.4	-118.6	103.1	2

VO (C), VANADIUM MONOXIDE

T(C)	DELTA H	DELTA F	ENTHALPY	ENTROPY	REF
25	-103.2	-96.6	-103.2	9.3	1
50	-103.2	-96.1	-102.9	10.2	2
75	-103.1	-95.5	-102.6	11.0	2
100	-103.1	-95.0	-102.4	11.8	2
150	-103.0	-93.9	-101.8	13.3	2
200	-102.9	-92.8	-101.2	14.7	2
250	-102.7	-91.8	-100.5	15.9	2
300	-102.6	-90.7	-99.9	17.1	2

VO++ (AQ)

T(C)	DELTA H	DELTA F	ENTHALPY	ENTROPY	REF
25	-116.3	-106.7	-116.3	-42.0	1
50	-116.2	-105.9	-115.0	-37.9	3
75	-116.1	-105.1	-113.3	-32.6	3
100	-116.0	-104.3	-111.1	-26.5	3
150	-115.9	-102.7	-107.4	-17.1	3
200	-115.6	-101.3	-102.9	-6.6	3
250	-115.7	-99.7	-97.9	3.4	3
300	-115.4	-98.2	-92.3	13.6	3

VO2 (C), VANADIUM DIOXIDE
TRANSITION PT FOR VO2 AT 72C

T(C)	DELTA H	DELTA F	ENTHALPY	ENTROPY	REF
25	-171.5	-158.5	-171.5	12.3	2
50	-171.5	-157.4	-171.1	13.5	2
72	-171.4	-156.5	-170.8	14.5	2
72	-170.4	-156.5	-169.8	17.5	2
75	-170.4	-156.3	-169.7	17.6	2
100	-170.3	-155.3	-169.3	18.7	2
150	-170.2	-153.3	-168.5	20.7	2
200	-170.0	-151.3	-167.7	22.5	2
250	-169.9	-149.4	-166.9	24.3	2
300	-169.7	-147.4	-166.0	25.9	2

VO2+ (AQ)

T(C)	DELTA H	DELTA F	ENTHALPY	ENTROPY	REF
25	-155.3	-140.3	-155.3	-15.1	1
50	-155.1	-139.0	-154.3	-11.8	3
75	-154.8	-137.8	-152.9	-7.7	3
100	-154.3	-136.6	-151.3	-2.9	3
150	-153.8	-134.3	-148.4	4.2	3
200	-152.7	-132.1	-144.9	12.6	3
250	-151.8	-130.0	-141.0	20.4	3
300	-150.5	-128.0	-136.6	28.3	3

VO4- (AQ)

T(C)	DELTA H	DELTA F	ENTHALPY	ENTROPY	REF
25	-210.9	-203.9	-210.9	102.1	5
50	-209.1	-203.4	-209.1	107.9	3
75	-207.0	-203.0	-207.1	113.7	3
100	-204.5	-202.9	-205.0	119.6	3
150	-201.3	-202.7	-202.4	125.8	3
200	-194.9	-203.8	-196.9	139.1	3
250	-188.1	-205.0	-191.3	150.3	3
300	-180.9	-207.0	-185.3	161.2	3

VOSCN+ (AQ)

T(C)	DELTA H	DELTA F	ENTHALPY	ENTROPY	REF
25	-113.2	-95.5	-113.2	3.0	1
50	-113.3	-94.0	-112.3	5.7	3
75	-113.3	-92.5	-111.2	9.1	3
100	-113.4	-91.0	-109.9	12.9	3
150	-113.7	-88.0	-107.6	18.6	3
200	-113.6	-85.0	-104.7	25.4	3
250	-113.9	-81.9	-101.5	31.8	3
300	-113.8	-78.9	-98.0	38.2	3

W (C), TUNGSTEN

T(C)	DELTA H	DELTA F	ENTHALPY	ENTROPY	REF
25	0.0	0.0	0.0	7.8	1
50	0.0	0.0	0.2	8.3	2
75	0.0	0.0	0.3	8.7	2
100	0.0	0.0	0.4	9.1	2
150	0.0	0.0	0.7	9.9	2
200	0.0	0.0	1.0	10.5	2
250	0.0	0.0	1.3	11.1	2
300	0.0	0.0	1.6	11.7	2

WBR5 (C,L) TUNGSTEN PENTABROMIDE
MELTING PT = 286C

T(C)	DELTA H	DELTA F	ENTHALPY	ENTROPY	REF
25	-74.5	-64.4	-74.5	65.0	2
50	-74.8	-63.6	-73.6	68.0	2
58	-74.9	-63.3	-73.3	69.0	2
58	-92.6	-63.3	-73.3	69.0	2
75	-92.4	-61.9	-72.6	70.9	2
100	-92.2	-59.7	-71.6	73.5	2
150	-91.6	-55.4	-69.7	78.5	2
200	-91.0	-51.1	-67.6	83.1	2
250	-90.3	-47.0	-65.5	87.3	2
286	-89.7	-44.0	-64.0	90.1	2
286	-85.6	-44.0	-59.9	97.4	2
300	-85.4	-43.0	-59.3	98.5	2

WBR6 (C), TUNGSTEN HEXABROMIDE

T(C)	DELTA H	DELTA F	ENTHALPY	ENTROPY	REF
25	-82.0	-69.5	-82.0	75.0	2
50	-82.3	-68.4	-80.9	78.5	2
58	-82.4	-68.1	-80.5	79.6	2
58	-103.8	-68.1	-80.5	79.6	2
75	-103.6	-66.4	-79.8	81.8	2
100	-103.2	-63.7	-78.7	84.9	2
150	-102.5	-58.4	-76.4	90.7	2
200	-101.8	-53.3	-74.0	96.0	2
250	-101.0	-48.2	-71.6	100.9	2
300	-100.1	-43.2	-69.1	105.4	2

WCL2 (C), TUNGSTEN DICHLORIDE

T(C)	DELTA H	DELTA F	ENTHALPY	ENTROPY	REF
25	-61.5	-52.6	-61.5	31.2	2
50	-61.4	-51.8	-61.0	32.7	2
75	-61.3	-51.1	-60.6	34.1	2
100	-61.2	-50.4	-60.1	35.4	2
150	-60.9	-49.0	-59.1	37.8	2
200	-60.7	-47.6	-58.2	40.0	2
250	-60.4	-46.2	-57.2	42.0	2
300	-60.2	-44.8	-56.2	43.8	2

WCL4 (C), TUNGSTEN TETRACHLORIDE

T(C)	DELTA H	DELTA F	ENTHALPY	ENTROPY	REF
25	-105.9	-54.2	-105.9	47.4	2
50	-106.1	-49.8	-105.1	49.9	2
75	-106.3	-45.4	-104.3	52.3	2
100	-106.4	-41.1	-103.5	54.5	2
150	-106.8	-32.3	-101.9	58.5	2
200	-107.2	-23.5	-100.3	62.2	2
250	-107.5	-14.6	-98.6	65.6	2
300	-107.8	-5.7	-96.9	68.8	2

WCL5 (C,L), TUNGSTEN PENTACHLORIDE
MELTING PT = 253C

T(C)	DELTA H	DELTA F	ENTHALPY	ENTROPY	REF
25	-122.6	-96.1	-122.6	52.0	2
50	-122.3	-93.8	-121.7	55.0	2
75	-122.0	-91.6	-120.7	57.9	2
100	-121.7	-89.5	-119.7	60.6	2
150	-121.1	-85.2	-117.7	65.6	2
200	-120.4	-81.0	-115.7	70.2	2
250	-119.6	-76.9	-113.5	74.5	2
253	-119.6	-76.7	-113.4	74.7	2
253	-114.6	-76.7	-108.5	84.1	2
300	-113.9	-73.3	-106.4	87.8	2

WCL6 (C,L), TUNGSTEN HEXACHLORIDE
TRANSITION PTS AT 177C AND 230C; MELT PT = 282C

T(C)	DELTA H	DELTA F	ENTHALPY	ENTROPY	REF
25	-141.9	-108.9	-141.9	57.0	2
50	-141.6	-106.1	-140.8	60.4	2
75	-141.3	-103.4	-139.8	63.7	2
100	-140.9	-100.7	-138.6	66.7	2
150	-140.2	-95.4	-136.4	72.5	2
177	-139.8	-92.5	-135.1	75.4	2
177	-138.8	-92.5	-134.1	77.6	2
200	-138.3	-90.2	-132.9	80.1	2
230	-137.8	-87.1	-131.4	83.2	2
230	-134.0	-87.1	-127.6	90.7	2
250	-133.8	-85.3	-126.8	92.5	2
282	-133.3	-82.3	-125.3	95.1	2
282	-131.7	-82.3	-123.7	98.0	2
300	-131.5	-80.7	-122.8	99.6	2

WO2 (C), TUNGSTEN DIOXIDE

T(C)	DELTA H	DELTA F	ENTHALPY	ENTROPY	REF
25	-140.9	-127.6	-140.9	12.1	1
50	-140.9	-126.5	-140.6	13.2	2
75	-140.9	-125.4	-140.2	14.2	2
100	-140.9	-124.3	-139.9	15.3	2
150	-140.7	-122.0	-139.1	17.2	2
200	-140.6	-119.8	-138.3	19.0	2
250	-140.5	-117.6	-137.5	20.6	2
300	-140.3	-115.5	-136.6	22.2	2

WO3 (C), TUNGSTEN TRIOXIDE

T(C)	DELTA H	DELTA F	ENTHALPY	ENTROPY	REF
25	-201.5	-182.6	-201.5	18.1	1
50	-201.4	-181.0	-201.0	19.6	2
75	-201.4	-179.5	-200.5	21.0	2
100	-201.3	-177.9	-200.1	22.3	2
150	-201.2	-174.8	-199.1	24.8	2
200	-201.0	-171.7	-198.0	27.1	2
250	-200.8	-168.6	-197.0	29.2	2
300	-200.6	-165.5	-195.9	31.2	2

WO4-- (AQ)

T(C)	DELTA H	DELTA F	ENTHALPY	ENTROPY	REF
25	-266.6	-220.0	-266.6	-9.3	5
50	-269.7	-216.0	-270.2	-20.7	3
75	-272.8	-211.7	-274.1	-32.6	3
100	-275.8	-207.2	-278.4	-44.7	3
150	-280.6	-197.9	-285.4	-62.1	3
200	-289.2	-187.3	-295.6	-85.7	3
250	-295.7	-176.3	-306.7	-108.0	3
300	-304.2	-164.4	-318.9	-130.3	3

Y (C), YTTRIUM

T(C)	DELTA H	DELTA F	ENTHALPY	ENTROPY	REF
25	0.0	0.0	0.0	10.6	1
50	0.0	0.0	0.2	11.1	2
75	0.0	0.0	0.3	11.6	2
100	0.0	0.0	0.5	12.1	2
150	0.0	0.0	0.8	12.9	2
200	0.0	0.0	1.1	13.6	2
250	0.0	0.0	1.5	14.3	2
300	0.0	0.0	1.8	14.9	2

Y+++ (AQ)

T(C)	DELTA H	DELTA F	ENTHALPY	ENTROPY	REF
25	-172.9	-165.8	-172.9	-75.0	1
50	-172.9	-165.2	-171.3	-69.8	3
75	-172.9	-164.6	-169.1	-63.1	3
100	-173.1	-164.0	-166.4	-55.4	3
150	-173.4	-162.7	-161.5	-43.2	3
200	-173.5	-161.5	-155.9	-30.0	3
250	-174.5	-160.2	-149.5	-17.3	3
300	-174.8	-158.8	-142.5	-4.5	3

YB (G), YTTERBIUM (GAS)

T(C)	DELTA H	DELTA F	ENTHALPY	ENTROPY	REF
25	0.0	0.0	0.0	41.3	1
50	0.0	0.0	0.1	41.8	2
75	0.0	0.0	0.3	42.1	2
100	0.0	0.0	0.4	42.5	2
150	0.0	0.0	0.6	43.1	2
200	0.0	0.0	0.9	43.7	2
250	0.0	0.0	1.1	44.2	2
300	0.0	0.0	1.4	44.6	2

YB+++ (AQ)

T(C)	DELTA H	DELTA F	ENTHALPY	ENTROPY	REF
25	-197.6	-182.2	-197.6	-72.0	1
50	-197.5	-180.9	-196.0	-66.9	3
75	-197.6	-179.6	-193.8	-60.3	3
100	-197.8	-178.3	-191.1	-52.8	3
150	-198.1	-175.7	-186.4	-40.8	3
200	-198.2	-173.0	-180.9	-27.9	3
250	-199.2	-170.3	-174.7	-15.5	3
300	-199.6	-167.5	-167.8	-2.9	3

ZN (C), ZINC

T(C)	DELTA H	DELTA F	ENTHALPY	ENTROPY	REF
25	0.0	0.0	0.0	10.0	1
50	0.0	0.0	0.2	10.4	2
75	0.0	0.0	0.3	10.9	2
100	0.0	0.0	0.5	11.3	2
150	0.0	0.0	0.8	12.1	2
200	0.0	0.0	1.1	12.8	2
250	0.0	0.0	1.4	13.5	2
300	0.0	0.0	1.8	14.1	2

ZN(CN)4-- (AQ)
(EXTRAPOLATED AS CRISS & COBBLE TYPE 3)

T(C)	DELTA H	DELTA F	ENTHALPY	ENTROPY	REF
25	82.0	100.4	82.0	86.5	5
50	83.3	101.9	83.1	89.9	3
75	85.1	103.2	84.2	93.3	3
100	87.3	104.4	85.4	96.6	3
150	90.2	106.6	86.6	99.5	3
200	95.6	107.8	90.0	107.7	3
250	101.7	108.8	93.2	114.1	3
300	108.0	109.2	96.7	120.4	3

ZN(NH3)4++ (AQ)

T(C)	DELTA H	DELTA F	ENTHALPY	ENTROPY	REF
25	-127.5	-72.2	-127.5	62.0	1
50	-129.7	-67.5	-127.3	62.8	3
75	-132.4	-62.5	-127.0	63.7	3
100	-135.4	-57.4	-126.6	64.6	3
150	-141.4	-46.4	-126.3	65.3	3
200	-147.3	-34.8	-125.5	67.4	3
250	-154.3	-22.6	-124.7	69.0	3
300	-161.2	-9.7	-123.8	70.6	3

ZN++ (AQ)

T(C)	DELTA H	DELTA F	ENTHALPY	ENTROPY	REF
25	-36.8	-35.1	-36.8	-36.8	1
50	-36.7	-35.0	-35.6	-32.8	3
75	-36.5	-34.9	-33.9	-27.8	3
100	-36.4	-34.8	-31.8	-21.9	3
150	-36.4	-34.5	-28.3	-13.0	3
200	-36.0	-34.4	-23.9	-2.9	3
250	-36.2	-34.2	-19.2	6.7	3
300	-35.9	-34.0	-13.8	16.4	3

ZNBR2 (C), ZINC BROMIDE

T(C)	DELTA H	DELTA F	ENTHALPY	ENTROPY	REF
25	-78.6	-74.6	-78.6	33.1	1
50	-78.7	-74.3	-78.1	34.4	2
58	-78.8	-74.1	-78.0	34.8	2
58	-85.9	-74.1	-78.0	34.8	2
75	-85.9	-73.6	-77.8	35.6	2
100	-85.8	-72.7	-77.3	36.7	2
150	-85.8	-70.9	-76.5	38.8	2
200	-85.7	-69.2	-75.6	40.7	2
250	-85.5	-67.5	-74.8	42.5	2
300	-85.4	-65.8	-73.8	44.2	2

ZNCL2 (C), ZINC CHLORIDE

T(C)	DELTA H	DELTA F	ENTHALPY	ENTROPY	REF
25	-99.2	-88.3	-99.2	26.6	1
50	-99.1	-87.4	-98.8	28.0	2
75	-99.1	-86.5	-98.4	29.2	2
100	-99.1	-85.6	-98.0	30.3	2
150	-98.9	-83.8	-97.1	32.4	2
200	-98.9	-82.0	-96.3	34.3	2
250	-98.8	-80.2	-95.4	36.0	2
300	-98.6	-78.4	-94.6	37.6	2

ZNCO3 (C), ZINC CARBONATE

T(C)	DELTA H	DELTA F	ENTHALPY	ENTROPY	REF
25	-194.3	-174.8	-194.3	19.7	1
50	-194.2	-173.2	-193.8	21.3	2
75	-194.2	-171.6	-193.3	22.8	2
100	-194.2	-170.0	-192.7	24.3	2
150	-194.1	-166.7	-191.6	27.1	2
200	-193.9	-163.5	-190.4	29.8	2

ZNF2 (C), ZINC FLUORIDE

T(C)	DELTA H	DELTA F	ENTHALPY	ENTROPY	REF
25	-182.7	-170.5	-182.7	17.6	1
50	-182.6	-169.5	-182.3	18.9	2
75	-182.6	-168.5	-181.9	20.0	2
100	-182.6	-167.5	-181.5	21.2	2
150	-182.5	-165.5	-180.7	23.2	2
200	-182.4	-163.5	-179.9	25.0	2
250	-182.3	-161.5	-179.1	26.6	2
300	-182.2	-159.5	-178.3	28.1	2

ZNI2 (C), ZINC IODIDE

T(C)	DELTA H	DELTA F	ENTHALPY	ENTROPY	REF
25	-49.7	-50.0	-49.7	38.5	1
50	-49.8	-50.0	-49.3	39.8	2
75	-49.9	-50.0	-48.9	41.0	2
100	-50.0	-50.0	-48.5	42.1	2
114	-50.1	-50.0	-48.3	42.7	2
114	-53.8	-50.0	-48.3	42.7	2
150	-54.1	-49.6	-47.7	44.2	2
185	-54.4	-49.2	-47.1	45.6	2
185	-64.4	-49.2	-47.1	45.6	2
200	-64.4	-48.7	-46.8	46.1	2
250	-64.3	-47.1	-45.9	47.9	2
300	-64.1	-45.4	-45.0	49.6	2

ZNO (C), ZINC OXIDE

T(C)	DELTA H	DELTA F	ENTHALPY	ENTROPY	REF
25	-83.2	-76.1	-83.2	10.4	1
50	-83.2	-75.5	-83.0	11.2	2
75	-83.2	-74.9	-82.7	12.0	2
100	-83.2	-74.3	-82.5	12.7	2
150	-83.2	-73.1	-81.9	14.1	2
200	-83.1	-71.9	-81.4	15.3	2
250	-83.1	-70.7	-80.8	16.5	2
300	-83.0	-69.5	-80.2	17.5	2

ZNS (C), ZINC SULFIDE

T(C)	DELTA H	DELTA F	ENTHALPY	ENTROPY	REF
25	-64.6	-57.6	-64.6	13.8	1
50	-64.5	-57.0	-64.3	14.7	2
75	-64.5	-56.4	-64.0	15.5	2
100	-64.5	-55.8	-63.7	16.3	2
150	-64.4	-54.7	-63.1	17.8	2
200	-64.3	-53.6	-62.5	19.2	2
250	-64.3	-52.4	-61.9	20.4	2
300	-64.2	-51.3	-61.3	21.5	2

ZNSO4 (C), ZINC SULFATE

T(C)	DELTA H	DELTA F	ENTHALPY	ENTROPY	REF
25	-250.2	-218.5	-250.2	28.6	1
50	-250.1	-215.8	-249.5	30.8	2
75	-250.1	-213.1	-248.9	32.9	2
100	-250.0	-210.5	-248.1	34.9	2
150	-249.8	-205.2	-246.7	38.5	2
200	-249.5	-200.0	-245.2	41.9	2
250	-249.3	-194.7	-243.6	45.0	2
300	-249.0	-189.5	-242.0	47.9	2

ZNSO4 (AQ), UNDISSOCIATED COMPLEX

T(C)	DELTA H	DELTA F	ENTHALPY	ENTROPY	REF
25	-265.7	-225.8	-265.7	1.2	1
50	-266.6	-222.4	-266.0	0.3	6
75	-267.1	-219.0	-265.9	0.6	6
100	-267.3	-215.5	-265.5	2.0	6
150	-266.1	-208.7	-263.2	8.2	6
200	-263.3	-201.9	-259.8	17.0	6

ZNSO4.H2O (C), ZINC SULFATE MONOHYDRATE

T(C)	DELTA H	DELTA F	ENTHALPY	ENTROPY	REF
25	-327.1	-280.1	-327.1	33.1	1
50	-327.1	-276.1	-326.3	35.9	2
75	-327.1	-272.2	-325.4	38.6	2
100	-327.1	-268.2	-324.5	41.1	2
150	-327.0	-260.3	-322.6	45.7	2
200	-326.9	-252.5	-320.7	49.9	2
250	-326.8	-244.6	-318.8	53.8	2
300	-326.6	-236.8	-316.8	57.4	2

ZNSO4.6H2O (C), ZINC SULFATE HEXAHYDRATE

T(C)	DELTA H	DELTA F	ENTHALPY	ENTROPY	REF
25	-679.2	-565.1	-679.2	86.9	1
50	-679.2	-555.5	-677.0	93.8	2
75	-679.2	-546.0	-674.9	100.3	2
100	-679.2	-536.4	-672.7	106.3	2
150	-679.2	-517.3	-668.3	117.4	2
200	-679.1	-498.1	-663.7	127.5	2
250	-679.0	-479.0	-659.1	136.8	2
300	-678.9	-459.9	-654.4	145.3	2

ZNSO4.7H2O (C), ZINC SULFATE HEPTAHYDRATE

T(C)	DELTA H	DELTA F	ENTHALPY	ENTROPY	REF
25	-750.9	-622.1	-750.9	92.9	1
50	-751.1	-611.2	-748.7	100.3	2
75	-751.2	-600.4	-746.3	107.1	2
100	-751.3	-589.6	-744.0	113.6	2
150	-751.6	-567.9	-739.3	125.5	2
200	-751.7	-546.2	-734.5	136.2	2
250	-751.9	-524.5	-729.6	146.1	2
300	-751.9	-502.7	-724.6	155.2	2

ZR (C), ZIRCONIUM

T(C)	DELTA H	DELTA F	ENTHALPY	ENTROPY	REF
25	0.0	0.0	0.0	9.3	1
50	0.0	0.0	0.2	9.8	2
75	0.0	0.0	0.3	10.3	2
100	0.0	0.0	0.5	10.7	2
150	0.0	0.0	0.8	11.5	2
200	0.0	0.0	1.1	12.2	2
250	0.0	0.0	1.4	12.9	2
300	0.0	0.0	1.8	13.5	2

ZRBR2 (C), ZIRCONIUM DIBROMIDE

T(C)	DELTA H	DELTA F	ENTHALPY	ENTROPY	REF
25	-100.0	-95.8	-100.0	31.5	2
50	-100.1	-95.4	-99.5	33.2	2
58	-100.1	-95.3	-99.3	33.7	2
58	-107.2	-95.3	-99.3	33.7	2
75	-107.1	-94.7	-99.0	34.7	2
100	-106.9	-93.8	-98.4	36.1	2
150	-106.7	-92.1	-97.4	38.8	2
200	-106.4	-90.4	-96.4	41.1	2
250	-106.1	-88.7	-95.3	43.2	2
300	-105.8	-87.1	-94.2	45.2	2

ZRBR3 (C), ZIRCONIUM TRIBROMIDE

T(C)	DELTA H	DELTA F	ENTHALPY	ENTROPY	REF
25	-152.0	-145.2	-152.0	41.1	2
50	-152.2	-144.6	-151.4	43.0	2
58	-152.3	-144.4	-151.2	43.7	2
58	-162.9	-144.4	-151.2	43.7	2
75	-162.8	-143.5	-150.8	44.8	2
100	-162.7	-142.2	-150.2	46.5	2
150	-162.5	-139.4	-149.0	49.6	2
200	-162.2	-136.7	-147.7	52.3	2
250	-162.0	-134.0	-146.5	54.8	2
300	-161.7	-131.4	-145.3	57.1	2

ZRBR4 (C), ZIRCONIUM TETRABROMIDE

T(C)	DELTA H	DELTA F	ENTHALPY	ENTROPY	REF
25	-181.6	-173.1	-181.6	53.7	2
50	-181.9	-172.4	-180.9	56.1	2
58	-181.9	-172.2	-180.6	56.9	2
58	-196.1	-172.2	-180.6	56.9	2
75	-196.0	-171.0	-180.1	58.4	2
100	-195.9	-169.2	-179.3	60.5	2
150	-195.5	-165.7	-177.8	64.4	2
200	-195.2	-162.2	-176.2	67.9	2
250	-194.8	-158.7	-174.6	71.1	2
300	-194.4	-155.3	-173.0	73.9	2

ZRCL2 (C), ZIRCONIUM DICHLORIDE

T(C)	DELTA H	DELTA F	ENTHALPY	ENTROPY	REF
25	-132.0	-121.4	-132.0	27.0	2
50	-131.9	-120.5	-131.5	28.5	2
75	-131.8	-119.6	-131.1	29.9	2
100	-131.7	-118.8	-130.6	31.2	2
150	-131.5	-117.0	-129.6	33.6	2
200	-131.3	-115.3	-128.7	35.7	2
250	-131.0	-113.7	-127.7	37.7	2
300	-130.8	-112.0	-126.7	39.5	2

ZRCL3 (C), ZIRCONIUM TRICHLORIDE

T(C)	DELTA H	DELTA F	ENTHALPY	ENTROPY	REF
25	-206.0	-189.7	-206.0	31.3	2
50	-205.9	-187.3	-205.4	33.2	2
75	-205.8	-185.8	-204.8	34.9	2
100	-205.6	-184.4	-204.3	36.5	2
150	-205.4	-181.6	-203.0	39.6	2
200	-205.1	-178.8	-201.8	42.3	2
250	-204.8	-176.0	-200.5	44.8	2
300	-204.5	-173.3	-199.3	47.2	2

ZRCL4 (C), ZIRCONIUM TETRACHLORIDE

T(C)	DELTA H	DELTA F	ENTHALPY	ENTROPY	REF
25	-234.4	-212.7	-234.4	43.4	1
50	-234.2	-210.9	-233.6	45.7	2
75	-234.0	-209.1	-232.9	47.9	2
100	-233.9	-207.4	-232.1	50.0	2
150	-233.5	-203.8	-230.6	53.8	2
200	-233.2	-200.3	-229.1	57.2	2
250	-232.8	-196.9	-227.6	60.2	2
300	-232.5	-193.5	-226.0	63.1	2

ZRF2 (C), ZIRCONIUM DIFLUORIDE

T(C)	DELTA H	DELTA F	ENTHALPY	ENTROPY	REF
25	-230.0	-219.0	-230.0	21.0	2
50	-229.8	-218.1	-229.4	22.8	2
75	-229.5	-217.2	-228.9	24.6	2
100	-229.3	-216.4	-228.3	26.2	2
150	-228.8	-214.7	-227.1	29.2	2
200	-228.3	-213.0	-225.9	31.9	2
250	-227.8	-211.4	-224.6	34.4	2
300	-227.3	-209.9	-223.3	36.8	2

ZRF3 (C), ZIRCONIUM TRIFLUORIDE

T(C)	DELTA H	DELTA F	ENTHALPY	ENTROPY	REF
25	-361.0	-341.4	-361.0	16.3	2
50	-361.0	-339.8	-360.5	17.9	2
75	-360.9	-338.2	-360.0	19.4	2
100	-360.8	-336.5	-359.5	20.8	2
150	-360.7	-333.3	-358.5	23.4	2
200	-360.5	-330.0	-357.4	25.9	2
250	-360.4	-326.8	-356.2	28.1	2
300	-360.2	-323.6	-355.1	30.3	2

ZRF4 (C), ZIRCONIUM TETRAFLUORIDE

T(C)	DELTA H	DELTA F	ENTHALPY	ENTROPY	REF
25	-456.8	-432.6	-456.8	25.0	1
50	-456.7	-430.6	-456.2	27.0	2
75	-456.6	-428.5	-455.5	29.0	2
100	-456.5	-426.5	-454.9	30.8	2
150	-456.2	-422.5	-453.5	34.2	2
200	-456.0	-418.6	-452.1	37.3	2
250	-455.7	-414.6	-450.7	40.2	2
300	-455.4	-410.7	-449.2	42.8	2

ZRI2 (C), ZIRCONIUM DIIODIDE

T(C)	DELTA H	DELTA F	ENTHALPY	ENTROPY	REF
25	-62.0	-61.7	-62.0	35.9	2
50	-61.9	-61.6	-61.4	37.7	2
75	-61.8	-61.6	-60.9	39.4	2
100	-61.8	-61.6	-60.3	41.0	2
114	-61.8	-61.6	-60.0	41.8	2
114	-65.5	-61.6	-60.0	41.8	2
150	-65.6	-61.2	-59.2	43.8	2
185	-65.7	-60.8	-58.4	45.6	2
185	-75.7	-60.8	-58.4	45.6	2
200	-75.6	-60.4	-58.0	46.4	2
250	-75.3	-58.8	-56.9	48.7	2
300	-74.9	-57.2	-55.7	50.8	2

ZRI3 (C), ZIRCONIUM TRIIODIDE

T(C)	DELTA H	DELTA F	ENTHALPY	ENTROPY	REF
25	-95.0	-94.4	-95.0	48.9	2
50	-95.0	-94.3	-94.4	50.9	2
75	-95.1	-94.3	-93.8	52.8	2
100	-95.1	-94.2	-93.1	54.5	2
114	-95.2	-94.2	-92.8	55.4	2
114	-100.7	-94.2	-92.8	55.4	2
150	-101.1	-93.6	-91.9	57.7	2
185	-101.5	-92.9	-91.0	59.7	2
185	-116.5	-92.9	-91.0	59.7	2
200	-116.4	-92.1	-90.6	60.5	2
250	-116.1	-89.6	-89.3	63.1	2
300	-115.9	-87.1	-88.1	65.4	2

ZRI4 (C), ZIRCONIUM TETRAIODIDE

T(C)	DELTA H	DELTA F	ENTHALPY	ENTROPY	REF
25	-115.9	-114.9	-115.9	61.4	2
50	-116.0	-114.8	-115.2	63.8	2
75	-116.1	-114.7	-114.4	66.0	2
100	-116.2	-114.6	-113.7	68.1	2
114	-116.3	-114.5	-113.3	69.2	2
114	-123.7	-114.5	-113.3	69.2	2
150	-124.2	-113.6	-112.1	71.9	2
185	-124.7	-112.8	-111.1	74.3	2
185	-144.8	-112.8	-111.1	74.3	2
200	-144.7	-111.7	-110.6	75.3	2
250	-144.4	-108.2	-109.1	78.4	2
300	-144.1	-104.8	-107.6	81.2	2

ZRO++ (AQ)

T(C)	DELTA H	DELTA F	ENTHALPY	ENTROPY	REF
25	-223.1	-200.9	-223.1	-81.9	5
50	-222.6	-199.1	-221.4	-76.4	3
75	-221.9	-197.3	-219.1	-69.4	3
100	-221.2	-195.5	-216.3	-61.4	3
150	-219.8	-192.2	-211.2	-48.6	3
200	-218.0	-189.1	-205.3	-34.9	3
250	-216.6	-186.1	-198.7	-21.7	3
300	-214.5	-183.3	-191.4	-8.3	3

ZRO2 (C), ZIRCONIUM DIOXIDE

T(C)	DELTA H	DELTA F	ENTHALPY	ENTROPY	REF
25	-263.0	-249.2	-263.0	12.0	1
50	-263.0	-248.1	-262.7	13.1	2
75	-263.0	-246.9	-262.3	14.2	2
100	-263.0	-245.8	-262.0	15.2	2
150	-262.9	-243.5	-261.2	17.1	2
200	-262.8	-241.2	-260.4	18.9	2
250	-262.7	-238.9	-259.6	20.5	2
300	-262.6	-236.6	-258.8	22.0	2

ZRSIO4 (C), ZIRCONIUM ORTHOSILICATE

T(C)	DELTA H	DELTA F	ENTHALPY	ENTROPY	REF
25	-486.0	-458.7	-486.0	20.1	1
50	-486.0	-456.4	-485.4	22.0	2
75	-486.0	-454.1	-484.8	23.9	2
100	-486.0	-451.8	-484.1	25.8	2
150	-485.9	-447.2	-482.7	29.3	2
200	-485.8	-442.6	-481.3	32.5	2
250	-485.7	-438.0	-479.7	35.6	2
300	-485.5	-433.5	-478.2	38.4	2

PREDICTION OF DATA FOR ADDITIONAL SUBSTANCES

Occasionally, those who use this compilation will need to generate thermo-dynamic data for substances not covered in Chapter 3. Therefore, the computer programs used to develop the data are described in this chapter. Specific examples of input information and the resulting tables of generated data are also presented.

Computer programs have been written in Fortran IV for Digital Equipment "DEC" System 10 (KI). Listings of the programs and subroutines are reproduced in the appendix.

INPUT DATA REQUIREMENTS AND TEMPERATURE LIMITATIONS

These input data requirements, calculation procedures and temperature limitations apply:

Compounds

The calculations are based on standard and rigorous methods. Heat capacity and phase transition data must be supplied over the temperature range of interest by the user. The heat of formation and entropy at 25°C must also be provided. There is no upper temperature limit.

Ions in Aqueous Solution

Temperature extrapolations are made by the method of Criss and Cobble.[1] The user must provide the entropy and free energy of formation for the ion at 25°C. The upper temperature limit is 300°C.

Neutral Complexes in Aqueous Solution

Temperature extrapolations are made by the method of Helgeson.[2] The user must provide the entropy and free energy of formation for the complex and its dissociation products (ions). The upper temperature limit is 200°C.

A convenient source for the 25°C input data is the National Bureau of Standard series of Technical Notes 270.[3]

COMPOUNDS

Program IONCOM calculates the heat and free energy of formation and the enthalpy and entropy of elements and compounds as a function of temperature. This program requires subroutines DELHS, DELHS1, and NNCODE. The data file (FOR21.DAT) containing reference state data for the elements must be available, and a second data file (CONVR.DAT) containing input data on the compound of interest must be created by the user. Output data are stored in file FOR24.DAT.

Reference Data for Elements

Reference data for the elements are stored in data file FOR21.DAT (see appendix). This file contains enthalpy, entropy, and heat capacity data for the elements in their reference states. The heat capacity data are stored in the form of coefficients to one of two heat capacity equations:

$$c_p^o = A + B10^{-3}T + C10^5 T^{-2} + D10^{-6}T^2 \tag{4.1}$$

$$c_p^o = A + B10^{-3}T + C10^5 T^{-2} + D10^8 T^{-3} \tag{4.2}$$

Provisions are made to accommodate phase transitions for the elements and to designate alternate sets of coefficients for heat capacity expressions in different temperature ranges. Reference state data for 71 elements are included in the present FOR21.DAT data file. This file can be expanded by users to cover additional elements. Data are stored in this manner:

Sequence of Element Data Stored in FOR21.DAT

TWO	H	S	R	T		
A	B	C	D	NCP	K	
TWO	H	S	R	T		
A	B	C	D	NCP	K	
	etc.					

Each pair of lines represents one file location of data for a given temperature range.

where

TWO = location of data for next temperature range for c_p^o (0 if none).

H = enthalpy (kcal/mole) at low temperature end (normally 298°K) of lowest temperature range for c_p^o; ΔH of transition (kcal/mole) for successively higher temperature ranges.

S = entropy (cal/deg-mole) at low temperature end of lowest temperature range for c_p^o; ΔS of transition (cal/deg-mole) for successively higher ranges.

R = lowest temperature (°K) in present temperature range.

T = highest temperature (°K) in present temperature range.

A = A coefficient in heat capacity equation (c_p^o in cal/deg K-mole; temperature in °K).

B = B coefficient in heat capacity equation.

C = C coefficient in heat capacity equation.

D = D coefficient in heat capacity equation.

K = location of present data set (use element code numbers for lowest range; see subroutine NNCODE).

NCP = 0 for c_p^o Eq. (4.1); 1 for Eq. (4.2).

Input Data for Compound

Input data for the compound of interest are stored in CONVR.DAT. These data consist of the number of temperature points to be listed in the output table, the temperature and data reference for each point, the enthalpy of formation and entropy at 298.15°K, the coefficients for the heat capacity expressions, the temperature extremes applicable to each heat capacity expression, the enthalpy and entropy changes applicable at each phase transition (if any), and the number of molecules of elemental species making up the compound.

The structure of data for CONVR.DAT is best illustrated by example. Input data for iron sulfide are specified below.

Example File (CONVR.DAT) for FeS, Iron(2) Sulfide

Input	Comments
FES	Compound name or identification (A20).
10 0	Number of temperature points (two for each phase transition); code for heat capacity expression: 0 for Eq. (4.1), 1 for Eq. (4.2).
25	Lowest temperature (^oC; do not go below 25^oC).
50	Next temperature (^oC)
75	Etc
100	
137.85	Temperature of phase transition for FeS in ^oC (= ^oK -273.15) (program generates data for compound before transition).
137.85	Temperature of phase transition for FeS in ^oC (program also generates data for compound after transition).
150	Next temperature (^oC).
200	Etc
250	
300	
1	Code to designate reference for data at 25^oC.
2	Code to designate reference for c_p^o data at 50^oC.
2	Etc
2	
2	
2	
2	
2	
2	
2	
2	No. of expressions used to define c_p^o.

-22.8 16.1   Enthalpy of formation (kcal/mole) and entropy (cal/deg K-mole) at lowest temperature (normally 25°C). [Use reference state from National Bureau of Standards Technical Note Series 270; for sulfur and phosphorus, program converts to reference states S_2 (gas) and to P_4 (gas) at the boiling point of phosphorus.]

5.19 26.40 A, B, C, and D for heat capacity equation in lowest temperature range. (C_p^O in cal/deg K-mole, T in °K.)

298.15 411   Temperature range (°K) for above heat capacity expression.

0.57 1.387   Enthalpy (kcal/mole) and entropy (cal/deg K-mole) of transition for FeS.

17.4 A, B, C, and D for heat capacity equation in next temperature range. (C_p^O in cal/deg K-mole, T in °K.)

411 598   Temperature range (°K) for above heat capacity expression.

2   Number of elements in compound.

FE 1.0   Symbol for iron and number of molecules of Fe in FeS.

S 0.5   Symbol for sulfur and number of molecules of S_2 in FeS.

Note: Spaces between lines are for explanations, and do not appear in real file.

Program IONCOM

Program IONCOM may be executed once the input data files CONVR.DAT and FOR21.DAT have been placed in the user area. Subroutines DELHS, DELHS1, and NNCODE must also be available. A brief outline of the sequence of calculations is now given.

Initially the program sets up all arrays. The Digital Equipment "DEC" System 10 (KI) initializes all locations to zero. Data from CONVR.DAT and FOR21.DAT are input, and element names are changed to code numbers by subroutine NNCODE at statement 5 plus two.

The temperature array is established beginning at statement 6 plus one. The data for each elemental species forming the compound are scanned to

determine whether a phase transition occurs in the temperature range of
interest, in which case two identical temperature levels are inserted consecutively into the array. These temperatures represent phase transitions
in the elements and are included to aid in the interpolation of the output
data. The adjusted temperature scale is complete at statement 106 plus
one. Temperature and element counters are initialized at statement 106
plus two. Enthalpy and entropy accumulators are set to zero at 11 plus
one.

The enthalpy and entropy for the specified compound and its consti-
tuent elemental species are calculated at the designated temperatures and
at the phase transition temperatures. These quantities are calculated
according to these relationships:

$$H_T^o = \Delta H_{298}^o + \Sigma \int_{298}^T C_p^o(T)\,dT + \Sigma \Delta H_{T_t}^t \tag{4.3}$$

$$S_T^o = S_{298}^o + \Sigma \int_{298}^T (C_p^o(T)/T)\,dT + \Sigma \Delta S_{T_t}^t \tag{4.4}$$

where

H_T^o is the enthalpy (compound or elemental species) at 1 atm. pressure
and temperature T.

ΔH_{298}^o is the heat of formation of the substance from the elements in
their reference states at 298.15°K. (The heat of formation of the
elements in their reference state is zero by definition.)

$C_p^o(T)$ is the heat capacity of the substance at constant pressure.
The summations include all phases between 298.15 and T°K.

$\Delta H_{T_t}^t$ and $\Delta S_{T_t}^t$ are the heat and entropy of transition of the substance,
respectively, at temperature T_t. The summations include all transfor-
mations between 298.15 and T°K.

S_T^o is the entropy of the substance at 1 atm. pressure and temperature
T.

S_{298}^o is the entropy of the substance at 1 atm. pressure and 298.15°K.

Two expressions are used for the heat capacity. Subroutine DELHS is
based on Eq. (4.1), and DELHS1 is based on Eq. (4.2).

The enthalpy and entropy for each element species forming the compound are calculated at each temperature beginning at statement 8. Beginning at 115 plus two, similar calculations are made for the compound.

Finally, the heat of formation, entropy of formation, and free energy of formation of the compound are calculated at each temperature using the relationships (statement 20 plus two)

$$\Delta H_T^O = H_T^O \; (compound) \; - \; \Sigma H_T^O \; (elements) \tag{4.5}$$

$$\Delta S_T^O = S_T^O \; (compound) \; - \; \Sigma S_T^O \; (elements) \tag{4.6}$$

$$\Delta F_T^O = \Delta H_T^O \; - \; T\Delta S_T^O \tag{4.7}$$

The summations in Eqs. (4.5) and (4.6) include, of course, the appropriate stoichiometric coefficients to form the compound from the elemental species.

Examples

Input data for lithium hydroxide (*c*)

Data File CONVR.DAT

```
LIOH
8 0
25
50
75
100
150
200
250
300
2
2
2
2
2
2
2
2
1
-115.84 10.2
11.99 8.24 -2.27 0
298.15 2740
3
LI 1
O 0.5
H 0.5
```

Execution of program

.EXECUTE IONCOM, DELHS, DELHS1, NNCODE
LOADING

IONCOM 6K CORE
EXECUTION

CRU´S: 67   ELAPSED TIME: 39.58
NO EXECUTION ERRORS DETECTED

EXIT

Output data for lithium hydroxide (c)

Output data are temperature ($^{\circ}$C), ΔH_T° (kcal/mole), ΔF_T° (kcal/mole), H_T° (kcal/mole), S_T° (cal/deg K-mole), mean heat capacity[*] (cal/deg K-mole) between temperature T and 25°C, and data reference code:

	T	ΔH_T°	ΔF_T°	H_T°	S_T°	$\overline{C}_p^{\circ}$	
LIOH	25	−115.84	−104.85	−115.84	10.20		2
LIOH	50	−115.86	−103.93	−115.54	11.18	12.19	2
LIOH	75	−115.87	−103.01	−115.22	12.13	12.47	2
LIOH	100	−115.87	−102.08	−114.89	13.05	12.72	2
LIOH	150	−115.86	−100.24	−114.19	14.79	13.16	2
LIOH	180.5	−115.85	−99.11	−113.75	15.79	13.41	2
LIOH	180.5	−116.57	−99.11	−113.75	15.79		2
LIOH	200	−116.56	−98.36	−113.47	16.41	14.75	2
LIOH	250	−116.53	−96.44	−112.71	17.93	15.18*	2
LIOH	300	−116.46	−94.52	−111.92	19.37	15.75*	2

Input data for iron sulfide (c)

Data file CONVR.DAT

```
FES
10 0
25
50
75
100
137.85
137.85
150
200
250
300
```

[*] Above 200°C the mean heat capacity is evaluated between temperature T and 200°C, and above 250°C between T and 250°C. These mean heat capacities are required for calculating the properties of ions by program IONIC.

ONIC calculates the heats and free energies of formation and the
nd entropy of ions as a function of temperature. The calcula-
ased on 25°C data provided by the user and on temperature extrap-
de by the Criss and Cobble[1] entropy correspondence principle.
 IONIC is an interactive program in which the user must provide
the ion of interest. The program also accesses file FOR21.DAT
 p. 94) for reference state data on the elements. Subroutines
N, DELHS, DELHS1, ELEMEN, NNCODE, and PRCENT must be available.
 output data are stored in file FOR24.DAT.

Ion

f the interactive program IONIC will bring on a series of
must be answered by the user. At the completion of the ques-
t table is generated and stored in FOR24.DAT.
ments for data input are now explained.

ed by IONIC	User Response
ie	A string of 20 characters that will iden-tify the ion in the output table.
	Enter reference code for 25°C data.
e at 25C	Enter charge on ion (include sign, e.g., -2 for SO_4^{2-}).
s in	Enter number of different elements in ion.
	Enter element symbol (A2), and then number of molecules of that element in the ion for each element (F3.1), allow-ing no spaces between data sets (e.g., for $HAsO_4^-$, enter H Ø.5AS1.ØO 2.Ø).
	Enter conventional ion entropy (cal/deg K-mole) at 25°C (program converts to "absolute" scale).

Ion type (2,3,4)	Enter Criss and Cobble ion type (appears only for anions):
	2 = Simple anions + OH^- (e.g., F^-, I^-, Cl^-, S^{2-}).
	3 = Oxy anions (e.g., SO_4^{2-}, PO_4^{2-}).
	4 = Acid oxy anions (e.g., HSO_4^-, $H_3P_2O_7^-$).
F of ion at 25	Enter free energy (kcal/mole) of ion at 25°C. When sulfur is present, the free energy of the ion must be based on the reference state of the ideal diatomic gas, $S_2(g)$. (Program IONIC does not make the conversion for the sulfur refer-ence state.)

Program IONIC

Program IONIC may be executed once all subroutines and the file of element data (FOR21.DAT) have been placed in the user area. A brief outline of the calculation sequence is now given.

Initially, the program sets up all arrays. The process of accepting data from the terminal begins at statement 2000. The loop at 2002 plus two assigns the (Criss and Cobble) reference code 3 to all temperatures above 25°C. Subroutine NNCODE is accessed to assign code numbers for the elemental species forming the ion.

Statement 10 plus two establishes whether an ion is positive or negative. If positive, control is transferred to assign the ion as type 1; if negative, more information on ion type is requested from the user.

The temperature array is established beginning at statement 12 plus one. The data for each elemental species forming the ion are scanned to determine whether a phase transition occurs in the temperature range of interest, in which case two identical temperature levels are inserted consecutively into the array. These temperatures represent phase transitions in the elements and are included to aid in interpolating the output data. The adjusted tem-perature scale is complete at statement 1061 plus two.

At this point provision is made to incorporate the hydrogen half-cell reaction into the defining chemical equation for the formation of the ion.

This eliminates ambiguities in defining the properties of electrons. As an example, the formation of $Al^{3+}(aq)$,

$$Al(c) \rightarrow Al^{3+}(aq) + 3e \qquad (4.8)$$

is combined with the hydrogen half-cell reaction

$$H^+(aq) + e \rightarrow \tfrac{1}{2}H_2(g) \qquad (4.9)$$

to give

$$Al(c) + 3H^+(aq) \rightarrow Al^{3+}(aq) + \frac{3}{2} H_2(g) \qquad (4.10)$$

Program IONIC uses the full-cell reaction, Eq. (4.10), to extrapolate the properties of ions to higher temperatures. This is treated in the program by the device of incrementing the "number of elements" by one at statement 1061 plus two.

Subroutine ELEMEN is called at statement 14 plus three to calculate the thermodynamic properties of the ion at $25^{\circ}C$. Since most of the data (at $25^{\circ}C$) were input, the only new quantities generated are ΔH, H and S. At this point, S is on the Criss and Cobble "absolute" entropy scale.

Subroutine ELEMEN is sequentially accessed to calculate the properties of the ion at each specified temperature. A brief description of this subroutine follows.

Subroutine ELEMEN

Subroutine ELEMEN is accessed at each temperature to calculate the properties of the specified ion. The main components of the calculation are summarized below, followed by a brief outline of the program statements in which key quantities are calculated.

The free energy of formation of an ion at temperature T is calculated [for a chemical reaction of the type represented by Eq. (4.10)] using the relationship

$$\Delta F_T^O = \Delta F_{T_r}^O - \Delta S_{T_r}^O \Delta T + \Delta \overline{C}_p^O \Big]_{T_r}^T [\Delta T - T \ln T/T_r] \qquad (4.11)$$

where

ΔF^O = the free energy of formation of the ion.

ΔS^O = the change in entropy for the ion formation reaction [chemical reaction of the type represented in Eq. (4.10)].

```
1
2
2
2
2
2
2
2
2
2
2
-23.9 14.41
5.19 26.4 0 0
298.15 411
.57 1.387
17.4 0 0 0
411 1500
2
FE 1
S 0.5
```

Execution of program

```
.EXECUTE IONCOM, DELHS, DELHS1, NNCODE
LOADING

IONCOM 6K CORE
EXECUTION


CRU´S: 67           ELAPSED TIME: 40.03
NO EXECUTION ERRORS DETECTED

EXIT
```

Output data for iron sulfide (c)

	T	ΔH_T^O	ΔF_T^O	H_T^O	S_T^O	$\overline{C}_p^O$
FES	25	-39.24	-33.47	-39.24	14.41	13.39
FES	50	-39.15	-32.99	-38.91	15.49	13.72
FES	75	-39.06	-32.51	-38.55	16.53	14.05
FES	100	-38.95	-32.05	-38.19	17.55	14.55
FES	137.9	-38.76	-31.36	-37.60	19.06	
FES	137.9	-38.19	-31.36	-37.03	20.44	
FES	150	-38.11	-31.15	-36.82	20.95	17.40
FES	200	-37.79	-30.35	-35.95	22.89	17.40
FES	250	-37.48	-29.58	-35.08	24.64	17.40*
FES	300	-37.19	-28.84	-34.21	26.23	17.40*

$\Delta \overline{C}_p^O \Big]_{T_r}^T$ = the change in average heat capacity (over ΔT) for the ion formation reaction.

ΔT = $T - T_r$.

The subscripts T and T_r designate the temperature and reference temperature, respectively.

The entropy of an ion is calculated on the "absolute" scale from the Criss and Cobble correspondence principle. The "absolute" entropy of an ion at 25°C is given by

$$\overline{S}_{298}^O(absolute) = \overline{S}_{298}^O(conventional) - 5.0\,z \qquad (4.12)$$

where z is the ionic charge of the ion.

The average heat capacity for ions is computed by the relationship

$$\overline{C}_p^O \Big]_{T_r}^T = (\overline{S}_T^O - \overline{S}_{T_r}^O)/\ln(T/T_r) \qquad (4.13)$$

where $\overline{S}_T^O$ and $\overline{S}_{T_r}^O$ are the entropies of the ion (Criss and Cobble) at temperatures T and T_r. The average heat capacities for the elements are obtained by integration of the heat capacity expression [either Eq. (4.1) or (4.2)].

The heat of formation of an ion at temperature T is calculated by the relationship

$$\Delta H_T^O = \Delta F_T^O + T\Delta S_T^O \qquad (4.14)$$

Finally, the enthalpy of an ion is computed from

$$H_T^O = \Delta H_{298}^O + \overline{C}_p^O(ion)\Big]_{T_r}^T [T - 298.15] \qquad (4.15)$$

where

ΔH_{298}^O = the enthalpy of formation of the ion at 298.15°K.

$\overline{C}_p^O(ion)\Big]_{T_r}^T$ = the mean heat capacity of the ion between temperatures 298.15°K and T.

Twenty-five degrees centigrade is generally used as the reference temperature (T_r) below 200°C. Because the linear ΔT and $\Delta \ln T$ averages of $\overline{C}_p^O$ are not identical over a large temperature interval, the calculations

above 200°C are carried out in 50° intervals (see Criss and Cobble, ref. 1, p. 5393). Therefore, above 200°C, a reference temperature of 200°C is used, and above 250°C a reference temperature of 250°C is used. Furthermore, whenever a phase transition in the forming element (for example, 58°C for bromine) is exceeded, the transition temperature is used as the reference temperature.*

Counters and arrays are initialized in subroutine ELEMEN at statement 11. Subroutines DELHS or DELHS1 [depending on the form of the heat capacity expression, i.e., Eq. (4.1) or (4.2)] are accessed to calculate the enthalpy and entropy for each elemental species appearing in the whole-cell reaction [including $H_2(g)$]. Statements 33 through 33 plus three establish the initial conditions for temperature extrapolations above 200°C.

At statements 101 and 101 plus one the subroutines CPMEAN and CPHYDR are accessed to determine the entropy and mean heat capacities for H^+ and the ion of interest, respectively. The subroutine PRCENT interpolates the temperature scale linearly as part of the process in obtaining the entropies of the ions from the Criss and Cobble correlation.

The entropy of reaction at the given temperature along with $\Delta \bar{C}_p^{\text{o}}\big]_{T_r}^{T}$ are subsequently calculated. Following this, ΔF_T^{o}, ΔH_T^{o}, and H_T^{o} are calculated according to Eqs. (4.11), (4.14) and (4.15).

The statements beginning with 10 relate to handling the data when a phase transition for one of the elements is encountered.

Examples

Input data for $H_2PO_3^-$ (execution of program)

```
.EXECUTE IONIC,CPHYDR,CPMEAN,DELHS,DELHS1,ELEMEN,NNCODE,PRCENT
LOADING

IONIC 6K CORE
EXECUTION
```

---

* Above either 200°C or a phase transition temperature for one of the forming elements, the enthalpy is calculated from

$$H_T^{\text{o}} = H_{T_r}^{\text{o}} + \bar{C}_p^{\text{o}}(ion)\big]_{T_r}^{T}[T - T_r]$$

where T_r is the reference temperature and $\bar{C}_p^{\text{o}}(ion)\big]_{T_r}^{T}$ the mean heat capacity between T and T_r.

ENTER ION NAME... H2PO3-

ENTER REFERENCE AT 25C ... 5

IONIC CHARGE?... -1

HOW MANY ELEMENTS IN ANION?... 3

ION NAME... H 1.0P 1.0O 1.5

S OF ION AT 25... 19.0

ION TYPE (2,3,4)... 4

F OF ION AT 25... -202.4

CRU´S: 70   ELAPSED TIME: 1:43.30
NO EXECUTION ERRORS DETECTED

EXIT

Output data for $H_2PO_3^-$

Output data are temperature (^oC), ΔH_T^o (kcal/mole), ΔF_T^o (kcal/mole), H_T^o (kcal/mole), $\overline{S}_T^o$ (cal/deg K-mole), mean heat capacity* (cal/deg K-mole) between temperature T and 25^oC, and data-reference code:

	T	ΔH_T^o	ΔF_T^o	H_T^o	$\overline{S}_T^o$	$\overline{C}_p^o$	
H2PO3-	25	-235.54	-202.40	-235.54	24.00		5
H2PO3-	44.2	-236.36	-200.25	-236.27	21.60	-38.50	2
H2PO3-	44.2	-236.52	-200.25	-236.27	21.60	-38.50	2
H2PO3-	50	-236.78	-199.57	-236.51	20.87	-38.85	3
H2PO3-	75	-237.64	-196.67	-237.49	17.95	-39.05	3
H2PO3-	100	-238.33	-193.71	-238.49	15.16	-39.41	3
H2PO3-	150	-241.02	-187.47	-241.55	7.14	-48.14	3
H2PO3-	200	-243.09	-181.03	-244.24	1.04	-49.72	3
H2PO3-	250	-245.14	-174.37	-247.38	-5.28	-62.91*	3
H2PO3-	273.9	-246.25	-171.11	-248.99	-8.29	-67.62*	2
H2PO3-	273.9	-249.14	-171.11	-248.99	-8.29	-67.62*	2
H2PO3-	300	-250.48	-167.49	-250.85	-11.60	-69.27*	3

Input data for $S_5O_6^=$ (execution of program)

.EXECUTE IONIC,CPHYDR,CPMEAN,DELHS,DELHS1,ELEMEN,NNCODE,PRCENT
LOADING

IONIC 6K CORE
EXECUTION

* Above 200^oC the mean heat capacity is evaluated between temperature T and 200^oC, and above 250^oC between T and 250^oC.

ENTER ION NAME... S5O6--

ENTER REFERENCE AT 25C ... 5

IONIC CHARGE?... -2

HOW MANY ELEMENTS IN ANION?... 2

ION NAME... S 2.5O 3.0

S OF ION AT 25... 40.1

ION TYPE (2,3,4)... 3

F OF ION AT 25... -275.9

CRU´S: 70   ELAPSED TIME: 1:26.58
NO EXECUTION ERRORS DETECTED

EXIT

Output data for $S_5O_6^=$

	T	ΔH_T^O	ΔF_T^O	H_T^O	$\bar{S}_T^O$	$\bar{C}_p^O$	
S5O6--	25	-357.71	-275.90	-357.71	50.10		5
S5O6--	50	-358.48	-269.01	-358.40	47.87	-27.75	3
S5O6--	75	-358.90	-262.09	-359.21	45.46	-29.91	3
S5O6--	100	-359.03	-255.17	-360.10	42.95	-31.88	3
S5O6--	150	-359.83	-241.21	-361.99	38.12	-34.22	3
S5O6--	200	-360.07	-227.29	-363.73	34.20	-34.42	3
S5O6--	250	-359.83	-213.27	-365.96	29.73	-44.50*	3
S5O6--	300	-359.84	-199.26	-368.46	25.16	-50.07*	3

UNDISSOCIATED NEUTRAL COMPLEXES

Program UNDIS1 has been written to calculate the thermodynamic properties of undissociated neutral complexes, such as $AgCl(aq)$. The calculations are based on 25^OC data provided by the user and on temperature extrapolations made by the generalized correlations of Helgeson[2] and Criss and Cobble.[1] This program has a temperature range of $25\text{-}200^O$C.

Program UNDIS1 is an interactive program in which the user must provide 25^OC data on the complex of interest and on its ionic species of dissociation. The program also accesses file FOR21.DAT (described on p. 94) for reference state data on the elements. Subroutines CPHYDR, CPMEAN, DELHS, DELHS1, ELEMEN, HELGE, HELGES, NNCODE, PRCENT, and SUBION must be available. The generated output data appears in file FOR24.DAT.

Input Data for UNDIS1

The execution of the interactive program UNDIS1 will bring on a series of
questions that must be answered by the user. Data on the neutral complex
and its dissociation products must be supplied. For example, for $AgCl(aq) \rightleftharpoons$
$Ag^+(aq) + Cl^-(aq)$, data on all three species must be entered. At the completion of the questions, the output table is generated and stored in
FOR24.DAT.

The requirements for data input are now explained.

Question Posed by UNDIS1	User Response
Enter ident	A string of 20 characters that will identify the complex in output table.
Enter reference at 25C	Enter reference code for 25°C data.
How many elements in undiss. species?	No. of elements in complex.
Name of species	Enter element symbol (A2) and the number of molecules of that element in complex for each element, allowing no spaces between data sets (e.g., for H_2CO_3 enter H 1.∅C 1.∅O 1.5).
Entropy of species	Enter entropy (cal/deg K-mole) of complex at 25°C.
Free energy of species	Enter free energy of formation of (kcal/mole) complex at 25°C. When sulfur is present, the free energy of the complex must be based on the reference state of the ideal diatomic gas, $S_2(g)$. (Program does <u>not</u> make the conversion for the sulfur reference state.)
Number of ions in reaction	Enter number of distinct ionic species into which complex dissociates. (The Helgeson correlation was formulated for neutral complexes dissociating to one positive and one negative ion.)

Ionic charge?   Enter ionic charge of first ionic
                                       species.

How many elements in   Enter number of elements in ion.
{cation}?
{anion }

Ion name   List elements and number of molecules
                                       of elements (same format as in "name of
                                       species").

S of ion at 25   Enter conventional ion entropy at $25^{\circ}C$
                                       (program converts to "absolute" scale).

Ion type (2,3,4)   Enter Criss and Cobble ion type (appears
                                       only for anions):

                                       2 = Simple anions + OH^{+}

                                       3 = Oxy anions (e.g., SO_4^{2-}, PO_4^{2-}).

                                       4 = Acid oxy anions (e.g., HSO_4^{-}, $H_3P_2O_7^{-}$).

F of ion at 25   Enter free energy (kcal/mole) of ion at
                                       $25^{\circ}C$. When sulfur is present, the free
                                       energy of the ion must be based on the
                                       reference state of the ideal diatomic
                                       gas, $S_2(g)$. [Program does not make the
                                       conversion for the sulfur reference
                                       state.]

Stoich. coef. for above   Enter stoichiometric coefficient for
ion   ion in dissociation reaction.

Ionic charge?   Enter ionic charge of second ionic
                                       species.

Etc.   Similar input as last five questions for
                                       second ionic species. Sequence is
                                       repeated for additional ionic species
                                       as needed.

Program UNDIS1

Program UNDIS1 may be executed once all subroutines and the file of element data (FOR21.DAT) have been placed in the user area. A brief outline of the bases of the calculations and their sequence is given below.

The program initially sets up the temperature array. Data for the neutral complex are entered between statements 10 plus one and 9 plus one. At 9 plus two the program scans the elements contained in the neutral complex and adds two identical temperature levels at consecutive locations in the temperature array at each elemental transition point. These additional temperatures are included as an aid to interpolating the output data. The adjusted temperature scale is complete at statement 1061.

The temperature extrapolations in UNDIS1 are based on the Helgeson[2] and Criss and Cobble[1] correlations. Helgeson's correlation is used to estimate the temperature dependence of ΔF^O and ΔS^O for the dissociation reactions of neutral complexes, as represented by the example

$$AgCl(aq) \rightleftharpoons Ag^+(aq) + Cl^-(aq) \tag{4.16}$$

The Criss and Cobble correlation, on the other hand, is used to estimate the thermodynamic properties of the dissociation product ions, $Ag^+(aq)$ and $Cl^-(aq)$ in the above example. The properties of the complex [$AgCl(aq)$] are then obtained by difference.

The thermodynamic properties of the ions are summed starting at statement 1061 plus one. The properties of the ions are calculated in subroutine SUBION. This subroutine is similar to program IONIC (with the exception of output data), which is described in the section on ions. After the calculations for the ions have been completed (statement 15), the program computes all the 25^OC output data for the neutral complex.

The loop at statement 15 plus eight begins calculating the higher temperature data for the neutral complex. Subroutine HELGE is accessed to compute the free energy of dissociation for the complex at temperature T using the expression (Helgeson, ref. 2)

$$\Delta F_T^O (dissociation) =$$

$$- \Delta S_{T_r}^O (dissociation) \left[T_r - \frac{\theta}{\omega} \{1 - \exp[\exp(b - aT) - c + (T - T_r)/\theta]\} \right] +$$

$$\Delta H_{T_r}^O (dissociation) \tag{4.17}$$

where

a = 0.01875

b = -12.741

c = exp $(b+aT_r)$

θ = 219.0

ω = 1 + $ac\theta$.

Subroutine HELGES calculates the entropy of dissociation for the complex at temperature T using the expression (Helgeson, ref. 2)

$$\Delta S_T^O(dissociation) =$$

$$\frac{\Delta S_{T_r}^O(dissociation)\{\exp[\exp(b+aT)-c + (T-T_r)/\theta]\}[1+\phi \exp(b+aT)]}{\omega} \qquad (4.18)$$

where ϕ = 4.106. The reference temperature T_r in the above equation is taken as 298.15°K.

The free energy of formation and the entropy of the neutral complex are calculated by difference, for example,

$$\Delta F_T^O(complex) = \Sigma\Delta F_T^O(ion) - \Delta F_T^O(dissociation) \qquad (4.19)$$

The mean heat capacity of the complex is calculated at statement 15 plus fifteen by the relation

$$\overline{C}_p^O\Big]_{T_r}^T = (S_T^O-S_{298}^O)/\ln(T/298.15) \qquad (4.20)$$

At statement 15 plus sixteen, the enthalpy of reaction for the dissociation reaction is calculated from

$$\Delta H_T^O(dissociation) = \Delta F_T^O(dissociation) + T\Delta S_T^O(dissociation) \qquad (4.21)$$

The enthalpy of formation of the complex is then calculated by difference. The enthalpy of the complex is computed at 15 plus eighteen from

$$H_T^O = \Delta H_{298}^O(complex) + \overline{C}_p^O(complex)\Big]_{298}^T[T-298.15] \qquad (4.22)$$

Examples

Input data for $HIO_3(aq)$ (execution of program)

```
.EXECUTE UNDIS1,CPHYDR,CPMEAN,DELHS,DELHS1,ELEMEN,HELGE,HELGES, NNCODE,
PRCENT,SUBION
LOADING

UNDIS1 7K CORE
EXECUTION

ENTER IDENT... HIO3(AQ)

ENTER REFERENCE AT 25C... 1

HOW MANY ELEMENTS IN UNDISS. SPECIES?... 3

NAME OF SPECIES... H 0.5I 0.5O 1.5

ENTROPY OF SPECIES... 39.9

FREE ENERGY OF SPECIES... -31.7

#OF IONS IN REACTION... 2

IONIC CHARGE?... 1

HOW MANY ELEMENTS IN CATION?... 1

ION NAME... H 0.5

S OF ION AT 25... 0

F OF ION AT 25... 0

STOIC. COEF. FOR ABOVE ION... 1

IONIC CHARGE?... -1

HOW MANY ELEMENTS IN ANION?... 2

ION NAME... I 0.5O 1.5

S OF ION AT 25... 28.3

ION TYPE (2,3,4)... 3

F OF ION AT 25... -30.6

STOIC. COEF. FOR ABOVE ION... 1
CRU´S: 79        ELAPSED TIME: 6:12.07
NO EXECUTION ERRORS DETECTED

EXIT
```

Output data for $HIO_3(aq)$

Output data are temperature (^{O}C), ΔH_T^O (kcal/mole), ΔF_T^O (kcal/mole), H_T^O (kcal/mole), S_T^O (cal/deg K-mole), mean heat capacity (cal/deg K-mole) between temperature T and 25°C, and data-reference code:

	T	ΔH_T^O	ΔF_T^O	H_T^O	S_T^O	$\overline{C}_p^O$	
HIO3(AQ)	25	-50.51	-31.70	-50.51	39.90	----	1
HIO3(AQ)	50	-51.53	-30.08	-51.01	38.28	-20.11	6
HIO3(AQ)	75	-52.40	-28.39	-51.37	37.24	-17.17	6
HIO3(AQ)	100	-53.14	-26.64	-51.58	36.69	-14.31	6
HIO3(AQ)	113.9	-53.41	-25.66	-51.57	36.78	-11.96	2
HIO3(AQ)	113.9	-55.27	-25.66	-51.57	36.78	-11.96	2
HIO3(AQ)	150	-55.81	-22.87	-51.34	37.58	-6.62	6
HIO3(AQ)	184.9	-56.10	-20.12	-50.93	38.78	-2.62	2
HIO3(AQ)	184.9	-61.12	-20.12	-50.93	38.78	-2.62	2
HIO3(AQ)	200	-60.91	-18.76	-50.53	39.85	-0.12	6

Input data for $H_2SO_3(aq)$ (execution of program)

```
.EXECUTE UNDIS1,CPHYDR,CPMEAN,DELHS,DELHS1,ELEMEN,HELGE,HELGES, NNCODE,
PRCENT,SUBION
LOADING

UNDIS1 7K CORE
EXECUTION

ENTER IDENT... H2SO3(AQ)

ENTER REFERENCE AT 25C... 1

HOW MANY ELEMENTS IN UNDISS. SPECIES?... 3

NAME OF SPECIES... H 1.0S 0.5O 1.5

ENTROPY OF SPECIES... 55.5

FREE ENERGY OF SPECIES... -138.04

#OF IONS IN REACTION... 2

IONIC CHARGE?... 1

HOW MANY ELEMENTS IN CATION?... 1

ION NAME... H 0.5

S OF ION AT 25... 0

F OF ION AT 25... 0

STOIC. COEF. FOR ABOVE ION... 1
```

IONIC CHARGE?... -1

HOW MANY ELEMENTS IN ANION?... 3

ION NAME... H 0.5S 0.5O 1.5

S OF ION AT 25... 33.4

ION TYPE (2,3,4)... 4

F OF ION AT 25... -135.63

STOIC. COEF. FOR ABOVE ION... 1

CRU´S: 79   ELAPSED TIME: 3:8.40
NO EXECUTION ERRORS DETECTED

EXIT

Output data for $H_2SO_3(aq)$

	T	ΔH_T^O	ΔF_T^O	H_T^O	S_T^O	$\overline{C}_p^O$	
H2SO3(AQ)	25	-160.84	-138.04	-160.84	55.50		1
H2SO3(AQ)	50	-159.73	-136.17	-159.19	60.80	65.79	6
H2SO3(AQ)	75	-157.96	-134.43	-156.90	67.72	78.83	6
H2SO3(AQ)	100	-155.59	-132.88	-154.01	75.94	91.09	6
H2SO3(AQ)	150	-151.55	-129.95	-149.05	88.52	94.31	6
H2SO3(AQ)	200	-143.59	-127.97	-140.56	109.02	115.90	6

REFERENCES

1. C. M. Criss and J. W. Cobble, *J. Am. Chem. Soc.*, 86, 5385 (1964).

2. H. C. Helgeson, *J. Phys. Chem.*, 71, 3121 (1967).

3. *Selected Values of Chemical Thermodynamic Properties*, National Bureau of Standards, Technical Notes 270-3 (1968), 270-4 (1969), 270-5 (1971), 270-6 (1971), and 270-7 (1973).

SYMBOLS

A, B, $\overset{\circ}{a}_i$	Parameters in Debye-Hückel equation.
a	Activity
a, b, c, θ, ω, ϕ	Parameters in Helgeson's correlations for neutral complexes.
$a(T)$, $b(T)$	Constants dependent on class of ions and on temperature used with "correspondence principle."
B	Denotes substance B.
C_p^o	Standard heat capacity at constant pressure.
ΔC_p^o	Change in C_p^o during reaction.
$\Delta \overline{C}_p^o \Big]_{T_r}^{T}$	Average value of C_p^o between T and T_r.
e	Electron.
ΔF_T^o	Standard Gibbs free energy of formation of substance, or standard Gibbs free energy change of reaction, at temperature T.
f	Fugacity of gas.
f^o	Fugacity of gas in standard state.
f^*	Fugacity of pure liquid.
H_T^o	Enthalpy of substance at temperature T.
ΔH_T^o	Standard heat of formation of substance, or standard enthalpy change of reaction, at temperature. T.
$\Delta H_{T_t}^t$	Enthalpy change of substance due to phase transition at temperature T_t.
$\Delta H \text{(mixing)}$	Enthalpy change due to mixing.
I	Ionic strength.

117

i	Subscript denotes substance i.
K	Equilibrium constant.
m	Concentration on a molality basis.
n	Moles.
p	Pressure.
p^*	Vapor pressure of pure liquid.
Q	Heat input to system.
R	Gas constant.
S_T^O	Entropy of substance in standard state at temperature T.
ΔS_T^O	Standard entropy change of reaction at temperature T.
T	Absolute temperature.
V	Molar volume of gas.
x	Liquid concentration on mole fraction basis.
z	Ionic charge.
γ	Activity coefficient.
ν	Stoichiometric coefficient.

APPENDIX
FORTRAN LISTINGS
OF COMPUTER PROGRAMS

CPHYDR   120
CPMEAN   120
DELHS   121
DELHS1   121
ELEMEN   122
FOR21.DAT   125
HELGE   133
HELGES   134
IONCOM   134
IONIC   141
NNCODE   145
PRCENT   146
SUBION   147
UNDIS1   149

Glossary

    Dimensioned variables   154
    Other variables   155

```
SUBROUTINE CPHYDR

C THIS SUBROUTINE CALCULATES MEAN HEAT CAPACITY FOR H+
C
      SUBROUTINE CPHYDR(KTEM)
      COMMON /RICK3/ ENTROI, ENTROC, ION, Z, ENTRO, TEMP(16)
      COMMON /RICK4/ ENTH1,NTEMP1,DELF(16),DELH(16),CP,CP2,CPH
      DIMENSION ENTROH(7)
      DATA ENTROH/-5.,-2.5,2.,6.5,11.1,16.1,20.7/
C
C CALCULATE ENTROPY OF H+ AT TEMP(KTEM)
C
      CALL PRCENT(IRANGE,PERCEN,KTEM)
      ENTRO=PERCEN*(ENTROH(IRANGE+1)-ENTROH(IRANGE))+ENTROH(IRANGE)
C
C CALCULATE ENTROPY OF H+ AT TEMP(NTEMP1)
C
      CALL PRCENT(IRANGE,PERCEN,NTEMP1)
      ENTRH1=PERCEN*(ENTROH(IRANGE+1)-ENTROH(IRANGE))+ENTROH(IRANGE)
      IF(ABS(TEMP(KTEM)-TEMP(NTEMP1)).LT..001) GO TO 1
C
C CALCULATE MEAN CP OF H+ FROM TEMP(NTEMP1) TO TEMP(KTEM)
C
      CPH=(ENTRO-ENTRH1)/ALOG(TEMP(KTEM)/TEMP(NTEMP1))
    1 RETURN
      END


SUBROUTINE CPMEAN

C THIS SUBROUTINE CALCULATES THE MEAN HEAT CAPACITY OF ANY ION, EXCEPT
C H+.  THE CP'S ARE TAKEN FROM TEMP(NTEMP1) AND 25C TO TEMP(KTEM).
C THE ENTROPY OF THE ION AT TEMP(KTEM) IS ALSO FOUND.
C
      SUBROUTINE CPMEAN(KTEM)
      COMMON /RICK3/ ENTROI, ENTROC, ION, Z, ENTRO, TEMP(16)
      COMMON /RICK4/ ENTH1,NTEMP1,DELF(16),DELH(16),CP,CP2,CPH
      DIMENSION CONSTA(4,7), CONSTB(4,7)
      DATA ((CONSTA(I,J),J=1,7),I=1,4)/0.,3.9,10.3,16.2,23.3,29.9,
     1 36.6,0.,-5.1,-13.,-21.3,-30.2,-38.7,-49.2,0.,-14.,-31.,-46.4,
     2 -67.,-86.5,-106.,0.,-13.5,-30.3,-50.,-70.,-90.,-110./
      DATA ((CONSTB(I,J),J=1,7),I=1,4)/1.,.955,.876,.792,.711,.63,
     1 .548,1.,.969,1.,.989,.981,.978,.972,1.,1.217,1.476,1.687,
     2 2.02,2.32,2.618,1.,1.38,1.894,2.381,2.96,3.53,4.1/
C
C CALCULATE THE CRISS AND COBBLE CONSTANTS FOR ENTROPY AT TEMP(KTEM)
C
      CALL PRCENT(IRANGE,PERCEN,KTEM)
      A=PERCEN*(CONSTA(ION,IRANGE+1)-CONSTA(ION,IRANGE))
     1 +CONSTA(ION,IRANGE)
      B=PERCEN*(CONSTB(ION,IRANGE+1)-CONSTB(ION,IRANGE))
     1 +CONSTB(ION,IRANGE)
C
```

```
C CALCULATE THE ABSOLUTE ENTROPIES AT 25C AND TEMP(KTEM)
C
      ENTROA=ENTROI-5*Z
      ENTROC=A+B*ENTROA
C
C CALCULATE THE MEAN CP FROM 25C TO TEMP(KTEM)
C
      IF(ABS(TEMP(KTEM)-298.15).LT..001) GO TO 1
      CP=(ENTROC-ENTROA)/ALOG((TEMP(KTEM))/298.15)
C
C CALCULATE THE CRISS AND COBBLE CONSTANTS FOR ENTROPY AT TEMP(NTEMP1)
C
      CALL PRCENT(IRANGE,PERCEN,NTEMP1)
      A=PERCEN*(CONSTA(ION,IRANGE+1)-CONSTA(ION,IRANGE))
    1 +CONSTA(ION,IRANGE)
      B=PERCEN*(CONSTB(ION,IRANGE+1)-CONSTB(ION,IRANGE))
    1 +CONSTB(ION,IRANGE)
C
C CALCULATE THE ABSOLUTE ENTROPY AT TEMP(NTEMP1)
C
      ENTROB=A+B*ENTROA
C
C CALCULATE THE MEAN CP FROM TEMP(NTEMP1) TO TEMP(KTEM)
C
      IF(ABS(TEMP(KTEM)-TEMP(NTEMP1)).LT..001) GO TO 1
      CP2=(ENTROC-ENTROB)/ALOG(TEMP(KTEM)/TEMP(NTEMP1))
    1 RETURN
      END
```

SUBROUTINE DELHS

```
      SUBROUTINE DELHS(A,B,C,D,T1,T2,DELH,DELS)
      IF(ABS(T2-T1).LT..001) GO TO 1
      DELH=(A*(T2-T1)+(B/2.)*.001*(T2**2-T1**2)-C*100000.*(1./T2-1./T1
    1)+(D/3.)*.000001*(T2**3-T1**3))*.001
      DELS=(A*ALOG(T2/T1)+B*.001*(T2-T1)-(C/2.)*100000.*(1./T2**2-1./T
    11**2)+(D/2.)*.000001*(T2**2-T1**2))*.001
      GO TO 2
    1 DELH=0.
      DELS=0.
    2 RETURN
      END
```

SUBROUTINE DELHS1

```
      SUBROUTINE DELHS1(A,B,C,D,T1,T2,DELH,DELS)
      IF(ABS(T2-T1).LT..001) GO TO 1
      DELH=(A*(T2-T1)+(B/2.)*.001*(T2**2-T1**2)-C*100000.*(1./T2-1./T1
    1)-(D/2.)*(1.0E+08)*(1./T2**2-1./T1**2))*.001
      DELS=(A*ALOG(T2/T1)+B*.001*(T2-T1)-(C/2.)*100000.*(1./T2**2-1./T
    11**2)-(D/3.)*(1.0E+08)*(1./T2**3-1./T1**3))*.001
      GO TO 2
    1 DELH=0.
      DELS=0.
    2 RETURN
      END
```

```
SUBROUTINE ELEMEN

C THIS SUBROUTINE CALCULATES HEAT AND FREE ENERGY OF FORMATION, ENTHALPY,
C ENTROPY, AND MEAN HEAT CAPACITY FOR IONS AT TEMP(KTEM)
C
        SUBROUTINE ELEMEN(KTEM)
        COMMON /RICK1/DHEL(16),DSEL(16),ENTH(16), DELS
        COMMON /RICK2/ NELE(8), FNUM(8), NEL(8), NELS
        COMMON /RICK3/ ENTROI, ENTROC, ION, Z, ENTRO, TEMP(16)
        COMMON /RICK4/ ENTH1,NTEMP1,DELF(16),DELH(16),CP,CP2,CPH
        COMMON /RICK5/ DELST, DELCP(16), DELSI(16)
        COMMON /RICK7/ ATWO(250), AH(250), AS(250), ATL(250), ATH(250)
        COMMON /RICK8/ AAA(250), AAB(250), AAC(250), AAD(250), ANCP(250)
C
C INITIALIZE ACCOUNTING
C
        CPELEM=0.
   11   KELEM=1
        DHEL(KTEM)=0.
        DSEL(KTEM)=0.
C
C START SUMMATION OF ELEMENTAL DATA
C
    7   KREAD=NEL(KELEM)
    8   A=AAA(KREAD)
        B=AAB(KREAD)
        C=AAC(KREAD)
        D=AAD(KREAD)
        TWO=ATWO(KREAD)
        H=AH(KREAD)
        S=AS(KREAD)
        TL=ATL(KREAD)
        TH=ATH(KREAD)
        NCP=ANCP(KREAD)
C
C IF T IS LESS THAN THE HIGHEST T IN THIS CP RANGE, GO TO 12
C IF NOT, SUM THE ENTHALPY AND ENTROPY FOR THE ENTIRE RANGE
C
        IF(TEMP(KTEM).LE.TH+.001) GO TO 12
        IF(NCP.EQ.0) CALL DELHS(A,B,C,D,TL,TH,DH,DS)
        IF(NCP.NE.0) CALL DELHS1(A,B,C,D,TL,TH,DH,DS)
        DHEL(KTEM)=DHEL(KTEM)+(DH+H)*FNUM(KELEM)
        DSEL(KTEM)=DSEL(KTEM)+(DS+S*.001)*FNUM(KELEM)
        KREAD=TWO+.01
        GO TO 8
C
C IF T IS A TRANSITION POINT, GO TO 13
C IF NOT, SUM THE ENTHALPY AND ENTROPY UP TO T AND GO ON TO THE
C NEXT ELEMENT IF ANY
C
   12   IF(ABS(TEMP(KTEM)-TH).LT..001) GO TO 13
        T2=TEMP(KTEM)
        IF(NCP.EQ.0) CALL DELHS(A,B,C,D,TL,T2,DH,DS)
        IF(NCP.NE.0) CALL DELHS1(A,B,C,D,TL,T2,DH,DS)
        DHEL(KTEM)=DHEL(KTEM)+(DH+H)*FNUM(KELEM)
```

```
          DSEL(KTEM)=DSEL(KTEM)+(DS+S*.001)*FNUM(KELEM)
          KELEM=KELEM+1
          IF(KELEM-NELS) 7, 7, 15
C
C IF THE LAST T WAS A TRANSITION POINT GO TO 14
C IF NOT, SUM THE ENTHALPY AND ENTROPY FOR THE ENTIRE RANGE
C
   13     IF(ABS(TEMP(KTEM)-TEMP(KTEM-1)).LT..001) GO TO 14
          IF(NCP.EQ.0) CALL DELHS(A,B,C,D,TL,TH,DH,DS)
          IF(NCP.NE.0) CALL DELHS1(A,B,C,D,TL,TH,DH,DS)
          DHEL(KTEM)=DHEL(KTEM)+(DH+H)*FNUM(KELEM)
          DSEL(KTEM)=DSEL(KTEM)+(DS+S*.001)*FNUM(KELEM)
          KELEM=KELEM+1
          IF(KELEM-NELS) 7, 7, 15
C
C T IS THE TEMPERATURE POINT AFTER A PHASE TRANSITION, SO JUST ADD
C THE ENTHALPY AND ENTROPY OF TRANSITION TO THE SUMMATION
C
   14     KREAD=TWO+.01
          H=AH(KREAD)
          S=AS(KREAD)
          DHEL(KTEM)=DHEL(KTEM-1)+H*FNUM(KELEM)
          DSEL(KTEM)=DSEL(KTEM-1)+S*.001*FNUM(KELEM)
C
C CHECK OTHER ELEMENTS FOR TRANSITIONS AT THIS T, AND IF THERE ARE
C ANY ADD H & S
C
 1501     KELEM=KELEM+1
          IF(KELEM.GT.NELS) GO TO 15
          KREAD=NEL(KELEM)
 1502     TWO=ATWO(KREAD)
          H=AH(KREAD)
          S=AS(KREAD)
          TH=ATH(KREAD)
          IF(ABS(TEMP(KTEM)-TH).LT..001) GO TO 1500
          IF(ABS(TWO).LT..001) GO TO 1501
          KREAD=TWO+.01
          IF(KREAD) 110, 110, 1502
 1500     DHEL(KTEM)=(DHEL(KTEM)+H*FNUM(KELEM))
          DSEL(KTEM)=(DSEL(KTEM)+S*.001*FNUM(KELEM))
          GO TO 1501
   15     CONTINUE
C
C IF T IS OVER 200C START NEW RANGES DUE TO CRISS AND COBBLE FORMULA
C
   33     IF(TEMP(KTEM).GT.473.15) ENTH1=DHEL(KTEM-1)
          IF(TEMP(KTEM).GT.473.15) NTEMP1=KTEM-1
          IF(TEMP(KTEM).GT.473.15) DELS=DELST
C
C IF T IS SECOND TRANSITION TEMPERATURE, GO TO 10
C
          IF(ABS(TEMP(KTEM)-TEMP(KTEM-1)).LT..001) GO TO 10
C
C IF TEMP(KTEM) = TEMP(NTEMP1) DO NOT CALCULATE CP MEAN
C
```

```
          IF(KTEM.EQ.NTEMP1) GO TO 102
          CPELEM=(DHEL(KTEM)-ENTH1)/(TEMP(KTEM)-TEMP(NTEMP1))
C
C IF T IS ABOVE 25C, DO NOT CALCULATE DELTA S
C
  102     IF(ABS(TEMP(KTEM)-298.15).GT..001) GO TO 101
          DELS=ENTROI*.001+DSEL(1)
C
C CALCULATE DELTA H & F, S, AND H, ALONG WITH CP MEAN FOR ION
C
  101     CALL CPMEAN(KTEM)
          CALL CPHYDR(KTEM)

          DELST=DSEL(KTEM)+(ENTROC-Z*ENTRO)*.001
          DELTAC=CPELEM+(CP2-Z*CPH)*.001
          DELTAT=TEMP(KTEM)-TEMP(NTEMP1)
          DELF(KTEM)=DELF(NTEMP1)-DELS*DELTAT+DELTAC*(DELTAT-
     1    TEMP(KTEM)*ALOG(TEMP(KTEM)/TEMP(NTEMP1)))
          DELH(KTEM)=DELF(KTEM)+TEMP(KTEM)*DELST
          ENTH(KTEM)=CP*.001*(TEMP(KTEM)-TEMP(1))+DELH(1)
C
C IF T IS ABOVE 200C, CALCULATE H FROM THE CP MEAN CALCULATED FROM
C THE LAST TEMPERATURE INTERVAL AS PER CRISS & COBBLE
C
          IF(TEMP(KTEM).GT.473.15) ENTH(KTEM)=DELTAT*.001*CP2+ENTH(KTEM-1)
          DELSI(KTEM)=ENTROC
          DELCP(KTEM)=CP
C
C USE ABOVE 200C METHODS FOR CALCULATING CP MEAN IF T IS ABOVE 200C
C
          IF(ABS(TEMP(KTEM)-473.15).LT..001) ENTH2=ENTH(KTEM)
          IF(ABS(TEMP(KTEM)-523.15).LT..001) ENTH3=ENTH(KTEM)
          IF(TEMP(KTEM).GT.473.16) DELCP(KTEM)=(ENTH(KTEM)-ENTH2)*1000.
     1    /(TEMP(KTEM)-473.15)
          IF(TEMP(KTEM).GT.523.16) DELCP(KTEM)=(ENTH(KTEM)-ENTH3)*1000.
     1    /(TEMP(KTEM)-523.15)
          GO TO 100
C
C AFTER TRANSITION CALCULATE A NEW DELTA S OF REACTION AND USE THIS
C TO GET THE REST OF THE DATA AT THIS TEMPERATURE POINT
C
   10     DELS=(ENTROC-Z*ENTRO)*.001+DSEL(KTEM)
          DELF(KTEM)=DELF(KTEM-1)
          DELCP(KTEM)=DELCP(KTEM-1)
          DELSI(KTEM)=ENTROC
          ENTH(KTEM)=ENTH(KTEM-1)
          DELH(KTEM)=DELF(KTEM)+TEMP(KTEM)*DELS
          NTEMP1=KTEM
          ENTH1=DHEL(KTEM)
  100     RETURN
  110     STOP
          END
```

FILE FOR21.DAT (ELEMENT REFERENCE STATE DATA)

```
100.00000      0.00000     10.17000    298.15000    1234.00000 AG (C)
  5.09000      2.04000      0.36000      0.00000       0     1
102.00000      0.00000      6.77000    298.15000     932.00000 AL (C)
  4.94000      2.96000      0.00000      0.00000       0     2
  0.00000      0.00000      8.40000    298.15000     885.00000 AS (ALPHA)
  5.23000      2.22000      0.00000      0.00000       0     3
104.00000      0.00000     11.33000    298.15000    1336.00000 AU (C)
  5.66000      1.22200      0.03400      0.00000       0     4
106.00000      0.00000      1.40000    298.15000    2450.00000 B  (BETA)
  4.73500      1.38000     -2.20100      0.00000       0     5
107.00000      0.00000     15.00000    298.15000     643.00000 BA (ALPHA)
  5.43200      3.15100     -0.06800      0.00000       0     6
110.00000      0.00000      2.27000    298.15000    1556.00000 BE (C)
  4.54000      2.05000     -0.80000      0.00000       0     7
112.00000      0.00000     13.56000    298.15000     544.50000 BI (C)
  5.48100      2.42000      0.00000      0.00000       0     8
113.00000      0.00000     36.38400    298.15000     331.40000 BR2 (L)
 17.10000      0.00000      0.00000      0.00000       0     9
114.00000      0.00000      1.37200    298.15000    1100.00000 C  (GRAPHITE)
  0.02600      9.30700     -0.35400     -4.15500       0    10
115.00000      0.00000      9.90000    298.15000     737.00000 CA (ALPHA)
  5.24000      3.50000      0.00000      0.00000       0    11
118.00000      0.00000     12.37000    298.15000     594.00000 CD (GAMMA)
  5.31000      2.94000      0.00000      0.00000       0    12
120.00000      0.00000     15.30000    298.15000    1003.00000 298 REF #2
  5.65000      2.30000      1.20000      0.00000       0    13 CE (BETA)
  0.00000      0.00000     53.28800    298.15000    3000.00000 CL2 (G)
  8.82000      0.06000     -0.68000      0.00000       0    14
122.00000      0.00000      7.18000    298.15000     700.00000 CO (ALPHA)
  4.74000      4.00000      0.00000      0.00000       0    15
127.00000      0.00000      5.68000    298.15000    2176.00000 CR (C)
  4.73000      3.07000     -0.06200      0.00000       0    16
129.00000      0.00000     20.16000    298.15000     301.80000 298 REF #2
  1.93200     19.17800      0.00000      0.00000       0    17 CS (C)
130.00000      0.00000      7.92300    298.15000    1357.00000 CU (C)
  5.41000      1.50000      0.00000      0.00000       0    18
  0.00000      0.00000     48.44000    298.15000    2000.00000 F2 (G)
  8.29000      0.44000     -0.80000      0.00000       0    19
132.00000      0.00000      6.52000    298.15000    1033.00000 FE (ALPHA)
  4.18000      5.92000      0.00000      0.00000       0    20
136.00000      0.00000      9.77000    298.15000     302.90000 GA (C)
  6.19000      0.00000      0.00000      0.00000       0    21
  0.00000      0.00000     16.27000    298.15000    1623.00000 GD (C)
  9.22100     -1.21500      0.00000      0.00000       0    22
138.00000      0.00000      7.43000    298.15000    1213.00000 GE (C)
  5.98000      0.82000     -0.56000      0.00000       0    23
  0.00000      0.00000     31.20800    298.15000    3000.00000 H2 (G)
  6.52000      0.78000      0.12000      0.00000       0    24
139.00000      0.00000     10.41000    298.15000    1000.00000 HF (ALPHA)
  5.60700      1.82000      0.00000      0.00000       0    25
143.00000      0.00000     18.17000    298.15000     630.00000 HG (L)
  7.26000     -2.74000      0.00000      2.42700       0    26
```

144.00000	0.00000	13.82000	298.15000	429.30000	IN (C)
5.14000	4.20000	0.00000	0.00000	0	27
146.00000	0.00000	8.48000	298.15000	2727.00000	IR (C)
5.58000	1.38000	0.00000	0.00000	0	28
148.00000	0.00000	27.75700	298.15000	387.00000	I2 (C)
-12.10500	59.01200	6.68600	0.00000	0	29
150.00000	0.00000	15.46000	298.15000	336.35000	K (C)
1.87300	17.18800	0.00000	0.00000	0	30
152.00000	0.00000	13.60000	298.15000	1141.00000	298 REF #2
6.17000	1.60000	0.00000	0.00000	0	31 LA (BETA)
154.00000	0.00000	6.95000	298.15000	453.69000	298 REF #2
3.33200	8.21200	0.00000	0.00000	0	32 LI (C)
157.00000	0.00000	7.81000	298.15000	923.00000	MG (C)
5.33000	2.45000	-0.10300	0.00000	0	33
159.00000	0.00000	7.65000	298.15000	990.00000	MN (ALPHA)
5.70000	3.38000	-0.37500	0.00000	0	34
164.00000	0.00000	6.85000	298.15000	2890.00000	MO (C)
5.19000	1.65800	0.00000	0.00000	0	35
0.00000	0.00000	45.77000	298.15000	2500.00000	N2 (G)
6.66000	1.02000	0.00000	0.00000	0	36
165.00000	0.00000	12.23000	298.15000	370.98000	298 REF #2
3.53500	10.57100	0.00000	0.00000	0	37 NA (C)
168.00000	0.00000	8.70000	298.15000	2740.00000	NB (C)
5.67000	0.96000	0.00000	0.00000	0	38
170.00000	0.00000	17.10000	298.15000	1135.00000	ND (ALPHA)
3.50300	6.43400	1.07100	0.00000	0	39
172.00000	0.00000	7.14000	298.15000	630.00000	NI (ALPHA)
7.80000	-0.43000	-1.33500	0.00000	0	40
0.00000	0.00000	49.00300	298.15000	3000.00000	O2 (G)
7.16000	1.00000	-0.40000	0.00000	0	41
174.00000	0.00000	7.80000	298.15000	3300.00000	OS (C)
5.63000	0.92000	0.00000	0.00000	0	42
175.00000	0.00000	9.82000	298.15000	317.30000	P (ALPHA,WHITE)
4.57000	3.78000	0.00000	0.00000	0	43
177.00000	0.00000	15.49000	298.15000	600.60000	PB (C)
5.62900	2.32800	0.00000	0.00000	0	44
180.00000	0.00000	8.98000	298.15000	1823.00000	PD (C)
5.79000	1.37400	0.00000	0.00000	0	45
181.00000	0.00000	9.95000	298.15000	2043.00000	PT (C)
5.79600	1.28500	0.00000	0.00000	0	46
182.00000	0.00000	12.30000	298.15000	395.00000	298 REF #2
3.92000	12.50000	0.00000	0.00000	0	47 PU (ALPHA)
188.00000	0.00000	18.10000	298.15000	312.00000	298 REF #2
3.27000	13.71800	0.00000	0.00000	0	48 RB (C)
189.00000	0.00000	8.81000	298.15000	3453.00000	RE (C)
5.59000	1.34400	0.14900	0.00000	0	49
190.00000	0.00000	7.53000	298.15000	2239.00000	RH (C)
5.25000	2.40000	0.00000	0.00000	0	50
191.00000	0.00000	6.82000	298.15000	1308.00000	RU (ALPHA)
5.25000	1.50000	0.00000	0.00000	0	51
0.00000	30.68000	54.51000	298.15000	2000.00000	S2 (G)
8.54000	0.28000	-0.79000	0.00000	0	52
195.00000	0.00000	10.92000	298.15000	903.00000	SB (C)
5.34000	2.14000	0.00000	0.00000	0	53

198.00000	0.00000	10.14400	298.15000	494.00000	SE (BLACK)
3.82100	7.21800	0.00000	0.00000	0 54	
200.00000	0.00000	4.50000	298.15000	1685.00000	SI (C)
5.45500	0.92200	-0.84600	0.00000	0 55	
201.00000	0.00000	16.63000	298.15000	1190.00000	SM (ALPHA)
6.00000	5.84000	-0.61000	0.00000	0 56	
204.00000	0.00000	12.32000	298.15000	505.00000	SN (WHITE)
5.16000	4.34000	0.00000	0.00000	0 57	
206.00000	0.00000	12.50000	298.15000	862.00000	SR (ALPHA)
5.31000	3.32000	0.00000	0.00000	0 58	
209.00000	0.00000	9.92000	298.15000	3269.00000	TA (C)
5.98000	0.59400	0.00000	0.00000	0 59	
210.00000	0.00000	8.00000	298.15000	2473.00000	298 REF #2
5.20000	2.00000	0.00000	0.00000	0 60	TC (C)
211.00000	0.00000	11.88000	298.15000	723.00000	TE (C)
4.58000	5.25000	0.00000	0.00000	0 61	
213.00000	0.00000	12.76000	298.15000	800.00000	298 REF #2
5.77300	2.54800	0.00000	0.00000	0 62	TH (ALPHA)
216.00000	0.00000	7.32000	298.15000	1155.00000	TI (ALPHA)
5.29600	2.45800	0.00000	0.00000	0 63	
218.00000	0.00000	15.34000	298.15000	507.00000	TL (ALPHA)
3.74000	6.04000	0.66700	0.00000	0 64	
221.00000	0.00000	12.03000	298.15000	941.00000	298 REF #2
2.61000	8.95000	1.17000	0.00000	0 65	U (ALPHA)
224.00000	0.00000	6.91000	298.15000	2190.00000	V (C)
4.90000	2.58000	0.20000	0.00000	0 66	
225.00000	0.00000	7.80000	298.15000	2500.00000	W (C)
5.47600	1.12000	0.00000	0.00000	0 67	
228.00000	0.00000	10.62000	298.15000	1758.00000	Y (ALPHA)
5.59000	1.90000	0.29000	0.00000	0 68	
0.00000	36.40000	41.35200	298.15000	2000.00000	YB (G)
4.97000	0.00000	0.00000	0.00000	0 69	
230.00000	0.00000	9.95000	298.15000	692.70000	ZN (C)
5.35000	2.40000	0.00000	0.00000	0 70	
232.00000	0.00000	9.32000	298.15000	1135.00000	ZR (ALPHA)
5.25200	2.78000	0.00000	0.00000	0 71	
0.00000	0.00000	0.00000	0.00000	0.00000	
0.00000	0.00000	0.00000	0.00000	0 72	
0.00000	0.00000	0.00000	0.00000	0.00000	
0.00000	0.00000	0.00000	0.00000	0 73	
0.00000	0.00000	0.00000	0.00000	0.00000	
0.00000	0.00000	0.00000	0.00000	0 74	
0.00000	0.00000	0.00000	0.00000	0.00000	
0.00000	0.00000	0.00000	0.00000	0 75	
0.00000	0.00000	0.00000	0.00000	0.00000	
0.00000	0.00000	0.00000	0.00000	0 76	
0.00000	0.00000	0.00000	0.00000	0.00000	
0.00000	0.00000	0.00000	0.00000	0 77	
0.00000	0.00000	0.00000	0.00000	0.00000	
0.00000	0.00000	0.00000	0.00000	0 78	
0.00000	0.00000	0.00000	0.00000	0.00000	
0.00000	0.00000	0.00000	0.00000	0 79	
0.00000	0.00000	0.00000	0.00000	0.00000	
0.00000	0.00000	0.00000	0.00000	0 80	
0.00000	0.00000	0.00000	0.00000	0.00000	

0.00000	0.00000	0.00000	0.00000	0 81
0.00000	0.00000	0.00000	0.00000	0.00000
0.00000	0.00000	0.00000	0.00000	0 82
0.00000	0.00000	0.00000	0.00000	0.00000
0.00000	0.00000	0.00000	0.00000	0 83
0.00000	0.00000	0.00000	0.00000	0.00000
0.00000	0.00000	0.00000	0.00000	0 84
0.00000	0.00000	0.00000	0.00000	0.00000
0.00000	0.00000	0.00000	0.00000	0 85
0.00000	0.00000	0.00000	0.00000	0.00000
0.00000	0.00000	0.00000	0.00000	0 86
0.00000	0.00000	0.00000	0.00000	0.00000
0.00000	0.00000	0.00000	0.00000	0 87
0.00000	0.00000	0.00000	0.00000	0.00000
0.00000	0.00000	0.00000	0.00000	0 88
0.00000	0.00000	0.00000	0.00000	0.00000
0.00000	0.00000	0.00000	0.00000	0 89
0.00000	0.00000	0.00000	0.00000	0.00000
0.00000	0.00000	0.00000	0.00000	0 90
0.00000	0.00000	0.00000	0.00000	0.00000
0.00000	0.00000	0.00000	0.00000	0 91
0.00000	0.00000	0.00000	0.00000	0.00000
0.00000	0.00000	0.00000	0.00000	0 92
0.00000	0.00000	0.00000	0.00000	0.00000
0.00000	0.00000	0.00000	0.00000	0 93
0.00000	0.00000	0.00000	0.00000	0.00000
0.00000	0.00000	0.00000	0.00000	0 94
0.00000	0.00000	0.00000	0.00000	0.00000
0.00000	0.00000	0.00000	0.00000	0 95
0.00000	0.00000	0.00000	0.00000	0.00000
0.00000	0.00000	0.00000	0.00000	0 96
0.00000	0.00000	0.00000	0.00000	0.00000
0.00000	0.00000	0.00000	0.00000	0 97
0.00000	0.00000	0.00000	0.00000	0.00000
0.00000	0.00000	0.00000	0.00000	0 98
0.00000	0.00000	0.00000	0.00000	0.00000
0.00000	0.00000	0.00000	0.00000	0 99
101.00000	2.85500	2.31400	1234.00000	2437.00000 AG (L)
7.30000	0.00000	0.00000	0.00000	0 100
0.00000	61.67700	25.30900	2437.00000	2500.00000 AG (G)
4.96800	0.00000	0.00000	0.00000	0 101
103.00000	2.60000	2.79000	932.00000	2723.00000 AL (L)
7.60000	0.00000	0.00000	0.00000	0 102
0.00000	69.50000	25.52300	2723.00000	3000.00000 AL (G)
4.97000	0.00000	0.00000	0.00000	0 103
105.00000	2.95500	2.21200	1336.00000	3081.00000 AU (L)
7.00000	0.00000	0.00000	0.00000	0 104
0.00000	80.07500	25.99000	3081.00000	3100.00000 AU (G)
2.99600	1.06100	9.53300	0.00000	0 105
0.00000	5.40000	2.20400	2450.00000	3600.00000 B (L)
7.50000	0.00000	0.00000	0.00000	0 106
108.00000	0.15000	0.23300	643.00000	983.00000 BA (BETA)
-1.36000	19.20000	0.00000	0.00000	0 107
109.00000	1.83000	1.86200	983.00000	1895.00000 BA (L)

```
 11.50000      0.00000      0.00000      0.00000        0    108
  0.00000     36.70000     19.36700   1895.00000   2200.00000  BA (G)
 -1.03000      3.93000      0.00000      0.00000        0    109
111.00000      3.50000      2.24900   1556.00000   2744.00000  BE (L)
  6.08000      0.51500      0.00000      0.00000        0    110
  0.00000     71.13700     25.92500   2744.00000   3000.00000  BE (G)
  5.28300     -0.39800     -0.20000      0.10800        0    111
  0.00000      2.60000      4.77500    544.50000   1852.00000  BI (L)
  7.60000      0.00000      0.00000      0.00000        0    112
  0.00000      7.10565     21.36662    331.40000   2000.00000  BR2 (G)
  8.92900      0.11100     -0.30900      0.00000        0    113
  0.00000      0.00000      0.00000   1100.00000   4073.00000  C  (GRAPHITE)
  5.84100      0.10400     -7.55900      0.00000        0    114
116.00000      0.06000      0.08100    737.00000   1123.00000  CA (BETA)
  2.59000      6.66000      0.00000      0.00000        0    115
117.00000      2.00000      1.78100   1123.00000   1762.00000  CA (L)
  7.40000      0.00000      0.00000      0.00000        0    116
  0.00000     36.50000     20.71500   1762.00000   2000.00000  CA (G)
  4.97000      0.00000      0.00000      0.00000        0    117
119.00000      1.53000      2.57600    594.00000   1043.00000  CD (L)
  7.10000      0.00000      0.00000      0.00000        0    118
  0.00000     23.90000     22.91500   1043.00000   2000.00000  CD (G)
  4.97000      0.00000      0.00000      0.00000        0    119
121.00000      0.72000      0.71800   1003.00000   1077.00000  CE(GAMMA)
  9.05000      0.00000      0.00000      0.00000        0    120
  0.00000      1.30000      1.20700   1077.00000   4083.00000  CE (L)
  9.35000      0.00000      0.00000      0.00000        0    121
123.00000      0.11000      0.15700    700.00000   1393.00000  CO (BETA)
  7.35000     -3.26000      0.00000      4.86000        0    122  CURIE PT = 1393
124.00000      0.00000      0.00000   1393.00000   1425.00000  CO (BETA)
127.50000    -82.20000      0.00000      0.00000        0    123
125.00000      0.00000      0.00000   1425.00000   1768.00000  CO (BETA)
 13.06800     -2.19200      0.00000      0.00000        0    124
126.00000      4.10000      2.31900   1768.00000   3174.00000  CO (L)
  9.00000      0.00000      0.00000      0.00000        0    125
  0.00000     89.76100     28.28000   3174.00000   4000.00000  CO (G)
  6.45600     -0.05100     -0.85700      0.00000        0    126
128.00000      5.00000      2.29800   2176.00000   2938.00000  CR (L)
  9.40000      0.00000      0.00000      0.00000        0    127
  0.00000     81.29400     27.67000   2938.00000   3200.00000  CR (G)
  2.32000      1.65300      9.53800      0.00000        0    128
  0.00000      0.52000      1.72300    301.80000    955.00000  CS (L)
  7.60000      0.00000      0.00000      0.00000        0    129
131.00000      3.12000      2.29900   1357.00000   2846.00000  CU (L)
  7.50000      0.00000      0.00000      0.00000        0    130
  0.00000     72.60000     25.50900   2846.00000   3000.00000  CU (G)
  5.36900     -0.72000     -0.20500      0.30700        0    131
133.00000      1.20000      1.16200   1033.00000   1183.00000  FE (BETA)
  9.00000      0.00000      0.00000      0.00000        0    132
134.00000      0.22000      0.18600   1183.00000   1673.00000  FE (GAMMA)
  1.84000      4.66000      0.00000      0.00000        0    133
135.00000      0.21000      0.12600   1673.00000   1809.00000  FE (DELTA)
 10.50000      0.00000      0.00000      0.00000        0    134
197.00000      3.30000      1.82400   1809.00000   3148.00000  FE (L)
  9.78000      0.40000      0.00000      0.00000        0    135
```

137.00000	1.33500	4.40700	302.90000	2520.00000	GA (L)
6.65000	0.00000	0.00000	0.00000	0	136
0.00000	61.46000	24.38900	2520.00000	3000.00000	GA (G)
5.07000	0.00000	0.00000	0.00000	0	137
0.00000	8.80000	7.25500	1213.00000	3125.00000	GE (L)
6.80000	0.00000	0.00000	0.00000	0	138
140.00000	0.00000	0.00000	1000.00000	2023.00000	HF (ALPHA)
6.31500	1.69600	0.00000	0.00000	0	139
141.00000	1.65000	0.81600	2023.00000	2495.00000	HF (BETA)
6.31500	1.69600	0.00000	0.00000	0	140
142.00000	5.75000	2.30500	2495.00000	4723.00000	HF (L)
8.00000	0.00000	0.00000	0.00000	0	141
0.00000	136.40000	28.88000	4723.00000	6000.00000	HF (G)
4.76000	0.83100	0.00000	0.00000	0	142
0.00000	14.13000	22.43800	630.00000	3000.00000	HG (G)
4.96800	0.00000	0.00000	0.00000	0	143
145.00000	0.78000	1.81700	429.30000	2343.00000	IN (L)
7.19000	0.00000	0.00000	0.00000	0	144
0.00000	55.41200	23.65000	2343.00000	2400.00000	IN (G)
7.21800	-0.60800	0.00000	0.00000	0	145
147.00000	6.30000	2.31000	2727.00000	4662.00000	IR (L)
9.35000	0.00000	0.00000	0.00000	0	146
0.00000	146.30000	31.38100	4662.00000	5000.00000	IR (G)
6.79500	0.16200	0.00000	0.00000	0	147
149.00000	3.71000	9.58700	387.00000	458.00000	I2 (L)
19.28100	0.00000	0.00000	0.00000	0	148
0.00000	10.02795	21.89147	458.00000	2000.00000	I2 (G)
8.94000	0.13600	-0.14800	0.00000	0	149
151.00000	0.56000	1.66500	336.35000	1037.00000	K (L)
8.88600	-4.57000	0.00000	2.94400	0	150
0.00000	19.04588	18.26266	1037.00000	2000.00000	K (G)
4.95500	0.01800	0.00000	0.00000	0	151
153.00000	0.68000	0.59600	1141.00000	1193.00000	LA (GAMMA)
8.10000	0.00000	0.00000	0.00000	0	152
235.00000	2.03000	1.70200	1193.00000	3693.00000	LA (L)
8.30000	0.00000	0.00000	0.00000	0	153
155.00000	0.71700	1.58000	453.69000	1200.00000	LI (L)
5.85000	1.30800	2.06700	-0.46700	0	154
156.00000	0.00000	0.00000	1200.00000	1620.00000	LI (L)
7.12000	-0.20000	0.00000	0.00000	0	155
0.00000	35.19027	21.48192	1620.00000	2000.00000	LI (G)
4.96400	0.00600	0.00000	0.00000	0	156
158.00000	2.14000	2.31900	923.00000	1378.00000	MG (L)
7.60000	0.00000	0.00000	0.00000	0	157
0.00000	30.50000	22.13400	1378.00000	2000.00000	MG (G)
4.96800	0.00000	0.00000	0.00000	0	158
160.00000	0.53500	0.54000	990.00000	1360.00000	MN (BETA)
8.33000	0.66000	0.00000	0.00000	0	159
161.00000	0.52500	0.38600	1360.00000	1410.00000	MN (GAMMA)
6.03000	3.56000	-0.44300	0.00000	0	160
162.00000	0.43000	0.30500	1410.00000	1517.00000	MN (DELTA)
11.30000	0.00000	0.00000	0.00000	0	161
163.00000	3.50000	2.30700	1517.00000	2324.00000	MN (L)
11.00000	0.00000	0.00000	0.00000	0	162

0.00000	52.80000	22.71900	2324.00000	3000.00000	MN (G)
4.15400	0.35700	0.00000	0.00000	0 163	
0.00000	6.65000	2.30100	2890.00000	4924.00000	MO (L)
10.00000	0.00000	0.00000	0.00000	0 164	
166.00000	0.62000	1.67100	370.98000	1156.00000	NA (L)
8.95500	-4.57800	0.00000	2.54200	0 165	
167.00000	23.32980	19.89526	1156.00000	1600.00000	NA (G)
4.96800	0.00000	0.00000	0.00000	0 166	
0.00000	0.00000	0.00000	1600.00000	2000.00000	NA (G)
4.95200	0.01000	0.00000	0.00000	0 167	
169.00000	6.30000	2.29900	2740.00000	5007.00000	NB (L)
8.00000	0.00000	0.00000	0.00000	0 168	
0.00000	163.30000	32.61400	5007.00000	5500.00000	NB (G)
4.25000	0.86500	0.00000	0.00000	0 169	
171.00000	0.71300	0.62800	1135.00000	1297.00000	ND (GAMMA)
10.65400	0.00000	0.00000	0.00000	0 170	
0.00000	1.70500	1.31500	1297.00000	3384.00000	ND (L)
11.66000	0.00000	0.00000	0.00000	0 171	
173.00000	0.14000	0.22200	630.00000	1728.00000	NI (BETA)
7.10000	1.00000	-2.23000	0.00000	0 172	
0.00000	4.21000	2.43600	1728.00000	3193.00000	NI (L)
9.30000	0.00000	0.00000	0.00000	0 173	
0.00000	7.59000	2.30000	3300.00000	4500.00000	OS (L)
8.60000	0.00000	0.00000	0.00000	0 174	
176.00000	0.15700	0.49500	317.30000	547.00000	P (L)
6.29200	0.00000	0.00000	0.00000	0 175	
0.00000	2.90995	5.28381	547.00000	2000.00000	(P4)/4 (G)
4.89050	0.04050	-0.80325	0.00000	0 176	
178.00000	1.14100	1.90000	600.60000	1400.00000	PB (L)
7.69900	-0.65900	0.00000	0.00000	0 177	
179.00000	0.00000	0.00000	1400.00000	2016.00000	PB (L)
6.10800	0.48500	0.00000	0.00000	0 178	
0.00000	42.36000	21.01200	2016.00000	2500.00000	PB (G)
5.64000	-1.43900	-0.29300	0.77800	0 179	
0.00000	4.20000	2.30400	1823.00000	3213.00000	PD (L)
8.30000	0.00000	0.00000	0.00000	0 180	
0.00000	4.70000	2.30100	2043.00000	4097.00000	PT (L)
8.30000	0.00000	0.00000	0.00000	0 181	
183.00000	0.96000	2.43000	395.00000	479.00000	PU (BETA)
4.97000	11.20000	0.00000	0.00000	0 182	
184.00000	0.14000	0.29200	479.00000	592.00000	PU (GAMMA)
4.97000	11.20000	0.00000	0.00000	0 183	
185.00000	0.16000	0.27000	592.00000	724.00000	PU (DELTA)
11.20000	0.00000	0.00000	0.00000	0 184	
186.00000	0.02000	0.02800	724.00000	749.00000	PU (DELTA 1)
11.20000	0.00000	0.00000	0.00000	0 185	
187.00000	0.47000	0.62800	749.00000	913.00000	PU (EPSILON)
9.70000	0.00000	0.00000	0.00000	0 186	
0.00000	0.94000	1.03000	913.00000	3508.00000	PU (L)
9.70000	0.00000	0.00000	0.00000	0 187	
0.00000	0.54000	1.73100	312.00000	967.00000	RB (L)
7.50000	0.00000	0.00000	0.00000	0 188	
0.00000	8.00000	2.31700	3453.00000	5960.00000	RE (L)
10.80000	0.00000	0.00000	0.00000	0 189	

0.00000	5.15000	2.30000	2239.00000	4000.00000	RH (L)
10.00000	0.00000	0.00000	0.00000	0	190
192.00000	0.03000	0.02300	1308.00000	1773.00000	RU (BETA & GAMMA)
7.20000	0.00000	0.00000	0.00000	0	191 BETA = GAMMA @1473
193.00000	0.23000	0.13000	1773.00000	2700.00000	RU (DELTA)
7.13100	1.06300	0.00000	0.00000	0	192
194.00000	6.21000	2.30000	2700.00000	4392.00000	RU (L)
10.00000	0.00000	0.00000	0.00000	0	193
0.00000	141.40000	32.19500	4392.00000	4500.00000	RU (G)
4.57200	0.87500	8.76000	0.00000	0	194
196.00000	4.69000	5.19400	903.00000	1908.00000	SB (L)
7.50000	0.00000	0.00000	0.00000	0	195
0.00000	54.43089	23.42721	1908.00000	2000.00000	SB (G)
4.46600	0.38100	0.00000	0.00000	0	196
0.00000	83.71000	26.59100	3148.00000	4000.00000	FE (G)
3.78400	0.79600	9.40300	-2.24300	1	197
0.00000	1.30000	2.63200	494.00000	958.00000	SE (L)
8.40000	0.00000	0.00000	0.00000	0	198
0.00000	0.00000	0.00000	0.00000	0.00000	
0.00000	0.00000	0.00000	0.00000	0	199
0.00000	12.00000	7.12200	1685.00000	3492.00000	SI (L)
6.50000	0.00000	0.00000	0.00000	0	200
202.00000	0.74400	0.62500	1190.00000	1345.00000	SM (GAMMA)
11.22000	0.00000	0.00000	0.00000	0	201
203.00000	2.13000	1.58400	1345.00000	2076.00000	SM (L)
12.57000	0.00000	0.00000	0.00000	0	202
0.00000	39.38000	18.96900	2076.00000	2200.00000	SM (G)
8.20800	-0.91500	-0.68300	0.00000	0	203
205.00000	1.67000	3.30700	505.00000	700.00000	SN (L)
6.70000	0.00000	0.72300	0.00000	0	204
0.00000	0.00000	0.00000	700.00000	2896.00000	SN (L)
6.85000	0.00000	0.00000	0.00000	0	205
207.00000	0.20000	0.23200	862.00000	1043.00000	SR (GAMMA)
3.03000	6.40000	0.00000	0.00000	0	206
208.00000	2.40000	2.30100	1043.00000	1648.00000	SR (L)
7.40000	0.00000	0.00000	0.00000	0	207
0.00000	33.20700	20.15000	1648.00000	2000.00000	SR (G)
4.90700	0.04500	0.00000	0.00000	0	208
0.00000	5.90000	1.80500	3269.00000	5513.00000	TA (L)
8.50000	0.00000	0.00000	0.00000	0	209
0.00000	5.68800	2.30000	2473.00000	4840.00000	TC (L)
10.00000	0.00000	0.00000	0.00000	0	210
212.00000	4.18000	5.78100	723.00000	1282.40000	TE (L)
9.00000	0.00000	0.00000	0.00000	0	211
0.00000	39.50082	21.84266	1282.40000	2000.00000	TE (G)
4.96400	0.01800	0.00000	0.00000	0	212
214.00000	0.00000	0.00000	800.00000	1633.00000	TH (ALPHA)
-3.21100	8.83400	26.09800	0.00000	0	213
215.00000	0.65300	0.40000	1633.00000	2028.00000	TH (BETA)
11.00000	0.00000	0.00000	0.00000	0	214
0.00000	3.85300	1.90000	2028.00000	5060.00000	TH (L)
11.00000	0.00000	0.00000	0.00000	0	215
217.00000	0.99000	0.85700	1155.00000	1933.00000	TI (BETA)
4.73900	1.89400	0.00000	0.00000	0	216

0.00000	4.45000	2.30200	1933.00000	3575.00000	TI (L)
8.50000	0.00000	0.00000	0.00000	0 217	
219.00000	0.09000	0.17800	507.00000	577.00000	TL (BETA)
5.00000	5.00000	0.00000	0.00000	0 218	
220.00000	0.97500	1.69000	577.00000	1760.00000	TL (L)
7.20000	0.00000	0.00000	0.00000	0 219	
0.00000	39.41000	22.39200	1760.00000	2000.00000	TL (G)
4.55400	0.38200	0.00000	0.00000	0 220	
222.00000	0.70000	0.74400	941.00000	1048.00000	U (BETA)
10.00000	0.00000	0.00000	0.00000	0 221	
223.00000	1.15000	1.09700	1048.00000	1403.00000	U (GAMMA)
9.58000	0.00000	0.00000	0.00000	0 222	
0.00000	3.00000	2.13800	1403.00000	3000.00000	U (L)
11.45000	0.00000	0.00000	0.00000	0 223	
0.00000	5.05000	2.30600	2190.00000	3652.00000	V (L)
9.50000	0.00000	0.00000	0.00000	0 224	
226.00000	0.00000	0.00000	2500.00000	3680.00000	W (C)
-50.66800	15.35300	1292.81799	0.00000	0 225	
227.00000	8.46000	2.29900	3680.00000	5936.00000	W (L)
8.50000	0.00000	0.00000	0.00000	0 226	
0.00000	192.82400	32.48400	5936.00000	6000.00000	W (G)
4.38600	0.90300	0.00000	0.00000	0 227	
229.00000	1.18900	0.67600	1758.00000	1803.00000	Y (GAMMA)
8.37000	0.00000	0.00000	0.00000	0 228	
0.00000	2.73200	1.51500	1803.00000	3000.00000	Y (L)
10.30000	0.00000	0.00000	0.00000	0 229	
231.00000	1.76500	2.54800	692.70000	1184.00000	ZN (L)
7.50000	0.00000	0.00000	0.00000	0 230	
0.00000	27.62000	23.32800	1184.00000	2000.00000	ZN (G)
4.96800	0.00000	0.00000	0.00000	0 231	
233.00000	0.96000	0.84600	1135.00000	2125.00000	ZR (BETA)
5.55400	1.11000	0.00000	0.00000	0 232	
0.00000	5.00000	2.35300	2125.00000	4777.00000	(L)
8.00000	0.00000	0.00000	0.00000	0 233	
0.00000	0.00000	0.00000	0.00000	0.00000	
0.00000	0.00000	0.00000	0.00000	0 234	
0.00000	96.10000	26.02200	3693.00000	4000.00000	LA (G)
7.74100	0.04600	-7.37100	1.75500	1 235	

SUBROUTINE HELGE

```
      SUBROUTINE HELGE(DG,KTEM)
      COMMON /RICK3/ ENTROI,ENTROC,ION,Z,ENTRO,TEMP(16)
      COMMON /RICK6/ DELHR,DELSR,NTEMP
      T=TEMP(KTEM)
      FOUR=EXP(-12.741+.01875*TEMP(1))
      ONE=1.-EXP(EXP(-12.741+.01875*T)-FOUR+(T-298.15)/219.)
      DG=-DELSR*(TEMP(1)-218.297*ONE)+DELHR
1     RETURN
      END
```

SUBROUTINE HELGES

```
SUBROUTINE HELGES(DS,KTEM)
COMMON /RICK3/ ENTROI, ENTROC, ION, Z, ENTRO, TEMP(16)
COMMON /RICK6/ DELHR, DELSR, NTEMP
T=TEMP(KTEM)
ONE=EXP(-12.741+.01875*T)
TWO=EXP(-12.741+.01875*298.15)
THREE=(T-298.15)/219.0
FOUR=1.+4.106*ONE
FIVE=(EXP(ONE-TWO+THREE))/1.00322
DS=DELSR*FIVE*FOUR
RETURN
END
```

PROGRAM IONCOM

```
C PROGRAM IONCOM CALCULATES THERMODYNAMIC DATA FOR COMPOUNDS AND ELEMENTS
C EXCLUDING PHOSPHORUS WHICH MUST BE HANDLED SEPARATELY DUE TO REFERENCE
C STATES CHOOSEN
C
C
C
        DIMENSION TEMP(40), HC(8), SC(8), AC(8), BC(8), CC(8), DC(8)
        DIMENSION RANGEL(8), RANGEH(8), NELE(8), FNUM(8), NEL(8)
        DIMENSION DHEL(40), DSEL(40), ENTH(40), DHCOM(40), DSCOM(40)
        DIMENSION DHREAC(40), DSREAC(40), DGREAC(40), IDENT(4), CP(40)
        DIMENSION NREF(40), CPARAY(2)
        DIMENSION AAA(250), AAB(250), AAC(250), AAD(250), ANCP(250)
        DIMENSION ATWO(250), AH(250), AS(250), ATL(250), ATH(250)
C
C SET UP INPUT FILE FOR DATA ON ELEMENT OR COMPOUND TO BE RUN
C
        CALL ASSDSK(1,4)
        CALL IFILE(1,'CONVR','DAT')
C
C GET INFORMATION ON ELEMENTS
C
        DO 1000 I=1,250
        READ(21,1001,END=1002) ATWO(I), AH(I), AS(I), ATL(I), ATH(I)
        READ(21,1001) AAA(I), AAB(I), AAC(I), AAD(I), ANCP(I)
 1000   CONTINUE
 1002   CONTINUE
 1001   FORMAT(5F)
C
C GET INFORMATION ON ELEMENT OR COMPOUD TO BE RUN
C
        READ(1,30) IDENT
 30     FORMAT(4A5)
        READ(1,1) NTEMPS, NCCP
 1      FORMAT(8I)
        DO 3 I=1,NTEMPS
        READ(1,2) TEMP(I)
 2      FORMAT(8F)
        TEMP(I)=TEMP(I)+273.15
```

```
      3         CONTINUE
                DO 31 I=1,NTEMPS
     31         READ(1,1) NREF(I)
                READ(1,1) NRANGE
                DO 4 I=1,NRANGE
                READ(1,2) HC(I), SC(I)
                READ(1,2) AC(I), BC(I), CC(I), DC(I)
                READ(1,2) RANGEL(I), RANGEH(I)
      4         CONTINUE
                READ(1,1) NELEM
                DO 6 I=1,NELEM
                READ(1,5) NELE(I), FNUM(I)
      5         FORMAT(A,F)
C
C CONVERT CHEMICAL SYMBOLS TO ELEMENT CODE NUMBERS
C
                NAME=NELE(I)
                CALL NNCODE(NAME,NCODE)
                NEL(I)=NCODE
      6         CONTINUE
C
C INITIALIZE ACCOUNTING
C
                KSETT=1
                KSETE=1
    100         KREAD=NEL(KSETE)
    104         TH=ATH(KREAD)
                TWO=ATWO(KREAD)
C
C IF HIGHEST TEMPERATURE IN SCALE IS LOWER THAN END OF CP RANGE
C NO TRANSITIONS OF THIS ELEMENT APPEAR IN THIS RANGE, GO ON TO
C NEXT ELEMENT
C
                IF(TEMP(NTEMPS).LT.TH) GO TO 105
C
C IF T IS GREATER THAN THE HIGHEST TEMPERATURE IN THE RANGE
C FOR CP, A TRANSITION OCCURED, SO INSERT TWO POINTS IN THE
C TEMPERATURE SCALE FOR THE TRANSITION
C IF NOT, INCREMENT THE TEMPERATURE POINT AND CHECK AGAIN
C
    101         IF(TEMP(KSETT).GT.TH) GO TO 102
                KSETT=KSETT+1
                IF(KSETT-NTEMPS) 101, 101, 105
    102         KOUNT=KSETT
                XTEMP=TEMP(KOUNT)
                NXREF=NREF(KOUNT)
                YTEMP=TEMP(KOUNT+1)
                NYREF=NREF(KOUNT+1)
                ZTEMP=TEMP(KOUNT+2)
                NZREF=NREF(KOUNT+2)
                TTEMP=TEMP(KOUNT+3)
                NTREF=NREF(KOUNT+3)
                TEMP(KOUNT)=TH
                NREF(KOUNT)=2
                TEMP(KOUNT+1)=TH
```

```
              NREF(KOUNT+1)=2
              NTEMPS=NTEMPS+2
   103        KOUNT=KOUNT+2
              TEMP(KOUNT)=XTEMP
              NREF(KOUNT)=NXREF
              TEMP(KOUNT+1)=YTEMP
              NREF(KOUNT+1)=NYREF
              XTEMP=ZTEMP
              NXREF=NZREF
              YTEMP=TTEMP
              NYREF=NTREF
              ZTEMP=TEMP(KOUNT+2)
              NZREF=NREF(KOUNT+2)
              TTEMP=TEMP(KOUNT+3)
              NTREF=NREF(KOUNT+3)
              IF(KOUNT+2.LE.NTEMPS) GO TO 103
              KREAD=TWO+.01
              KSETT=KSETT+2
              IF(KREAD) 110, 110, 104
   105        KSETE=KSETE+1
              KSETT=1
              IF(KSETE.LE.NELEM) GO TO 100
              NTEMP=NTEMPS
C
C CHECK TEMPERATURE SCALE TO SEE THAT THERE ARE NO MORE THAN TWO
C CONSECUTIVE LIKE TEMPERATURE POINTS
C
              DO 106 I=3,NTEMPS
   109        IF(ABS(TEMP(I)-TEMP(I-2)).GT..001) GO TO 106
              IF(TEMP(I-2).EQ.0) GO TO 106
              K=I
   107        TEMP(K)=TEMP(K+1)
              IF(K.GE.NTEMP) GO TO 108
              K=K+1
              GO TO 107
   108        NTEMP=NTEMP-1
              GO TO 109
   106        CONTINUE
              NTEMPS=NTEMP
C
C INITIALIZE ACCOUNTING
C
              KTEM=1
    11        KELEM=1
              DHEL(KTEM)=0.
              DSEL(KTEM)=0.
C
C START SUMMATION OF ELEMENTAL DATA
C
     7        KREAD=NEL(KELEM)
     8        A=AAA(KREAD)
              B=AAB(KREAD)
              C=AAC(KREAD)
              D=AAD(KREAD)
              TWO=ATWO(KREAD)
```

```
            H=AH(KREAD)
            S=AS(KREAD)
            TL=ATL(KREAD)
            TH=ATH(KREAD)
            NCP=ANCP(KREAD)
C
C IF T IS LESS THAN THE HIGHEST T IN THIS CP RANGE, GO TO 12
C IF NOT, SUM THE ENTHALPY AND ENTROPY FOR THE ENTIRE RANGE
C
            IF(TEMP(KTEM).LE.TH+.001) GO TO 12
            IF(NCP.EQ.0) CALL DELHS(A,B,C,D,TL,TH,DH,DS)
            IF(NCP.NE.0) CALL DELHS1(A,B,C,D,TL,TH,DH,DS)
            DHEL(KTEM)=(DHEL(KTEM)+(DH+H)*FNUM(KELEM))
            DSEL(KTEM)=(DSEL(KTEM)+(DS+S*.001)*FNUM(KELEM))
            KREAD=TWO+.01
            IF(KREAD) 110, 110, 8
C
C IF T IS A TRANSITION POINT, GO TO 13
C IF NOT, SUM THE ENTHALPY AND ENTROPY UP TO T AND GO ON TO THE
C NEXT ELEMENT IF ANY
C
   12       IF(ABS(TEMP(KTEM)-TH).LT..001) GO TO 13
            T2=TEMP(KTEM)
            IF(NCP.EQ.0) CALL DELHS(A,B,C,D,TL,T2,DH,DS)
            IF(NCP.NE.0) CALL DELHS1(A,B,C,D,TL,T2,DH,DS)
            DHEL(KTEM)=(DHEL(KTEM)+(DH+H)*FNUM(KELEM))
            DSEL(KTEM)=(DSEL(KTEM)+(DS+S*.001)*FNUM(KELEM))
            KELEM=KELEM+1
            IF(KELEM-NELEM) 7, 7, 15
C
C IF THE LAST T WAS A TRANSITION POINT GO TO 14
C IF NOT, SUM THE ENTHALPY AND ENTROPY FOR THE ENTIRE RANGE
C
   13       IF(ABS(TEMP(KTEM)-TEMP(KTEM-1)).LT..001) GO TO 14
            IF(NCP.EQ.0) CALL DELHS(A,B,C,D,TL,TH,DH,DS)
            IF(NCP.NE.0) CALL DELHS1(A,B,C,D,TL,TH,DH,DS)
            DHEL(KTEM)=(DHEL(KTEM)+(DH+H)*FNUM(KELEM))
            DSEL(KTEM)=(DSEL(KTEM)+(DS+S*.001)*FNUM(KELEM))
            KELEM=KELEM+1
            IF(KELEM-NELEM) 7, 7, 15
C
C T IS THE TEMPERATURE POINT AFTER A PHASE TRANSITION, SO JUST ADD
C THE ENTHALPY AND ENTROPY OF TRANSITION TO THE SUMMATION
C
   14       KREAD=TWO+.01
            H=AH(KREAD)
            S=AS(KREAD)
            DHEL(KTEM)=(DHEL(KTEM-1)+H*FNUM(KELEM))
            DSEL(KTEM)=(DSEL(KTEM-1)+S*.001*FNUM(KELEM))
C
C CHECK OTHER ELEMENTS FOR TRANSITIONS AT THIS T, AND IF THERE ARE
C ANY ADD H & S
C
```

```
1501     KELEM=KELEM+1
         IF(KELEM.GT.NELEM) GO TO 15
         KREAD=NEL(KELEM)
1502     TWO=ATWO(KREAD)
         H=AH(KREAD)
         S=AS(KREAD)
         TH=ATH(KREAD)
         IF(ABS(TEMP(KTEM)-TH).LT..001) GO TO 1500
         IF(ABS(TWO).LT..001) GO TO 1501
         KREAD=TWO+.01
         IF(KREAD) 110, 110, 1502
1500     DHEL(KTEM)=(DHEL(KTEM)+H*FNUM(KELEM))
         DSEL(KTEM)=(DSEL(KTEM)+S*.001*FNUM(KELEM))
         GO TO 1501
15       KTEM=KTEM+1
         IF(KTEM.LE.NTEMPS) GO TO 11
C
C REINITIALIZE FOR COMPOUND CALCULATIONS
C
         KTEM=1
16       KRANGE=1
         DHCOM(KTEM)=0.
         DSCOM(KTEM)=0.
C
C IF T IS LESS THAN THE HIGHEST T IN THIS RANGE FOR CP THEN GO TO 18
C IF NOT, CALCULATE THE H & S FOR THE ENTIRE RANGE
C
17       IF(TEMP(KTEM).LE.RANGEH(KRANGE)) GO TO 18
         A=AC(KRANGE)
         B=BC(KRANGE)
         C=CC(KRANGE)
         D=DC(KRANGE)
         T1=RANGEL(KRANGE)
         T2=RANGEH(KRANGE)
         IF(NCCP.EQ.0) CALL DELHS(A,B,C,D,T1,T2,DH,DS)
         IF(NCCP.NE.0) CALL DELHS1(A,B,C,D,T1,T2,DH,DS)
         DHCOM(KTEM)=DHCOM(KTEM)+HC(KRANGE)+DH
         DSCOM(KTEM)=DSCOM(KTEM)+SC(KRANGE)*.001+DS
         KRANGE=KRANGE+1
         GO TO 17
C
C IF T IS A TRANSITION POINT GO TO 19, IF NOT, CALCULATE H & S
C UP TO T
C
18       IF(ABS(TEMP(KTEM)-RANGEH(KRANGE)).LT..001) GO TO 19
         A=AC(KRANGE)
         B=BC(KRANGE)
         C=CC(KRANGE)
         D=DC(KRANGE)
         T1=RANGEL(KRANGE)
         T2=TEMP(KTEM)
         IF(NCCP.EQ.0) CALL DELHS(A,B,C,D,T1,T2,DH,DS)
         IF(NCCP.NE.0) CALL DELHS1(A,B,C,D,T1,T2,DH,DS)
         DHCOM(KTEM)=DHCOM(KTEM)+HC(KRANGE)+DH
```

```
          DSCOM(KTEM)=DSCOM(KTEM)+SC(KRANGE)*.001+DS
          GO TO 20
C
C IF T IS AFTER THE TRANSITION GO TO 21, IF NOT, CALCULATE H & S UP TO
C THE END OF THE RANGE WHICH SHOULD = T
C
  19      IF(ABS(TEMP(KTEM)-TEMP(KTEM-1)).LT..001) GO TO 21
          A=AC(KRANGE)
          B=BC(KRANGE)
          C=CC(KRANGE)
          D=DC(KRANGE)
          T1=RANGEL(KRANGE)
          T2=RANGEH(KRANGE)
          IF(NCCP.EQ.0) CALL DELHS(A,B,C,D,T1,T2,DH,DS)
          IF(NCCP.NE.0) CALL DELHS1(A,B,C,D,T1,T2,DH,DS)
          DHCOM(KTEM)=DHCOM(KTEM)+HC(KRANGE)+DH
          DSCOM(KTEM)=DSCOM(KTEM)+SC(KRANGE)*.001+DS
          GO TO 20
C
C T IS AFTER THE TRANSITION, SO ADD THE H & S FOR TRANSITION
C
  21      KRANGE=KRANGE+1
          DHCOM(KTEM)=DHCOM(KTEM-1)+HC(KRANGE)
          DSCOM(KTEM)=DSCOM(KTEM-1)+SC(KRANGE)*.001
  20      KTEM=KTEM+1
          IF(KTEM.LE.NTEMPS) GO TO 16
C
C CALULATE DELTA H, F, AND S OF REACTION FOR COMPOUND ALONG WITH ENTHALPY
C
          DO 22 I=1,NTEMPS
          DHREAC(I)=DHCOM(I)-DHEL(I)
          DSREAC(I)=DSCOM(I)-DSEL(I)
          DGREAC(I)=DHREAC(I)-TEMP(I)*DSREAC(I)
C
C THE NEXT STATEMENT TAKES CARE OF REFERENCE STATE CONVERSION AS WELL AS
C THE CALCULATION FOR ENTHALPY
C
          ENTH(I)=DHREAC(1)+DHCOM(I)-DHCOM(1)
  22      CONTINUE
C
C CALCULATE CP MEAN FOR USE WITH CRISS AND COBBLE IONS
C
          KTEM=2
          ENTH1=ENTH(1)
          TEMP1=TEMP(1)
          KRANGE=1
  33      IF(TEMP(KTEM).GT.473.15) ENTH1=ENTH(KTEM-1)
          IF(TEMP(KTEM).GT.473.15) TEMP1=TEMP(KTEM-1)
1504      IF(ABS(TEMP(KTEM)-RANGEH(KRANGE)).LT..001) GO TO 32
          IF(ABS(TEMP(KTEM)-TEMP(KTEM-1)).GT.001) GO TO 1503
          CP(KTEM)=0.
          ENTH1=ENTH(KTEM)
          TEMP1=TEMP(KTEM)
          KTEM=KTEM+1
          GO TO 1504
```

```
 1503      CP(KTEM)=1000.*(ENTH(KTEM)-ENTH1)/(TEMP(KTEM)-TEMP1)
           KTEM=KTEM+1
           IF(KTEM.LE.NTEMPS) GO TO 33
           GO TO 34
   32      CP(KTEM)=1000.*(ENTH(KTEM)-ENTH1)/(TEMP(KTEM)-TEMP1)
           KTEM=KTEM+1
           ENTH1=ENTH(KTEM)
           TEMP1=TEMP(KTEM)
           KTEM=KTEM+1
           KRANGE=KRANGE+1
           IF(KTEM.LE.NTEMPS) GO TO 33
   34      NTEMP=TEMP(1)-273.15
C
C CHANGE UNITS AND OUTPUT
C
           DSCOM(1)=DSCOM(1)*1000.
           WRITE(24,35) IDENT,NTEMP,DHREAC(1),DGREAC(1),ENTH(1),DSCOM(1),
      1    NREF(1)
   35      FORMAT(1X,4A5,I3,5X,3(F8.2,2X),F7.2,4X,'----',3X,I3)
           DO 23 I=2,NTEMPS
           CPIND=' '
           IF(TEMP(I).GT.473.15) CPIND='*'
   50      IF(ABS(CP(I)).LT..001) GO TO 36
   51      IF(CP(I).LT.9.995) GO TO 56
   52      IF(CP(I).GT.-9.995) GO TO 57
   53      IF(CP(I).LT.99.995) GO TO 58
   54      IF(CP(I).GT.-99.995) GO TO 59
   55      IF(CP(I).LT.999.995) GO TO 60
   62      ENCODE(7,40,CPARAY) CP(I)
   40      FORMAT(F7.2)
           GO TO 41
   36      CPARAY(1)='   ---'
           CPARAY(2)='- '
           GO TO 41
   37      ENCODE(7,42,CPARAY) CP(I)
   42      FORMAT(3X,F4.2)
           GO TO 41
   38      ENCODE(7,43,CPARAY) CP(I)
   43      FORMAT(2X,F5.2)
           GO TO 41
   39      ENCODE(7,44,CPARAY) CP(I)
   44      FORMAT(1X,F6.2)
   41      IF(ABS(DHREAC(I)).LT..005) DHREAC(I)=0.
           GO TO 61
   56      IF(CP(I).LT..001) GO TO 52
           GO TO 37
   57      IF(CP(I).GT.-.001) GO TO 53
           GO TO 38
   58      IF(CP(I).LT..001) GO TO 54
           GO TO 38
   60      IF(CP(I).LT..001) GO TO 62
           GO TO 39
   59      IF(CP(I).GT.-.001) GO TO 55
           GO TO 39
```

```
61      CONTINUE
        IF(ABS(DGREAC(I)).LT..005) DGREAC(I)=0.
        DSCOM(I)=DSCOM(I)*1000.
        TEMP(I)=TEMP(I)-273.15
        IF(ABS(TEMP(I)-AINT(TEMP(I))).LT..001) GO TO 25
        IF(ABS(TEMP(I)-TEMP(I-1)).LT..001) GO TO 26
        WRITE(24,27) IDENT,TEMP(I),DHREAC(I),DGREAC(I),ENTH(I),
     1  DSCOM(I),CPARAY,CPIND,NREF(I)
27      FORMAT(1X,4A5,F5.1,3X,3(F8.2,2X),F7.2,2X,A5,A2,A1,1X,I3)
        GO TO 23
26      WRITE(24,27) IDENT,TEMP(I),DHREAC(I),DGREAC(I-1),ENTH(I),
     1  DSCOM(I),CPARAY,CPIND,NREF(I)
        GO TO 23
25      NTEMP=TEMP(I)
        IF(ABS(TEMP(I)-TEMP(I-1)).LT..001) GO TO 28
        WRITE(24,29) IDENT,NTEMP,DHREAC(I),DGREAC(I),ENTH(I),DSCOM(I),
     1  CPARAY,CPIND,NREF(I)
29      FORMAT(1X,4A5,I3,5X,3(F8.2,2X),F7.2,2X,A5,A2,A1,1X,I3)
        GO TO 23
28      WRITE(24,29) IDENT,NTEMP,DHREAC(I),DGREAC(I-1),ENTH(I),
     1DS        COM(I),CPARAY,CPIND,NREF(I)
23      CONTINUE
110     STOP
        END


PROGRAM IONIC
C       PROGRAM IONIC CALCULATES HEAT AND FREE ENERGIES OF FORMATION FOR
C       SINGLE IONS IN HIGH TEMPERATURE AQUEOUS SOLUTIONS BY THE CRISS
C       AND COBBLE METHOD
C
C
C
        COMMON /RICK1/ DHEL(16), DSEL(16), ENTH(16), DELS
        COMMON /RICK2/ NELE(8), FNUM(8), NEL(8), NELS
        COMMON /RICK3/ ENTROI, ENTROC, ION, Z, ENTRO, TEMP(16)
        COMMON /RICK4/ ENTH1,NTEMP1,DELF(16),DELH(16),CP,CP2,CPH
        COMMON /RICK5/ DELST, DELCP(16), DELSI(16)
        COMMON /RICK7/ ATWO(250), AH(250), AS(250), ATL(250), ATH(250)
        COMMON /RICK8/ AAA(250), AAB(250), AAC(250), AAD(250), ANCP(250)
        DIMENSION IDENT(4), NREF(16)
C
C SET UP TEMPERATURE SCALE
C
        TEMP(1)=298.15
        TEMP(2)=323.15
        TEMP(3)=348.15
        TEMP(4)=373.15
        TEMP(5)=423.15
        TEMP(6)=473.15
        TEMP(7)=523.15
        TEMP(8)=573.15
C
C GET DATA FROM THE USER
C
```

```
        TYPE 2000
2000    FORMAT(' ENTER ION NAME... ',$)
        ACCEPT 2001, IDENT
2001    FORMAT(4A5)
        TYPE 2002
2002    FORMAT(' ENTER REFERENCE AT 25C ... ',$)
        ACCEPT 9, NREF(1)
        DO 2003 ICC=2,8
2003    NREF(ICC)=3
        TYPE 3
3       FORMAT(' IONIC CHARGE?... ',$)
        ACCEPT 5, Z
5       FORMAT(F)
        IF(Z.LT.0.) GO TO 100
        TYPE 4
4       FORMAT(' HOW MANY ELEMENTS IN CATION?... ',$)
        GO TO 1000
100     TYPE 101
101     FORMAT(' HOW MANY ELEMENTS IN ANION?... ',$)
1000    ACCEPT 9, NELS
9       FORMAT(I)
        TYPE 6
6       FORMAT(' ION NAME... ',$)
        READ(5,7) ((NELE(I),FNUM(I)),I=1,NELS)
7       FORMAT(8(A2,F3.1))
C
C SET STOICHIOMETRIC AMOUNTS, AND CONVERT CHEMICAL SYMBOLS TO
C ELEMENTS CODE NUMBERS
C
        DO 8 I=1,NELS
        FNUM(I)=-FNUM(I)
        NAME=NELE(I)
        CALL NNCODE(NAME,NCODE)
8       NEL(I)=NCODE
        TYPE 10
10      FORMAT(' S OF ION AT 25... ',$)
        ACCEPT 5, ENTROI
C
C IF SPECIES IS ANIONIC, GET TYPE NUMBER, IF NOT ASSIGN TYPE 1
C
        IF(Z.GT.0.) GO TO 102
        TYPE 103
103     FORMAT(' ION TYPE (2,3,4)... ',$)
        ACCEPT 9, ION
        GO TO 104
102     ION=1
104     CONTINUE
C
C INITIALIZE ACCOUNTING
C
12      NTEMPS=8
        KSETT=1
        KSETE=1
C
```

```
C ENTER ELEMENTAL DATA
C
        DO 3000 I=1,250
        READ(21,3001,END=3002) ATWO(I), AH(I), AS(I), ATL(I), ATH(I)
        READ(21,3001) AAA(I), AAB(I), AAC(I), AAD(I), ANCP(I)
 3000   CONTINUE
 3002   CONTINUE
 3001   FORMAT(5F)
 1001   KREAD=NEL(KSETE)
 1041   TH=ATH(KREAD)
        TWO=ATWO(KREAD)
C
C IF HIGHEST TEMPERATURE IN SCALE IS LOWER THAN END OF CP RANGE
C NO TRANSITIONS OF THIS ELEMENT APPEAR IN THIS RANGE, GO ON TO
C NEXT ELEMENT
C
        IF(TEMP(NTEMPS).LT.TH) GO TO 1051
C
C IF T IS GREATER THAN THE HIGHEST TEMPERATURE IN THE RANGE
C FOR CP, A TRANSITION OCCURED, SO INSERT TWO POINTS IN THE
C TEMPERATURE SCALE FOR THE TRANSITION
C IF NOT, INCREMENT THE TEMPERATURE POINT AND CHECK AGAIN
C
 1011   IF(TEMP(KSETT).GT.TH) GO TO 1021
        KSETT=KSETT+1
        IF(KSETT-NTEMPS) 1011, 1011, 1051
 1021   KOUNT=KSETT
        XTEMP=TEMP(KOUNT)
        NXREF=NREF(KOUNT)
        YTEMP=TEMP(KOUNT+1)
        NYREF=NREF(KOUNT+1)
        ZTEMP=TEMP(KOUNT+2)
        NZREF=NREF(KOUNT+2)
        TTEMP=TEMP(KOUNT+3)
        NTREF=NREF(KOUNT+3)
        TEMP(KOUNT)=TH
        NREF(KOUNT)=2
        TEMP(KOUNT+1)=TH
        NREF(KOUNT+1)=2
        NTEMPS=NTEMPS+2
 1031   KOUNT=KOUNT+2
        TEMP(KOUNT)=XTEMP
        NREF(KOUNT)=NXREF
        TEMP(KOUNT+1)=YTEMP
        NREF(KOUNT+1)=NYREF
        XTEMP=ZTEMP
        NXREF=NZREF
        YTEMP=TTEMP
        NYREF=NTREF
        ZTEMP=TEMP(KOUNT+2)
        NZREF=NREF(KOUNT+2)
        TTEMP=TEMP(KOUNT+3)
        NTREF=NREF(KOUNT+3)
        IF(KOUNT+2.LE.NTEMPS) GO TO 1031
        KREAD=TWO+.01
```

```
        KSETT=KSETT+2
        IF(KREAD) 110, 110, 1041
 1051   KSETE=KSETE+1
        KSETT=1
        IF(KSETE.LE.NELS) GO TO 1001
        NTEMP=NTEMPS
C
C CHECK TEMPERATURE SCALE TO SEE THAT THERE ARE NO MORE THAN TWO
C CONSECUTIVE LIKE TEMPERATURE POINTS
C
        DO 1061 I=3,NTEMPS
 1091   IF(ABS(TEMP(I)-TEMP(I-2)).GT..001) GO TO 1061
        K=I
 1071   TEMP(K)=TEMP(K+1)
        IF(K.GE.NTEMP) GO TO 1081
        K=K+1
        GO TO 1071
 1081   NTEMP=NTEMP-1
        GO TO 1091
 1061   CONTINUE
        NTEMPS=NTEMP
C
C ADD HYDROGEN TO REACTION TO COMPLETE WHOLE CELL
C
        NELS=NELS+1
        FNUM(NELS)=Z/2.
        NEL(NELS)=24
        TYPE 14
 14     FORMAT(' F OF ION AT 25... ',$)
        ACCEPT 5, DELF(1)
        NTEMP1=1
C
C CALCULATE DELTA H AT 25C
C
        CALL ELEMEN(1)
        DELSI(1)=ENTROC
        ENTH1=DHEL(1)
C
C CALCULATE THE DATA FOR THE ION AT EACH TEMPERATURE
C
        DO 15 IJ=2,NTEMP
        CALL ELEMEN(IJ)
 15     CONTINUE
C
C OUTPUT THE DATA
C
        ITEMP=TEMP(1)-273.14999
        WRITE(24,35) IDENT,ITEMP,DELH(1),DELF(1),ENTH(1),
     1  DELSI(1),NREF(1)
 35     FORMAT(1X,4A5,I3,5X,3(F8.2,2X),F7.2,4X,'----',3X,I3)
        DO 23 IJF=2,NTEMPS
        CPIND=' '
        IF(TEMP(IJF).GT.473.15) CPIND='*'
        TEMP(IJF)=TEMP(IJF)-273.14999
```

```
             IF(ABS(TEMP(IJF)-AINT(TEMP(IJF))).LT..001) GO TO 25
             IF(ABS(TEMP(IJF)-TEMP(IJF-1)).LT..001) GO TO 26
             WRITE(24,27) IDENT,TEMP(IJF),DELH(IJF),DELF(IJF),ENTH(IJF),
      1 DELSI(IJF),DELCP(IJF),CPIND,NREF(IJF)
   27        FORMAT(1X,4A5,F5.1,3X,3(F8.2,2X),F7.2,2X,F7.2,A1,1X,I3)
             GO TO 23
   26        WRITE(24,27) IDENT,TEMP(IJF),DELH(IJF),DELF(IJF-1),ENTH(IJF),
      1 DELSI(IJF),DELCP(IJF),CPIND,NREF(IJF)
             GO TO 23
   25        ITEMP=TEMP(IJF)
             IF(ABS(TEMP(IJF)-TEMP(IJF-1)).LT..001) GO TO 28
             WRITE(24,29) IDENT,ITEMP,DELH(IJF),DELF(IJF),ENTH(IJF),
      1 DELSI(IJF),DELCP(IJF),CPIND,NREF(IJF)
   29        FORMAT(1X,4A5,I3,5X,3(F8.2,2X),F7.2,2X,F7.2,A1,1X,I3)
             GO TO 23
   28        WRITE(24,29) IDENT,ITEMP,DELH(IJF),DELF(IJF-1),ENTH(IJF),
      1 DELSI(IJF),DELCP(IJF),CPIND,NREF(IJF)
   23        CONTINUE
  110        STOP
             END


SUBROUTINE NNCODE

C THIS SUBROUTINE CHANGES CHEMICAL SYMBOLS TO ELEMENT CODE NUMBERS
C
C                         CODE NUMBERS
C   AG=01 BI=08 CO=15 GD=22 I =29 N =36 P =43 RH=50 SN=57 TL=64 ZR=71
C   AL=02 BR=09 CR=16 GE=23 K =30 NA=37 PB=44 RU=51 SR=58 U =65
C   AS=03 C =10 CS=17 H =24 LA=31 NB=38 PD=45 S =52 TA=59 V =66
C   AU=04 CA=11 CU=18 HF=25 LI=32 ND=39 PT=46 SB=53 TC=60 W =67
C   B =05 CD=12 F =19 HG=26 MG=33 NI=40 PU=47 SE=54 TE=61 Y =68
C   BA=06 CE=13 FE=20 IN=27 MN=34 O =41 RB=48 SI=55 TH=62 YB=69
C   BE=07 CL=14 GA=21 IR=28 MO=35 OS=42 RE=49 SM=56 TI=63 ZN=70
C
C NOTE THAT IODINE IS NOT IN APHABETICAL ORDER
C
      SUBROUTINE NNCODE(NAME,NCODE)
      DIMENSION N(71)
      DATA N/'AG','AL','AS','AU','B','BA','BE','BI','BR','C','CA','CD','
     1CE','CL','CO','CR','CS','CU','F','FE','GA','GD','GE','H','HF','HG'
     2,'IN','IR','I','K','LA','LI','MG','MN','MO','N','NA','NB','ND','NI
     3','O','OS','P','PB','PD','PT','PU','RB','RE','RH','RU','S','SB','S
     4E','SI','SM','SN','SR','TA','TC','TE','TH','TI','TL','U','V','W','
     5Y','YB','ZN','ZR'/
      NCODE=0
      DO 1 I=1,71
      IF(NAME.EQ.N(I)) NCODE=I
   1  CONTINUE
      IF(NCODE.NE.0) GO TO 3
      TYPE 2, NAME
   2  FORMAT(' COULD NOT CODE ',A)
      STOP
   3  RETURN
      END
```

```
SUBROUTINE PRCENT

C THIS SUBROUTINE IS USED TO INTERPOLATE THE CRISS AND COBBLE ENTROPY
C CORRESPONDENCE PRINCIPLE DATA
C
      SUBROUTINE PRCENT(IRANGE,PERCEN,KTEM)
      COMMON /RICK3/ ENTROI, ENTROC, ION, Z, ENTRO, TEMP(16)
C
C IF TEMPERATURE IS BETWEEN 25 & 60 DEGREES C SET IRANGE = 1 AND
C CALCULATE THE PERCENTAGE OF DISTANCE ALONG TEMPERATURE SCALE THAT THE
C ACTUAL TEMPERATURE LIES
C
          IF(TEMP(KTEM).GT.333.15) GO TO 1
          IRANGE=1
          PERCEN=(TEMP(KTEM)-298.15)/35.
          GO TO 6
C
C ETC. FOR T = 60 TO 100C    IRANGE = 2
C
  1       IF(TEMP(KTEM).GT.373.15) GO TO 2
          IRANGE=2
          PERCEN=(TEMP(KTEM)-333.15)/40.
          GO TO 6
C
C ETC. FOR T = 100 TO 150C   IRANGE = 3
C
  2       IF(TEMP(KTEM).GT.423.15) GO TO 3
          IRANGE=3
          PERCEN=(TEMP(KTEM)-373.15)/50.
          GO TO 6
C
C ETC. FOR T = 150 TO 200C   IRANGE = 4
C
  3       IF(TEMP(KTEM).GT.473.15) GO TO 4
          IRANGE=4
          PERCEN=(TEMP(KTEM)-423.15)/50.
          GO TO 6
C
C ETC. FOR T = 200 TO 250C   IRANGE = 5
C
  4       IF(TEMP(KTEM).GT.523.15) GO TO 9
          IRANGE=5
          PERCEN=(TEMP(KTEM)-473.15)/50.
          GO TO 6
C
C ETC. FOR T = 250 TO 300C   IRANGE = 6
C
  9       IF(TEMP(KTEM).GT.573.15) GO TO 10
          IRANGE=6
          PERCEN=(TEMP(KTEM)-523.15)/50.
          GO TO 6
C
C TEMPERATURE LIES OUTSIDE OF CRISS AND COBBLE RANGE, TERMINATE PGM
C
```

```
10        TYPE 5
5         FORMAT(' TEMPERATURE ABOVE 300')
          STOP
6         RETURN
          END


SUBROUTINE SUBION

C THIS SUBROUTINE CALCULATES THERMODYNAMIC DATA FOR IONS
C
          SUBROUTINE SUBION(FNUMI)
          COMMON /RICK1/ DHEL(16), DSEL(16), ENTH(16), DELS
          COMMON /RICK2/ NELE(8), FNUM(8), NEL(8), NELS
          COMMON /RICK3/ ENTROI, ENTROC, ION, Z, ENTRO, TEMP(16)
          COMMON /RICK4/ ENTH1,NTEMP1,DELF(16),DELH(16),CP,CP2,CPH
          COMMON /RICK5/ DELST, DELCP(16), DELSI(16)
          COMMON /RICK6/ DELHR,DELSR,NTEMP
C
C INITIALIZE ACCOUNTING
C
          NTEMPS=NTEMP
          DO 2 I=1,16
          DHEL(I)=0.
          DSEL(I)=0.
          ENTH(I)=0.
          DELF(I)=0.
          DELH(I)=0.
          DELCP(I)=0.
          DELSI(I)=0.
2         CONTINUE
C
C GET INFORMATION ON ION
C
          TYPE 3
3         FORMAT(' IONIC CHARGE?...  ',$)
          ACCEPT 5, Z
5         FORMAT(F)
          IF(Z.LT.0.) GO TO 100
          TYPE 4
4         FORMAT(' HOW MANY ELEMENTS IN CATION?...  ',$)
          GO TO 1000
100       TYPE 101
101       FORMAT(' HOW MANY ELEMENTS IN ANION?...  ',$)
1000      ACCEPT 9, NELS
9         FORMAT(I)
          TYPE 6
6         FORMAT(' ION NAME...  ',$)
          READ(5,7) ((NELE(I),FNUM(I)),I=1,NELS)
7         FORMAT(8(A2,F3.1))
C
C SET STOICHIOMETRIC AMOUNTS AND ELEMENT CODE NUMBERS
C
          DO 8 I=1,NELS
          FNUM(I)=-FNUM(I)
```

```
         NAME=NELE(I)
         CALL NNCODE(NAME,NCODE)
   8     NEL(I)=NCODE
         TYPE 10
  10     FORMAT(' S OF ION AT 25... ',$)
         ACCEPT 5, ENTROI
         IF(Z.GT.0.) GO TO 102
         TYPE 103
 103     FORMAT(' ION TYPE (2,3,4)... ',$)
         ACCEPT 9, ION
         GO TO 104
 102     ION=1
 104     CONTINUE
C
C ADD HYDROGEN TO GET WHOLE CELL REATION
C
         NELS=NELS+1
         FNUM(NELS)=Z/2.
         NEL(NELS)=24
         TYPE 14
  14     FORMAT(' F OF ION AT 25... ',$)
         ACCEPT 5, DELF(1)
         TYPE 16
  16     FORMAT(' STOIC. COEF. FOR ABOVE ION... ',$)
         ACCEPT 5, FNUMI
         NTEMP1=1
C
C CALCULATE DATA FOR ION AT 25C
C
         CALL ELEMEN(1)
         DELSI(1)=ENTROC
         ENTH1=DHEL(1)
C
C CALCULATE DATA FOR ION AT SUCCEEDING TEMPERATURES
C
         DO 15 IJ=2,NTEMPS
         CALL ELEMEN(IJ)
  15     CONTINUE
C
C IF ION IS H+, SET DATA FOR ALL T'S
C
         IF(NELS.NE.2) GO TO 17
         IF(ABS(Z-1.).GT..001) GO TO 17
         IF(NELE(1).NE.'H') GO TO 17
         NTEMP1=1
         DO 1 I=1,NTEMP
         DELF(I)=0.
         DELH(I)=0.
         IF(TEMP(I).GT.473.16) NTEMP1=I-1
         IF(ABS(TEMP(I)-TEMP(I-1)).LT..001) NTEMP1=I-1
         CALL CPHYDR(I)
         DELSI(I)=ENTRO
         DELCP(I)=CPH
         ENTH(1)=0.
```

```
            ENTH(I)=CPH*.001*(TEMP(I)-TEMP(NTEMP1))+ENTH(NTEMP1)
   1        CONTINUE
  17        RETURN
            END


PROGRAM UNDIS1

C PROGRAM UNDIS1 CALCULATES THERMODYNAMIC DATA FOR UNDISSOCIATED NEUTRAL
C COMPLEXES IN AQUEOUS SOLUTIONS BY USE OF THE HELGESON GENERALIZED
C CORRELATION FOR DISSOCIATION CONSTANTS AND THE CRISS AND COBBLE
C TECHNIQUE FOR IONIC SPECIES.
C
C
            COMMON /RICK1/ DHEL(16), DSEL(16), ENTH(16), DELS
            COMMON /RICK2/ NELE(8), FNUM(8), NEL(8), NELS
            COMMON /RICK3/ ENTROI, ENTROC, ION, Z, ENTRO, TEMP(16)
            COMMON /RICK4/ ENTH1,NTEMP1,DELF(16),DELH(16),CP,CP2,CPH
            COMMON /RICK5/ DELST, DELCP(16), DELSI(16)
            COMMON /RICK6/ DELHR, DELSR,NTEMP
            COMMON /RICK7/ ATWO(250), AH(250), AS(250), ATL(250), ATH(250)
            COMMON /RICK8/ AAA(250), AAB(250), AAC(250), AAD(250), ANCP(250)
            DIMENSION IDENT(4), NREF(16), CPARAY(2)
            DIMENSION UDELG(16), UCP(16), US(16), UDELH(16), UENTH(16)
C
C GET INFORMATION ON ELEMENTS
C
            DO 2000 I=1,250
            READ(21,2001,END=2002) ATWO(I), AH(I), AS(I), ATL(I), ATH(I)
            READ(21,2001) AAA(I), AAB(I), AAC(I), AAD(I), ANCP(I)
 2000       CONTINUE
 2002       CONTINUE
 2001       FORMAT(5F)
C
C SET UP TEMPERATURE SCALE
C
            TEMP(1)=298.15
            TEMP(2)=323.15
            TEMP(3)=348.15
            TEMP(4)=373.15
            TEMP(5)=423.15
            TEMP(6)=473.15
C
C INITIALIZE ACCOUNTING
C
            NTEMP=6
            NTEMPS=6
            NTEMP2=1
C
C GET INFORMATION ON UNDISSOCIATED SPECIES
C
            TYPE 10
  10        FORMAT(" ENTER IDENT... ",$)
            ACCEPT 11, IDENT
  11        FORMAT(4A5)
            TYPE 12
```

```
 12        FORMAT(' ENTER REFERENCE AT 25C... ',$)
           ACCEPT 2, NREF(1)
           DO 14 ICC=2,8
 14        NREF(ICC)=6
           TYPE 1
  1        FORMAT(' HOW MANY ELEMENTS IN UNDISS. SPECIES?... ',$)
           ACCEPT 2, NUMUND
  2        FORMAT(8I)
           TYPE 3
  3        FORMAT(' NAME OF SPECIES... ',$)
           ACCEPT4, ((NELE(I),FNUM(I)),I=1,NUMUND)
  4        FORMAT(8(A2,F3.1))
C
C CHANGE CHEMICAL SYMBOLS TO ELEMENT CODE NUMBERS
C
           DO 5 I=1,NUMUND
           NAME=NELE(I)
           CALL NNCODE(NAME,NCODE)
  5        NEL(I)=NCODE
           TYPE 6
  6        FORMAT(' ENTROPY OF SPECIES... ',$)
           ACCEPT 7, UENTRO
  7        FORMAT(8F)
           TYPE 8
  8        FORMAT(' FREE ENERGY OF SPECIES... ',$)
           ACCEPT 7, UNDG
           TYPE 9
  9        FORMAT(' #OF IONS IN REACTION... ',$)
           ACCEPT 2, NUMION
C
C  INITIALIZE ACCOUNTING
C
           KSETT=1
           KSETE=1
 1001      KREAD=NEL(KSETE)
 1041      TH=ATH(KREAD)
           TWO=ATWO(KREAD)
C
C IF HIGHEST TEMPERATURE IN SCALE IS LOWER THAN END FO CP RANGE
C NO TRANSITIONS OF THIS ELEMENT APPEAR IN THIS RANGE, GO ON TO
C NEXT ELEMENT
C
           IF(TEMP(NTEMPS).LT.TH) GO TO 1051
C
C IF T IS GREATER THAN THE HIGHEST TEMPERATURE IN THE RANGE
C FOR CP, A TRANSITION OCCURED, SO INSERT TWO POINTS IN THE
C TEMPERATURE SCALE FOR THE TRANSITION
C IF NOT, INCREMENT THE TEMPERATURE POINT AND CHECK AGAIN
C
 1011      IF(TEMP(KSETT).GT.TH) GO TO 1021
           KSETT=KSETT+1
           IF(KSETT.GT.NTEMPS) GO TO 1051
           GO TO 1011
```

```
    1021    KOUNT=KSETT
            XTEMP=TEMP(KOUNT)
            NXREF=NREF(KOUNT)
            YTEMP=TEMP(KOUNT+1)
            NYREF=NREF(KOUNT+1)
            ZTEMP=TEMP(KOUNT+2)
            NZREF=NREF(KOUNT+2)
            TTEMP=TEMP(KOUNT+3)
            NTREF=NREF(KOUNT+3)
            TEMP(KOUNT)=TH
            NREF(KOUNT)=2
            TEMP(KOUNT+1)=TH
            NREF(KOUNT+1)=2
            NTEMPS=NTEMPS+2
    1031    KOUNT=KOUNT+2
            TEMP(KOUNT)=XTEMP
            NREF(KOUNT)=NXREF
            TEMP(KOUNT+1)=YTEMP
            NREF(KOUNT+1)=NYREF
            XTEMP=ZTEMP
            NXREF=NZREF
            YTEMP=TTEMP
            NYREF=NTREF
            ZTEMP=TEMP(KOUNT+2)
            NZREF=NREF(KOUNT+2)
            TTEMP=TEMP(KOUNT+3)
            NTREF=NREF(KOUNT+3)
            IF(KOUNT+2.LE.NTEMPS) GO TO 1031
            KREAD=TWO+.01
            KSETT=KSETT+2
            IF(KREAD.EQ.0) GO TO 110
            GO TO 1041
    1051    KSETE=KSETE+1
            KSETT=1
            IF(KSETE.LE.NUMUND) GO TO 1001
            NTEMP=NTEMPS
C
C CHECK TEMPERATURE SCALE TO SEE THAT THERE ARE NO MORE THAN TWO
C CONSECUTIVE LIKE TEMPERATURE POINTS
C
            DO 1061 I=3,NTEMPS
    1091    IF(ABS(TEMP(I)-TEMP(I-2)).GT..001) GO TO 1061
            K=I
    1071    TEMP(K)=TEMP(K+1)
            IF(K.GE.NTEMP) GO TO 1081
            K=K+1
            GO TO 1071
    1081    NTEMP=NTEMP-1
            GO TO 1091
    1061    CONTINUE
C
C SUM THE THERMODYNAMIC DATA ON ALL IONS AT EACH TEMPERATURE
C
            DO 15 KION=1,NUMION
            CALL SUBION(FNUMI)
```

```
            DO 15 KKTEM=1,NTEMP
            UDELG(KKTEM)=DELF(KKTEM)*FNUMI+UDELG(KKTEM)
            UCP(KKTEM)=UCP(KKTEM)+DELCP(KKTEM)*FNUMI
            US(KKTEM)=US(KKTEM)+DELSI(KKTEM)*FNUMI
            UDELH(KKTEM)=UDELH(KKTEM)+DELH(KKTEM)*FNUMI
            UENTH(KKTEM)=UENTH(KKTEM)+ENTH(KKTEM)*FNUMI
   15       CONTINUE
C
C CALCULATE 25C DATA FOR NEUTRAL COMPLEX
C
            DELSR=US(1)-UENTRO
            DELGR=UDELG(1)-UNDG
            DELHR=DELGR*1000.+298.15*DELSR
            UDELG(1)=UNDG
            UDELH(1)=UDELH(1)-DELHR*.001
            US(1)=UENTRO
            UENTH1=UDELH(1)
C
C CALCULATE DATA FOR NEUTRAL COMPLEX AT REMAINING TEMPERATURES
C
            DO 16 I=2,NTEMP
            CALL HELGE(DG,I)
            DG=DG/1000.
            UDELG(I)=UDELG(I)-DG
            CALL HELGES(DS,I)
            DELT=TEMP(I)-298.15
            US(I)=US(I)-DS
            UCP(I)=(US(I)-UENTRO)/ALOG(TEMP(I)/298.15)
            DH=DG+TEMP(I)*DS*.001
            UDELH(I)=UDELH(I)-DH
            UENTH(I)=UENTH1+UCP(I)*DELT/1000.
   16       CONTINUE
            UENTH(1)=UDELH(1)
C
C OUTPUT DATA
C
            ITEMP=TEMP(1)-273.14999
            WRITE(24,35) IDENT,ITEMP,UDELH(1),UDELG(1),UENTH(1),
      1     US(1),NREF(1)
            NTEMPS=NTEMP
   35       FORMAT(1X,4A5,I3,5X,3(F8.2,2X),F7.2,4X,'----',3X,I3)
            DO 23 IJF=2,NTEMPS
            CPIND=' '
            IF(TEMP(IJF).GT.473.15) CPIND='*'
   50       IF(ABS(UCP(IJF)).LT..001) GO TO 36
   51       IF(UCP(IJF).LT.9.995) GO TO 56
   52       IF(UCP(IJF).GT.-9.995) GO TO 57
   53       IF(UCP(IJF).LT.99.995) GO TO 58
   54       IF(UCP(IJF).GT.-99.995) GO TO 59
   55       IF(UCP(IJF).LT.999.995) GO TO 60
   62       ENCODE(7,40,CPARAY) UCP(IJF)
   40       FORMAT(F7.2)
            GO TO 41
   36       CPARAY(1)='  ---'
            CPARAY(2)='- '
```

```
          GO TO 41
37        ENCODE(7,42,CPARAY) UCP(IJF)
42        FORMAT(3X,F4.2)
          GO TO 41
38        ENCODE(7,43,CPARAY) UCP(IJF)
43        FORMAT(2X,F5.2)
          GO TO 41
39        ENCODE(7,44,CPARAY) UCP(IJF)
44        FORMAT(1X,F6.2)
41        CONTINUE
          GO TO 61
56        IF(UCP(IJF).LT..001) GO TO 52
          GO TO 37
57        IF(UCP(IJF).GT.-.001) GO TO 53
          GO TO 38
58        IF(UCP(IJF).LT..001) GO TO 54
          GO TO 38
60        IF(UCP(IJF).LT..001) GO TO 62
          GO TO 39
59        IF(UCP(IJF).GT.-.001) GO TO 55
          GO TO 39
61        CONTINUE
          TEMP(IJF)=TEMP(IJF)-273.14999
          IF(ABS(TEMP(IJF)-AINT(TEMP(IJF))).LT..001) GO TO 25
          IF(ABS(TEMP(IJF)-TEMP(IJF-1)).LT..001) GO TO 26
          WRITE(24,27) IDENT,TEMP(IJF),UDELH(IJF),UDELG(IJF),UENTH(IJF),
     1    US(IJF),CPARAY,CPIND,NREF(IJF)
27        FORMAT(1X,4A5,F5.1,3X,3(F8.2,2X),F7.2,2X,A5,A2,A1,1X,I3)
          GO TO 23
26        WRITE(24,27) IDENT,TEMP(IJF),UDELH(IJF),UDELG(IJF-1),
     1    UENTH(IJF-1),US(IJF),CPARAY,CPIND,NREF(IJF)
          GO TO 23
25        ITEMP=TEMP(IJF)
          IF(ABS(TEMP(IJF)-TEMP(IJF-1)).LT..001) GO TO 28
          WRITE(24,29) IDENT,ITEMP,UDELH(IJF),UDELG(IJF),UENTH(IJF),
     1    US(IJF),CPARAY,CPIND,NREF(IJF)
29        FORMAT(1X,4A5,I3,5X,3(F8.2,2X),F7.2,2X,A5,A2,A1,1X,I3)
          GO TO 23
28        WRITE(24,29) IDENT,ITEMP,UDELH(IJF),UDELG(IJF-1),UENTH(IJF-1),
     1    US(IJF),CPARAY,CPIND,NREF(IJF)
23        CONTINUE
110       STOP
          END
```

GLOSSARY--DIMENSIONED VARIABLES

AAA   A COEFFICIENT IN HEAT CAPACITY EXPRESSION FOR ELEMENTS
AAB   B COEFFICIENT IN HEAT CAPACITY EXPRESSION FOR ELEMENTS
AAC   C COEFFICIENT IN HEAT CAPACITY EXPRESSION FOR ELEMENTS
AAD   D COEFFICIENT IN HEAT CAPACITY EXPRESSION FOR ELEMENTS
AC   A COEFFICIENT IN HEAT CAPACITY EXPRESSION FOR COMPOUND
AH   ENTHALPY OF ELEMENT AT 25C, OR ENTHALPY OF TRANSITION AT HIGHER T'S
ANCP   INDICATES WHICH FORM OF THE HEAT CAPACITY EXPRESSION TO USE
AS   ENTROPY OF ELEMENT AT 25C, OR ENTROPY OF TRANSITION AT HIGHER T'S
ATH   HIGH ENDPOINT OF TEMPERATURE RANGE FOR CP EXPRESSION FOR ELEMENTS
ATL   LOW ENDPOINT OF TEMPERATURE RANGE FOR CP EXPRESSION FOR ELEMENTS
ATWO   LOCATION OF NEXT DATA SET FOR SPECIFIC ELEMENT
BC   B COEFFICIENT IN HEAT CAPACITY EXPRESSION FOR COMPOUND
CC   C COEFFICIENT IN HEAT CAPACITY EXPRESSION FOR COMPOUND
CONSTA DATA FOR A CONSTANT FROM CORRESPONDENCE PRINCIPLE
CONSTB DATA FOR B CONSTANT FROM CORRESPONDENCE PRINCIPLE
CP   (IONCOM ONLY) AVERAGE CP FROM 25, 200, 250C, OR LAST TRANSITION TO T
CPARAY USED TO CHANGE CP DATA TO A FORMAT FOR PAGE SPACING
DC   D COEFFICIENT IN HEAT CAPACITY EXPRESSION FOR COMPOUND
DELCP   MEAN HEAT CAPACITY OF ION FROM 25, 200, 250C, OR LAST TRANS. TO T
DELF   FREE ENERGY OF FORMATION OF ION RELATIVE TO H+
DELH   HEAT OF FORMATION OF ION RELATIVE TO H+
DELSI   CRISS AND COBBLE "ABSOLUTE" ENTROPY OF ION
DGREAC FREE ENERGY OF FORMATION FOR COMPOUND AT EACH TEMPERATURE
DHCOM   SUMMED ENTHALPY FOR COMPOUND AT EACH TEMPERATURE
DHEL   SUMMED ENTHALPIES OF FORMATION OF ALL ELEMENTS IN REACTION
DHREAC HEAT OF FORMATION FOR COMPOUND AT EACH TEMPERATURE
DSCOM   SUMMED ENTROPY FOR COMPOUND AT EACH TEMPERATURE
DSEL   SUMMED ENTROPIES OF ALL ELEMENTS IN REACTION
DSREAC ENTROPY OF FORMATION FOR COMPOUND AT EACH TEMPERATURE
ENTH   HEAT OF FORMATION AT 25C + ENTHALPY DUE TO HEAT FOR SPECIES
ENTROH DATA FOR "ABSOLUTE" ENTROPY OF H+
FNUM   STOICHIOMETRIC COEFFICIENTS OF ELEMENTS PARTICIPATING IN REACTION
HC   (1)=ENTHALPY OF COMPOUND AT 25C, (2) & UP =HEAT OF TRANSITION FOR COMP.
IDENT   IDENTIFICATION FOR SPECIES OF INTEREST
NEL   CODE NUMBER FOR EACH ELEMENT
NELE   CHEMICAL SYMBOL FOR EACH ELEMENT
NREF   REFERENCE CODE NUMBER FOR DATA AT EACH TEMPERATURE
RANGEH HIGH ENDPOINT OF TEMPERATURE RANGE FOR CP EXPRESSION FOR COMPOUND
RANGEL LOW ENDPOINT OF TEMPERATURE RANGE FOR CP EXPRESSION FOR COMPOUND
SC   (1)=ENTROPY OF COMPOUND AT 25C, (2) & UP =DELTA S OF TRANS. FOR COMP.
TEMP   TEMPERATURE SCALE
UCP   MEAN HEAT CAPACITY FOR NEU. COMP. FROM 25C OR LAST TRANSITION TO T
UDELG   FREE ENERGY OF FORMATION OF NEUTRAL COMPLEX
UDELH   ENTHALPY OF FORMATION OF NEUTRAL COMPLEX
UENTH   ENTHALPY OF NEUTRAL COMPLEX
US   ENTROPY OF NEUTRAL COMPLEX

GLOSSARY--OTHER VARIABLES

A   A COEFFICIENT IN HEAT CAPACITY EXPRESSION FOR ELEMENTS
        (CPMEAN ONLY) INTERPOLATED A CONSTANT FROM CRISS AND COBBLE DATA
B   B COEFFICIENT IN HEAT CAPACITY EXPRESSION FOR ELEMENTS
        (CPMEAN ONLY) INTERPOLATED B CONSTANT FROM CRISS AND COBBLE DATA
C   C COEFFICIENT IN HEAT CAPACITY EXPRESSION FOR ELEMENTS
CP   MEAN HEAT CAPACITY OF ION FROM 25C TO T
CP2   MEAN HEAT CAPACITY OF ION FROM 25, 200, 250C, OR LAST TRANSITION
        TO T
CPELEM SUMMED MEAN CP'S OF ELEMENTS FROM 25, 200, 250C, OR LAST TRANSITION
CPH   MEAN HEAT CAPACITY OF H+ FROM 25, 200, 250C, OR LAST TRANSITION
        TO T
CPIND   INDICATES THAT MEAN CP WAS TAKEN FROM LAST TEMPERATURE AT T'S
        ABOVE 200C
D   D COEFFICIENT IN HEAT CAPACITY EXPRESSION FOR ELEMENTS
DELGR   FREE ENERGY OF DISSOCIATION AT 25C
DELHR   ENTHALPY OF DISSOCIATION AT 25C
DELS   ENTROPY OF REACTION FOR WHOLE CELL AT 25, 200, 250C, OR LAST
        TRANSITION
DELSR   ENTROPY OF DISSOCIATION AT 25C
DELST   ENTROPY OF REACTION FOR WHOLE CELL AT T
DELT   DELTA T FROM 25C TO T
DELTAC DELTA CP FOR WHOLE CELL REACTION FROM 25, 200, 250C, OR LAST
        TRANSITION
DELTAT DELTA T FROM 25, 200, 250C, OR LAST TRANSITION TO T
DG   FREE ENERGY OF DISSOCIATION AT T
DH   CHANGE IN ENTHALPY DUE TO TEMPERATURE EFFECTS
        (UNDIS1 ONLY) ENTHALPY OF DISSOCIATION AT T
DS   CHANGE IN ENTROPY DUE TO TEMPERATURE EFFECTS
        (UNDIS1 ONLY) ENTROPY OF DISSOCIATION AT T
ENTH1   SUMMED ENTHALPIES OF ELEMENTS AT 25, 200, 250C, OR LAST TRANSITION
        (IONCOM) ENTHALPY OF COMPOUND AT LOW END OF SOME TEMPERATURE RANGE
ENTH2   ENTHALPY OF ION AT 200C
ENTH3   ENTHALPY OF ION AT 250C
ENTRH1 "ABSOLUTE" ENTROPY OF H+ AT TEMP(NTEMP1)
ENTRO   CRISS AND COBBLE "ABSOLUTE" ENTROPY OF H+ AT T
ENTROA "ABSOLUTE" ENTROPY OF ION AT 25C
ENTROB "ABSOLUTE" ENTROPY OF ION AT 25, 200, 250C, OR LAST TRANSITION
ENTROC CRISS AND COBBLE "ABSOLUTE" ENTROPY OF ION AT T
ENTROI CONVENTIONAL ENTROPY OF ION AT 25C
FNUMI   STOICHIOMETRIC NUMBER OF IONS IN DISSOCIATION REACTION
H   ENTHALPY OF ELEMENT AT 25C OR ENTHALPY OF TRANSITION AT HIGHER T'S
I   COUNTER
ICC   COUNTER
IJ   COUNTER
IJF   COUNTER
ION   CRISS AND COBBLE ION TYPE
IRANGE DENOTES WHICH SET OF DATA POINTS IS BEING INTERPOLATED BETWEEN
ITEMP   INTEGER TEMPERATURE FOR OUTPUT
K   COUNTER
KELEM   COUNTER FOR ELEMENTS
KION   COUNTER FOR NUMBER OF ION SPECIES IN DISSOCIATION REACTION
KKTEM   COUNTER FOR TEMPERATURE

KOUNT   COUNTER USED IN INSERTING TEMPERATURE POINTS IN TEMPERATURE SCALE
KRANGE   COUNTER FOR COMPOUND HEAT CAPACITY RANGES
KREAD   READING POINTER FOR USE IN ELEMENTAL DATA LOCATIONS
KSETE   COUNTER FOR ELEMENTS
KSETT   COUNTER FOR TEMPERATURES
KTEM   SUBSCRIPT FOR TEMPERATURE OF INTEREST
NAME   TEMPORARY LOCATION FOR CHEMICAL SYMBOL OF ELEMENT
NCCP   INDICATES WHICH FORM OF THE HEAT CAPACITY EXPRESSION TO USE FOR
         ELEMENTS
NCODE   TEMPORARY LOCATION FOR ELEMENT CODE NUMBER
NCP   INDICATES WHICH FORM OF THE HEAT CAPACITY EXPRESSION TO USE FOR
         COMPOUND
NELEM   NUMBER OF ELEMENTS IN COMPOUND
NELS   NUMBER OF ELEMENTAL SPECIES IN WHOLE CELL REACTION
NRANGE   NUMBER OF RANGES NEEDED TO DEFINE THE CP FUNCTION FOR COMPOUND
NTEMP   TEMPERARY STORAGE FOR NTEMPS OR INTEGER TEMPERATURE FOR OUTPUT
NTEMP1   SUBSCRIPT FOR TEMPERATURE AT 25, 200, 250C, OR LAST TRANSITION
NTEMPS   NUMBER OF TEMPERATURE POINTS
NTREF   TEMPORARY STORAGE FOR REFERENCE CODE NUMBER
NUMION   NUMBER OF IONIC SPECIES UPON DISSOCIATION
NUMUND   NUMBER OF ELEMENTS IN NEUTRAL COMPLEX
NXREF   TEMPORARY STORAGE FOR REFERENCE CODE NUMBER
NYREF   TEMPORARY STORAGE FOR REFERENCE CODE NUMBER
NZREF   TEMPORARY STORAGE FOR REFERENCE CODE NUMBER
PERCEN   LINEAR INTERPOLATION FACTOR
S   ENTROPY OF ELEMENT AT 25C OR ENTROPY OF TRANSITION AT HIGHER T'S
T1   TEMPERATURE AT LOW END OF HEAT CAPACITY RANGE
T2   TEMPERATURE AT HIGH END OF HEAT CAPACITY RANGE
         OR TEMPERATURE OF INTEREST
TEMP1   TEMPERATURE AT LOW END OF HEAT CAPACITY RANGE FOR COMPOUND
TH   TEMPERATURE AT HIGH END OF HEAT CAPACITY RANGE FOR ELEMENTS
TL   TEMPERATURE AT LOW END OF HEAT CAPACITY RANGE FOR ELEMENTS
TTEMP   TEMPORARY STORAGE FOR TEMPERATURE
TWO   LOCATION OF NEXT ELEMENTAL DATA SET
UENTH1   ENTHALPY OF FORMATION OF NEUTRAL COMPLEX AT 25C
UENTRO   ENTROPY OF NEUTRAL COMPLEX AT 25C
UNDG   FREE ENERGY OF FORMATION OF NEUTRAL COMPLEX AT 25C
XTEMP   TEMPORARY STORAGE FOR TEMPERATURE
YTEMP   TEMPORARY STORAGE FOR TEMPERATURE
Z   IONIC CHARGE
ZTEMP   TEMPORARY STORAGE FOR TEMPERATURE